T0203750

INSECT PEST MANAGEMENT
Techniques for Environmental Protection

AGRICULTURE AND ENVIRONMENT SERIES

Jack E. Rechcigl
Editor-in-Chief

Agriculture is an essential part of our economy on which we all depend for food, feed, and fiber. With the increased agricultural productivity in this country as well as abroad, the general public has taken agriculture for granted while voicing its concern and dismay over possible adverse effects of agriculture on the environment. The public debate that has ensued on the subject has been brought about, in part, by the indiscriminate use of agricultural chemicals and, in part, by misinformation, based largely on anecdotal evidence.

At the national level, recommendations have been made for increased research in this area by such bodies as the Office of Technology Assessment, the National Academy of Sciences, and the Carnegie Commission on Science, Technology, and Government. Specific issues identified for attention include: contamination of surface and groundwater by natural and chemical fertilizers, pesticides, and sediment; the continued abuse of fragile and nutrient-poor soils; and suitable disposal of industrial and agricultural waste.

Although a number of publications have appeared recently on specific environmental effects of some agricultural practices, no attempt has been made to approach the subject systematically and in a comprehensive manner. The aim of this series is to fill the gap by providing the synthesis and critical analysis of the state of the art in different areas of agriculture bearing on environment and vice versa. Efforts will also be made to review research in progress and comment on perspectives for the future. From time to time methodological treatises as well as compendia of important data in handbook form will also be included. The emphasis throughout the series will be on comprehensiveness, comparative aspects, alternative approaches, innovation, and worldwide orientation.

Specific topics will be selected by the Editor-in-Chief with the council of an international advisory board. Imaginative and timely suggestions for the inclusion in the series from individual scientists will be given serious consideration.

Published Titles

Biological and Biotechnological Control of Insect Pests
Environmentally Safe Approaches to Crop Disease Control
Insect Pest Management: Techniques for Environmental Protection
Soil Amendments and Environmental Quality
Soil Amendments: Impacts on Biotic Systems

INSECT PEST MANAGEMENT
Techniques for Environmental Protection

Jack E. Rechcigl
University of Florida
Soil and Water Science Department
Research and Education Center
Ona, Florida

Nancy A. Rechcigl
Yoder Brothers, Inc.
Parrish, Florida

CRC Press
Taylor & Francis Group
Boca Raton London New York

CRC Press is an imprint of the
Taylor & Francis Group, an **informa** business

CRC Press
Taylor & Francis Group
6000 Broken Sound Parkway NW, Suite 300
Boca Raton, FL 33487-2742

First issued in paperback 2019

ISBN-13: 978-1-56670-478-6 (hbk)
ISBN-13: 978-0-367-39937-5 (pbk)

Library of Congress Cataloging-in-Publication Data

Insect pest management: techniques for environmental protection /
[edited] by Jack E. Rechcigl and Nancy A. Rechcigl.
p. cm. (Agriculture & environment series)
Includes bibliographical references and index.
ISBN 1-56670-478-2 (alk. paper)
1. Insect pests — Biological control. I. Rechcigl, Jack E. II. Rechcigl, Nancy A. III. Series.
SB933.3.I53 1999
632'.9517—dc21   99-40543
                                                                                      CIP

Library of Congress Card Number 99-40543

Visit the Taylor & Francis Web site at
http://www.taylorandfrancis.com

and the CRC Press Web site at
http://www.crcpress.com

Preface

Insect pest management has always been and will continue to be a constant challenge to agricultural producers and researchers alike. As insect resistance to commonly used pesticides builds and the removal of more toxic pesticides from the market continues, controlling insect infestations will become increasingly more difficult. Growers are constantly faced with the dilemma of producing a high quality, pest-free crop within economical means, without endangering the environment and the workers' safety.

While the environmental concerns are in order, we have to, at the same time, guard against extreme points of view by zealous advocates, since, as the matters now stand, the traditional cultural practices are here to stay, and the use of chemical pesticides, in the foreseeable future, will remain playing an important role in pest management.

The purpose of this book is to present a balanced overview of environmentally safe and ecologically sound approaches applied to commonly used practices in insect pest management. The use of various biological agents and other physiological approaches, as well as of biotechnological techniques, is the subject of a separate publication and will therefore not be covered in this volume.

The first part of the book examines specific ecological measures which could be taken to prevent or lessen insect pest infestations of crops. The next section covers a variety of environmentally acceptable physical control measures to prevent insect pests from entering greenhouses and infesting specific plants. Judicious use of chemical pesticides forms the third section which encompasses a choice of recommended formulations and applications of preferable insecticides, including those that could be targeted against the destructive pests, without harming beneficial insects and the surrounding environment. The fourth section presents a detailed account of various agronomic and other cultural practices, which have proven to be effective against insect pests, yet which are considered environmentally sound, as well as other potential measures that could, with suitable modifications, be made environmentally acceptable. A separate chapter is devoted to integrated pest management based on the current state-of-the-art. Section five deals with biological control of insects, including biological control by Bti. The last section of the book is devoted to regulatory and legislative aspects, covering both the inspection and quarantine of plant materials, and the other regulations relating to insecticide control.

Each chapter has been written by an expert in the respective field. This publication should be a useful source of information to students and professionals in the field of entomology, agronomy, horticulture, ecology, and environmental sciences, as well as to agricultural practitioners, extension workers and industrial chemists, and last but not least, to all persons concerned with regulatory and legislative matters.

The editors wish to thank the individual contributors for the time and effort they put into the preparation of their chapters. In addition, special thanks are due to the Ann Arbor Press and CRC Press Staff and Editorial Board.

Jack E. Rechcigl
Nancy A. Rechcigl

The Editors

Jack E. Rechcigl is a Professor of Soil and Environmental Sciences at the University of Florida and is located at the Research and Education Center in Ona, FL. He received his B.S. degree (1982) in Agriculture from the University of Delaware, Newark, DE and his M.S. (1983) and Ph.D. (1986) degrees in Soil Science from Virginia Polytechnic Institute and State University, Blacksburg, VA. He joined the faculty of the University of Florida in 1986 as Assistant Professor, in 1991 was promoted to Associate Professor, and in 1996 attained Full Professorship. In 1999, he was named a University of Florida Research Foundation Professor.

Dr. Rechcigl has authored over 200 publications, including contributions to books, monographs, and articles in periodicals in the fields of soil fertility, environmental quality, and water pollution. His research has been supported by research grants totaling over $3 million from both private sources and government agencies. Dr. Rechcigl has been a frequent speaker at national and international workshops and conferences and has consulted in various countries, including Canada, Brazil, Nicaragua, Venezuela, Australia, New Zealand, Taiwan, Philippines, France, and the Czech Republic. He also serves on a number of national and international boards, including the University of Cukurova Mediterranean International Center for Soils and Environment Research in Turkey.

He is currently Editor-in-Chief of the *Agriculture and Environment Book Series*, Associate Editor of the Soil and Crop Science Society Proceedings, and until recently Associate Editor of the *Journal of Environmental Quality*. Most recently he has edited *Insect Pest Management: Techniques for Environmental Protection* (Lewis Publishers, 2000), *Environmentally Safe Approaches to Crop Disease Control* (Lewis Publishers and CRC Press, 1997), *Soil Amendments: Impacts on Biotic Systems* (Lewis Publishers and CRC Press, 1995), and *Use of By-Products and Wastes in Agriculture* (American Chemical Society, 1997). He is also serving as an invitational reviewer of manuscripts and grant proposals for scientific journals and granting agencies.

Dr. Rechcigl is a member of the American Chemical Society, Soil Science Society of America, American Society of Agronomy, International Soil Science Society, Czechoslovak Society of Arts and Sciences, various trade organizations, and the honorary societies of Sigma Xi, Gamma Sigma Delta, Phi Sigma, and Gamma Beta Phi.

Dr. Rechcigl has been the recipient of numerous awards, including the Sigma Xi Research Award, University of Philippines Research Award, University of Florida Research Honor Award, University of Florida Research Achievement Award, and University of Delaware Presidential Citation for Outstanding Achievement Award. Most recently he was elected a Fellow of the American Society of Agronomy, Fellow of the Soil Science Society of America, and the recipient of Honorary Professorship from the Czech Agricultural University in Prague.

Nancy A. Rechcigl holds the position of entomologist with Yoder Bros. Inc., Parrish, FL, specializing in plant disease and entomological problems of floricultural crops. Prior to joining Yoder Bros., Nancy worked for the University of Florida (1989–1994) as a County Horticultural Agent, providing diagnostic services and information on cultural practices and pest management to horticultural, landscape. and pest control industries. As an Extension Agent she was also responsible for supervising the County Master Gardener Program, providing instructional classes and operating a Plant Clinic that was popular with the urban community. From 1986 to 1989, she worked for Ball PanAm Inc., Parrish, FL as a Plant Pathologist responsible for the disease certification program of ornamental plants.

Over the past 12 years, Ms. Rechcigl has given numerous lectures on the identification and control of disease and pest problems of turf and ornamentals. In addition to writing a weekly gardening column "Suncoast Gardening" for the urban community, she frequently contributes articles to local trade and professional journals. Most recently she has co-edited the books *Environmentally Safe Approaches to Crop Disease Control* (Lewis Publishers and CRC Press, 1997), and *Insect Pest Management: Techniques for Environmental Protection* (Lewis Publishers, 2000).

Ms. Rechcigl received her B.S. degree (1983) in Plant Pathology from the University of Delaware, Newark, DE. She did her graduate work at Virginia Polytechnic Institute & State University, Blacksburg, VA, receiving her M.S. degree in 1986, specializing in Plant Virology.

Ms. Rechcigl is an active member of the American Phytopathological Society, Entomological Society of America, Florida Nurserymen and Growers Association, Czechoslovak Society of Arts and Sciences, and the Honorary Society of Phi Kappa Phi.

Contributors

D. A. Andow
Department of Entomology
University of Minnesota
St. Paul, Minnesota

Eitan Ben-Dov
Center for Biological Control
Department of Life Sciences
Ben-Gurion University of the Negev
Be'er-Sheva, Israel

Gary L. Cave
Animal and Plant Health Inspection
  Service Plant Protection and Quarantine
U.S.D.A./A.P.H.I.S.
Riverdale, Maryland

G. W. Cuperus
Department of Entomology and Plant
  Pathology
Oklahoma State University
Stillwater, Oklahoma

Clive A. Edwards
Department of Entomology
The Ohio State University
Columbus, Ohio

Jorge Fernandez-Cornejo
Economic Research Service
U.S.D.A./E.R.S.
Washington, D.C.

John K. Greifer
Animal and Plant Health Inspection
  Service Trade Support Team
U.S.D.A./A.P.H.I.S.
Washington, D.C.

James Robert Hagler
Western Cotton Research Laboratory
U.S.D.A./A.R.S.
Phoenix, Arizona

J. P. Harmon
Department of Entomology
University of Minnesota
St. Paul, Minnesota

David J. Horn
Department of Entomology
The Ohio State University
Columbus, Ohio

Edwin Imai
Animal and Plant Health Inspection
  Service Plant Protection and
  Quarantine
U.S.D.A./A.P.H.I.S.
Riverdale, Maryland

Robert P. Kahn
Plant Protection and Quarantine
Rockville, Maryland

Yoel Margalith
Center for Biological Control
Department of Life Sciences
Ben-Gurion University of the Negev
Be'er-Sheva, Israel

P. G. Mulder
Department of Entomology and Plant
  Pathology
Oklahoma State University
Stillwater, Oklahoma

Michael Ollinger
Economic Research Service
U.S.D.A./E.R.S.
Washington, D.C.

Christian Y. Oseto
Department of Entomology
Purdue University
West Lafayette, Indiana

Douglas G. Pfeiffer
Department of Entomology
Virginia Polytechnic Institute & State
 University
Blacksburg, Virginia

T. A. Royer
Department of Entomology and Plant
 Pathology
Oklahoma State University
Stillwater, Oklahoma

N. A. Schellhorn
Department of Entomology
University of Minnesota
St. Paul, Minnesota

Dedication

To our parents and our family for their love and support.

Contents

Section VI   Regulatory Aspects

SECTION **I**

Ecological Measures

CHAPTER 1

Ecological Control of Insects

David J. Horn

CONTENTS

1.1   INTRODUCTION

In a sense, when intended to reduce pest numbers, *any* manipulation of the environment might be considered as "ecological control," for any environmental factor that impinges on an insect pest is by definition "ecological." In a narrower view, ecological control is manipulation or adjustment of the environment surrounding an insect pest in order to enhance its control with minimal disruption of ecosystem function. Ecological control is therefore similar to what Frisbie and Smith

(1991) termed "biointensive" control, i.e., pest management that relies heavily on natural and biological controls, with a prescriptive chemical input only as a last resort. For effective ecological control, there needs to be an understanding of a pest's interaction with its environment, along with a fundamental understanding of the interconnections within an ecosystem. The past few decades have witnessed general acceptance of the necessity for considering ecology in developing pest management systems, yet there is little agreement as to what components of ecological theory are most applicable to pest management systems (Kogan, 1995). This is partly because ecology is a synthetic science, drawing on ideas and data from other fields in biology, and ecological theory is therefore in a continual state of flux. The lack of agreement among ecologists on such issues as the reality of equilibrium in population regulation and the relationship (if any) between species diversity and community stability can be frustrating to designers of pest management systems. This frustration is exacerbated by a number of differences between "natural" ecosystems (such as forests and abandoned fields) and managed, artificial ecosystems (such as crop fields or manicured landscaping); ecological theory generated from studies of natural ecosystems may not be applicable to artificial ecosystems. Also, even the most localized ecosystems are enormously complex and variable, and ecological experiments when performed in the field are subject to widely varying outputs. Results are not always easy to interpret and experiments are not easily replicated.

The ecosystem consists of the pest population and the surrounding interactive biotic and physical environment. The interactions between a single pest species and its environment are enormously complex, and all too frequently we are also faced with the necessity to manage a number of pests forming a "pest complex" associated with a single plant species. In an agricultural landscape there are usually several crops grown simultaneously, such as corn, soybeans, alfalfa, and wheat on farms in the midwestern U.S., or beans, squash, tomatoes, peppers, lettuce, and radishes in my own backyard garden. These plants coexist within a matrix of surrounding ecosystems each with its typical flora and fauna: abandoned weedy fields, hedgerows, forests, and so forth. Ecological processes within these surrounding habitats influence events within adjacent agricultural or landscaped ecosystems. In agricultural production we may cast aside the complexity and unpredictability of these ecological processes, and we may oversimplify, ignore, or override these ecological processes as best we can, with the appropriate goal of maintaining or increasing yields with minimal (financial) input in order to make a profit. However, our efforts to manage pests often disrupt whatever naturally occurring pest population regulation or "equilibrium" there may be, and we may be forced to commit additional environmental disruption to achieve economic goals. Even very successful integrated pest management (IPM) programs often display little attention to or appreciation of ecosystem functions (Kogan, 1986, 1995).

A recent report of the National Research Council (1996) has called for development of "ecologically based IPM," with the following components: (1) safety (to the environment, the crop, the producer, fish and wildlife, etc.); (2) cost effectiveness; (3) long-term sustainability; and (4) consideration of the ecosystem as a central focus. The implication is that to manage pests most effectively with minimal disruption, they must be considered within the context of the ecosystem in which they occur. Ecological control seeks to achieve successful pest management through an understanding of the complexities of ecosystem interactions, followed by application of this understanding to effectively achieve relative stability of pest populations below damaging levels without resorting to exclusive use of interventive and disruptive techniques. This is the ideal toward which to work in applying ecological control. This chapter explores some fundamentals of pest ecology in relation to natural and anthropogenic ecosystems, and how an understanding of these fundamentals can enhance pest management with minimal disruption of ecosystem processes.

1.2 LIFE SYSTEMS

The "life system" concept was initially conceived by Clark et al. (1967) to reinforce the idea that a population cannot be considered apart from the ecosystem with which it interacts. The life system consists of the pest population plus its "effective environment." Every insect (or other) population is surrounded by environmental factors that may impact it positively or negatively. The effective environment thus includes food supply, predators, pathogens, competitors, hiding places — in short, anything that may enhance or limit survival, reproduction, and/or dispersal of a pest species. A limitation to the life system concept is that the scale of the surrounding ecosystem is defined arbitrarily, and the intensity of environmental impacts is likely to vary depending upon whether one views the ecosystem as bounded by a single crop field, an entire farm, or the local or regional landscape beyond individual farms. Most ecological pest management concentrates on the agroecosystem, defined as the effective environment at the crop level (Altieri, 1987, 1994). Rabb (1978) suggested that the definition of agroecosystem be expanded to include natural (or unmanaged) habitats surrounding crops. Increasingly, ecological pest management needs to consider environmental interactions at least to the level of the local landscape (Collins and Qualset, 1999; Duelli, 1997). At any scale, the implication of the life system concept is that human-caused manipulations (such as tilling, harvesting, etc.) of an ecosystem can either disrupt or ameliorate the favorableness of the local environment to an insect, resulting in an increase or a decrease of its population. These manipulations can have a direct or indirect impact on the most carefully designed IPM systems when these have not considered the agroecosystem on a large enough scale.

1.3 ECONOMIC INJURY LEVEL

The economic injury level (EIL) is the determination of when an insect (or any other organism) becomes a "pest," so that management (ecological or otherwise) needs to be undertaken. Stern et al. (1959) pioneered the current concept of EIL and their view remains a useful, simplified way of illustrating when an insect becomes a pest. Upon introduction to a favorable environment, any population increases for a while, but eventually the combined negative impacts of dwindling food supply,

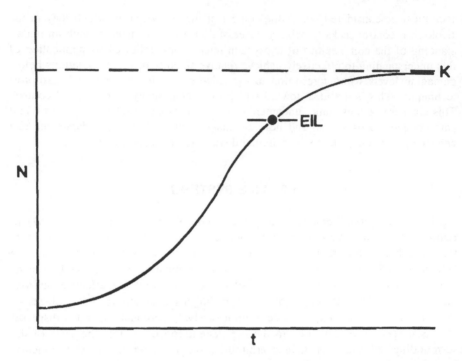

Figure 1.1   Relationship of Economic Injury Level (EIL) to general equilibrium population (K). When K exceeds EIL, an insect becomes a pest.

increased predation, parasitism, and perhaps factors intrinsic to the population (e.g., depressed reproduction due to crowding) at high densities limit further increase and the population density will no longer increase but oscillate around a "general equilibrium position" (or "carrying capacity" — Figure 1.1). If this general equilibrium position exceeds an arbitrary density (the EIL) above which the insect interferes with health, comfort, convenience, or profit, then the insect is considered a pest, and management efforts are undertaken.

Determination of EILs is increasingly sophisticated and has developed well beyond the simple model illustrated here (e.g., Higley and Wintersteen, 1992; Higley and Pedigo, 1996). In all such models it is assumed that an EIL can be measured, and this is central to the development of IPM programs. In a fundamental way, the goal of IPM is to reduce pest numbers below the EIL, and ecological insect control seeks to do this within the context of the life system without major environmental disruption. Ideally, ecological insect control seeks to adjust the ecosystem so that a new general equilibrium position is established permanently below the EIL.

One difficulty in attaining this goal is that in many instances the EIL cannot be estimated with precision. The arbitrariness of the EIL is especially evident in case of the so-called "aesthetic injury level," in which perception of damage is a factor varying from one person or group of persons to another. For example, as an entomologist I am both appreciative and tolerant of spiders in my house due to the beneficial impact of these agents of biological control. (They eat the flies that are attracted to food odors.) I do not consider the spiders to be pests, but my enthusiasm for having spiders indoors is not shared by other members of my household for whom more than one spider is cause for concern. Developing ecological control programs for such "nuisance" pests as indoor spiders can be complex and problematical; for instance, traditional biological control may not be suitable if it involves importation of more and larger spiders. A desire for high quality, blemish, and insect-free produce (such as in fresh fruit or cut flower production) may lead to extremely low EILs that are impossible to achieve through ecological management; the "general equilibrium position" for such a pest population within its complex environment may *always* exceed the EIL, at least until humans accept low levels of insect impact as inevitable and harmless.

The distinction between *injury* and *damage* is not universally appreciated. Injury is interference with optimal physiological function, whereas damage is actual or potential economic loss. To illustrate this distinction, most deciduous trees, if well watered and well fertilized, can lose up to 30% of their foliage before they are physiologically stressed, so they are not "injured" at low levels of defoliation. However, 30% defoliation is quite visible and is often seen as "damage" by landscapers and homeowners who insist on taking corrective action. (It is perhaps unfortunate that we use the term "Economic Injury Level" rather than "Economic Damage Level" to denote pest status, but the meaning of "Economic Injury Level" as it is currently used has been accepted and generally understood for many years.)

Assessing the impact of vectors of pathogens presents a special case, in that the presence or absence of the appropriate pathogen(s) may change the effective EIL. For example, in most of North America, mosquitoes are primarily a nuisance and low densities are tolerated, especially away from areas of high-density human habitation. There is thus some flexibility in the potential for ecological control. However, where malaria, yellow fever, dengue, and other mosquito-transmitted diseases are prevalent, the consequences of mosquito bite become severe; the EIL is much lower; and the range of pest management options is reduced.

The model of Stern et al. (1959) depends on a simplistic notion of population dynamics rooted in elegant but greatly simplified mathematical models of equilibrium developed early in the 20th century. These models are readily understandable, mathematically tractable, and intuitively satisfying, but in real populations there may not be a general equilibrium position for density of many, perhaps most insect species. The simplistic concept of EIL may need to be reconsidered in the light of novel approaches to theoretical population dynamics.

1.4   PEST POPULATION DYNAMICS

As noted, the interaction between a pest population and its effective environment is complex, and we may resort to simple population models to provide insight into ecological processes. Conceptually simplified population models can provide an array of outputs illustrating general principles of IPM. In simple population models, for instance, we often denote numbers with a single value "N" and (temporarily) suspend knowledge that individuals in a population vary widely in regard to an array of genetic and behavioral traits. As an illustration of how this simplification can mislead, consider that reproductive females alone contribute to population growth, so that a population consisting exclusively of fertile females is likely to increase at a much higher rate than a population dominated by nonreproductive ones. Although we use a single term "N" for convenience to denote population density, we must remember that it represents a range of individuals assumed identical only for study and preliminary analysis. For greater realism, we need to consider the following general characteristics of populations (Ehrlich et al., 1975): (1) Populations and their effective environments are changing constantly in space and time, and a description of a population at one location and time interval may not adequately represent events in the same population at another time and place. (This idea is the basis of the "metapopulation" concept discussed below.) (2) For practical reasons, it is necessary to consider management of local populations, although ideal management should give attention to the pest over its entire geographic range, so far as this is practicable. (3) Variation within a local population may equal or exceed variation among adjacent or distant populations of the same species. (4) Immigration does not always guarantee gene flow and changes in gene frequency do not necessarily follow after immigration. For instance, corn earworms migrating into the midwestern U.S. from the southern U.S. may not necessarily carry genes for insecticide resistance due to selection by heavy insecticide use at their point of origin.

A simple mathematical model to illustrate the role of equilibrium in population dynamics is the Lotka-Volterra "logistic" model of population growth (Lotka, 1920), standard fare in all basic ecology courses. This model recognizes the tendency of populations to be regulated about an equilibrium set by the effective environment. In the simplest form of the logistic model, K (the environmental carrying capacity) acts as a brake on population growth according to the following relationship (expressed as a difference equation):

$$N_{t+1} - N_t = N_t(b-d)(K-N_t)/K$$

where N = population density
    t = time interval
    b = birthrate
    d = death rate
    K = carrying capacity

This equation gives the familiar, intuitively satisfying sigmoid curve (Figures 1.1 and 1.2). In discussions of ecological models, the equation is usually presented in differential form, integrated to:

$$N = K/(1-e^{-rt})$$

where r = (b − d) and e = base of natural logarithms.

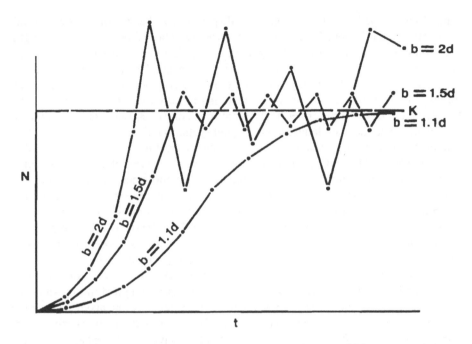

Figure 1.2   Results of simulations for logistic equation as birthrate (b) increases relative to deathrate(d). When b = 1.1d, the population increases according to a smooth sigmoid curve and levels off at K. When b = 1.5d, the population increases beyond K and a stable cycle results. When b = 2d, the population increases well beyond K, declines, and increases again with unstable cycles of great magnitude. (When b = 2.5d, the population increases exponentially so far above K that it declines to extinction in the subsequent time interval.)

Despite its simplicity, the difference equation form of the model is capable of a large array of outputs due to the built-in presence of time lags (Horn, 1988b; Figure 1.2). Most importantly for insect pest control, if b is large relative to d (characterizing a population with a high intrinsic rate of increase), there is a tendency for greater oscillations about K, along with greater instability. A population with high "r" may approach K with such speed that it exceeds K and the population then will decline. Computer simulations of this simple model result in everything from low amplitude cyclic oscillations about K, to stable limit cycles, to cycles whose periodicity cannot be distinguished from random, or overpopulation followed by crash and local extinction (Horn, 1988b). These different outcomes are simple functions of the ratio of b to d and/or the relationship of the initial N to K. Such outputs mirror observations from the real world on aphids, spider mites, and other arthropods with high fecundity and short generation time. The model predicts that insect populations with short generation time and high fecundity may fluctuate wildly and unpredictability and may never appear to be in equilibrium while exhibiting spectacular local instability. These populations also reach the EIL much more quickly than do those with lower r. This suite of adaptations (high fecundity, short generation time, low competitive ability, and high dispersal) has been termed "r-selection" (MacArthur and Wilson, 1967), and results from selection in environments favoring maximum growth, such as temporary habitats that occur early in ecological succession. Spider mites and many aphid species are examples of r-selected species, reproducing rapidly due to high fecundity and short generation time. They often quickly overexploit their environment, resulting in local extinction (as anyone who has had these pests on his or her house plants can attest). "K-selection," by contrast, is typical of habitats with longer temporal and spatial stability, favoring species with longer generation times, lower fecundity, higher competitive ability, and lower dispersal tendency. The codling moth and the corn earworm are (relatively) K-selected species and there is rarely more than one larva per apple core or ear of corn. Of course, characteristics of both r- and K-selection may occur in the same species and these may vary seasonally. During the growing season, saltmarsh planthoppers may be short winged (limiting dispersal) and display high fecundity. As the growing season ends and their food supply dwindles, they develop long-winged forms with lower fecundity and might be considered K-selected (Denno, 1994).

Adaptations of many agricultural pests are consistent with r-selection. Seasonal agricultural crops are periodically disrupted due to harvesting and tilling, and ecological succession (the orderly replacement of ecosystems by one another over time until or if a steady, sustained state is reached) may be reset to its starting point annually (or more often). This is likely to select in favor of phytophagous insects that can locate and exploit a resource quickly and efficiently. The initial colonizing species of plants and insects have adaptations consistent with r-selection; i.e., rapid dispersal and an ability to increase numbers quickly when suitable habitat is located. Many crop plants (or their ancestors) are typical of early successional stages, as are their associated insect pests. Conventional agriculture including soil tillage thus invites early-successional species that are very likely to undergo outbreaks simply due to their r-selected lifestyles. Populations of such pests may not display equilibrium at all; especially at the local level, there simply is not enough time for the population to increase to the carrying capacity. The model describes this situation with high r, i.e., birthrate greatly exceeds death rate (until harvest, when the insects all emigrate or die). The ephemeral nature of annual crops may mean that insufficient time is available for any equilibrium to be reached before harvest and subsequent crop destruction. Equilibrium might be more likely to occur in longer-lasting systems such as orchards and forests. Additionally, population fluctuations in these more complex ecosystems are partly buffered by the complex interactions within food webs, so there is less likelihood of outbreak of any particular pest species. (This is discussed further below in Section 1.5, on species diversity.)

The logistic model above describes so-called "density-dependent" population regulation, which (by definition) is the major way to regulate a population about an equilibrium. The impact of a density-dependent regulating factor is a function of the numbers within a population; at low density the impact is light or moderate, while at high density the impact is severe. Predation, parasitism and competition are examples of density-dependent factors. Density-*independent* factors, such as weather, volcanoes, and earthquakes (and chemical insecticides), may *control* a population but do not *regulate,* by definition. In most insect populations, both density-dependent and density independent factors exert impacts on the population, and the relative importance of each may vary, leading to the impression that one or the other is the dominant or exclusive influence in determining population density (Horn, 1968). The density-dependent model assumes that there *is* an equilibrium and that one among many factors is the one that regulates (Hunter, 1991). It has been argued that density-dependence may not be important in determining numbers of most populations (Strong, 1986; Stiling, 1988). Chesson (1981) argued that density-dependent regulation occurred mainly at extremes of abundance and that in most populations the influence of density-dependent regulating factors was indiscernible at medium ranges of density. This view is supported by many recent studies of natural populations (see Cappuccino and Price 1995 for examples).

A more realistic (although more complicated) characterization of actual population events assigns probability functions to birthrate, death rate, carrying capacity, and other components of the life system. The life system is thus described by functions that represent fluctuations about a mean. Models that incorporate probability functions (*stochastic* models) are less tractable mathematically and less intuitively understandable than are deterministic models, although such models supply greater realism in describing actual population events and are thus of greater utility in insect pest management. The use of computers has removed one major hurdle to application of stochastic models to pest management, although experimental verification of those models remains tedious (Pearl et al., 1989).

As the area of interest expands beyond a single crop to the landscape and regional levels, it is worthwhile to consider the behavior of populations of the same species in relation to one another by adding dispersal as a component of population regulation. The entire interactive system of local populations over its entire range can be considered a *metapopulation* (Gilpin and Hanski, 1991). The metapopulation occupies both favorable and unfavorable regions. Where the environment is favorable ("source" areas), the population is usually increasing (b > d) and the excess disperses to other regions, including "sink" areas where b < d but the population is supplemented by immigration. Movement among sources and sinks may create an impression that the resulting metapopulation is in equilibrium throughout its range, but there is no equilibrium evident in any localized area (Murdoch, 1994). Usually, the localized areas occupy the greatest interest when we deal with practical issues in pest management.

1.5 SPECIES DIVERSITY AND STABILITY

The effective environment includes all those components that impinge upon a particular species, and this may include a diverse array of other populations when one constructs food webs even for simple habitats. For example, Weires and Chiang (1973) exhaustively surveyed the invertebrate fauna associated with a single crop species (cabbage) in Minnesota and found 11 leaf feeders, 10 sap feeders, 4 root feeders, 21 feeders on decaying plant matter (saprobes), and 79 saccharophiles (feeding on sugar either from the plant or from Homoptera) for a total of 125 species of primary consumers (herbivores in the widest sense). Additionally, there were

85 carnivore species (mostly predatory and parasitic insects). Such variety within a local ecosystem is not unusual; I have found over 1000 species of insects and spiders in my small urban backyard. When considering ecosystem analysis, once again we often resort to simplified shorthand models for understanding.

Species diversity is formally measured as some combination of the numbers and proportion of each species present in an ecosystem. It is presumed that species diversity in turn reflects the number of links in a food web. The term *biodiversity* has become popular in discussions about complexity in agricultural and other human-dominated ecosystems (Stinner et al., 1997). An informal definition (Altieri and Nicholls, 1999) of biodiversity is "all species of plants, animals and microorganisms existing and interacting within an ecosystem." In agroecosystems, this includes phytophagous species (pests and non-pests), natural enemies, pollinators, and decomposers including earthworms and soil microbes. Great debate has raged among ecologists and pest managers as to whether there exists any direct relationship between species diversity and stability of individual populations within an ecosystem. In particular, it is often presumed that pest outbreaks are suppressed in more complex (and therefore more diverse) ecosystems. This so-called "diversity-stability hypothesis" holds that communities with a higher species diversity (or greater biodiversity) are more stable because outbreaks of pest species are ameliorated by the checks and balances and alternative pathways that exist within a large and integrated food web. Andow (1991) compiled an exhaustive list of studies addressing the diversity-stability hypothesis in agricultural systems, and found that in 52% of cases, herbivores were less abundant in diverse plantings (Table 1.1). Most of these studies mixed other plant species with the primary host of a specialist herbivore and this led to reduced populations of the specialist herbivore (Risch et al., 1983; Altieri, 1994). Root (1973) termed this the *Resource Concentration Hypothesis*: "herbivores are more likely to find and remain on hosts that are growing in dense or nearly pure stands; the most specialized species frequently attain higher densities in simple environments. As a result, biomass tends to become concentrated in a few species, causing a decrease in the diversity of herbivores in pure stands." The increases in herbivore populations in crop monocultures are generally due to higher rates of colonization and reproduction along with reductions in dispersal, predation, and parasitism. Other studies (e.g., Tilman et al., 1996) have shown experimentally that productivity increases and soil nutrients are more completely cycled in more diverse ecosystems, at least in grasslands. Altieri and Nicholls (1999) believed that as

Table 1.1   Population Changes of Arthropod Species in Response to Increased Plant Diversity in Polycultural versus Monocultural Agroecosystems

| | Population Response to Polyculture | | | |
	Increase	Decrease	No Change	Variable Response
Herbivores	44	149	36	58
Monophagous	17	130	31	42
Polyphagous	27	19	5	16
Predators	38	11	14	27
Parasitoids	30	1	3	6

From Andow, 1991.

biodiversity increases within an agroecosystem, more internal links develop within food webs and these links promote greater stability, resulting in fewer pest outbreaks. This presumes that most of the interconnecting trophic web is comprised of density-dependent links. The trophic structure in agricultural systems has rarely been analyzed at this daunting level of detail.

The relationship between diversity and stability is intuitively satisfying yet difficult to prove experimentally. Southwood and Way (1970) argued that stability of insect populations in agroecosystems depended on the "precision" of density-dependent responses within the food web, and that precision in turn depends on four major ecosystem parameters: (1) surrounding and within-crop vegetation; (2) permanence of the crop over space and time; (3) intensity of management, including frequency of disruptive events like tillage and chemical applications; and (4) degree of isolation of the agroecosystem from surrounding "natural" (i.e., unmanaged) vegetation. Overall, the stability of any ecosystem is a function of the sum of interactions among plant, pests, natural enemies, and pathogens. Structural diversity (Andow and Prokrym, 1990) is an important component; cropping systems with taller plants (such as corn among beans and squash) present more physical space to arthropods, and this enhances species diversity (Altieri, 1994).

Vandermeer (1995) suggested that biodiversity in agroecosystems has two components, planned and associated. *Planned* biodiversity is the portion of biodiversity consisting of cultivated crops, livestock, and associated organisms (such as agents of biological control) that are purposely included in an agroecosystem for direct economic benefit. Planned biodiversity is normally managed intensively to produce high yields. *Associated* biodiversity includes all the plants, herbivores, carnivores, and microbes that preexist in or immigrate into the agroecosystem, from surrounding habitats. Associated biodiversity persists to a greater or lesser degree within an agroecosystem depending on whether the ecological requirements of each organism are met. Vandermeer (1995) suggested that a high amount of associated biodiversity is essential to maintaining stability of arthropod populations likely to negatively impact planned biodiversity. Consideration of the relative amounts of planned versus associated biodiversity is useful in developing pest management practices that enhance overall biodiversity. This can lead to increased sustainability due to greater impact of biological control, enhanced on-site nutrient cycling, and reduced soil loss.

1.6   OPEN AND CLOSED ECOSYSTEMS

Ecosystems may be considered to be *open* (subsidized) or *closed* depending on the amount of nutrient and energy exchange with ecosystems outside themselves. Open ecosystems depend on periodic input of nutrients and energy, and there is periodic removal of a large proportion of nutrients. A cornfield in the midwestern U.S. is an example; there is a heavy importation of mineral fertilizer at planting and subsequent energy inputs associated with tilling, pesticides, and so forth. (Over 20 million tons of chemical fertilizer are used annually in the U.S.) Most of the nutrients in a cornfield are removed at harvest, either as yield or crop residue. Furthermore, the species assemblage in a cornfield is artificial and novel, with interspecific associations that are not longstanding. Corn is native to Mesoamerica, whereas many of its major insect pests (e.g., European corn borer) are of exotic origin, often European. Usually, it takes time for native natural enemies to expand their host or prey range to include exotic organisms. By contrast, in a closed ecosystem, such as a deciduous forest, most nutrient cycling occurs on-site with rather little of it resulting from importation. The nutrients and energy in the canopy fall to the ground either as leaves or as caterpillar frass, or are converted into caterpillars which are eaten by insectivorous birds, predatory insects, and parasitic wasps. Most of the flora and fauna do not leave the ecosystem. (Migratory birds do, of course, but many return.) Overall, there is rather little "leakage" of nutrients from the system. Moreover, many closed ecosystems (like the eastern deciduous forest) have existed as assemblages of the same species for millennia, resulting in many trophic links and close ecological associations among mostly native species. Such ecosystems are relatively "immune" to invasion by exotic species. (There are exceptions, as the gypsy moth has demonstrated in forests of the eastern U.S.)

Most man-made agroecosystems are artificial assemblages and open ecosystems, which make sense economically since our desire is to extract a usable product (Lowrance, Stinner, and House, 1984). Even in a landscaping ecosystem, although we may not harvest, we fertilize (providing input) and rake leaves and remove grass clippings (exporting productivity) to maintain a pleasing appearance. The species assemblage in planned landscapes often includes a preponderance of exotic species; for instance, from my front porch in Ohio I can view Norway maple, Siberian elm, Colorado blue spruce, English walnut and Chinese ginkgo trees.

"Open" and "closed" are arbitrary designations and the two types of ecosystems grade into one another, but there are distinct differences in the level of impact of pests and of management procedures (Altieri, 1987, 1994) Open, simplified agroecosystems are increasingly devoted to a single crop resulting in increased pest populations and lowered species diversity. Generally, the more modification in the direction of ecosystem simplification and subsidy, the more abundant are insect pests. These reductions in biodiversity can impact the normal functioning of surrounding ecosystems with further negative consequences for pest management (Flint and Roberts, 1988).

1.7 MONOCULTURE VERSUS POLYCULTURE

Monoculture, the planting of a single species of crop plant, often results in increased populations of specialist herbivores, as noted above (Altieri and Letourneau, 1982). On the other hand, polyculture may reduce impact of herbivorous pests through "associational resistance" (Tahvanainen and Root, 1972), in which the presence of a variety of plants disrupts orientation of specialist herbivores to their hosts. Cabbage flea beetles and cabbage aphids that locate their hosts via specific chemical cues (such as the alkaloid sinigrin) are less effective in locating these hosts against a variety of other plant species, resulting in lower populations. Local movement of cucumber beetles and coccinellid predators is enhanced when cucumbers are interplanted with corn and beans when compared with these insects' movement

Table 1.2   Examples of Agroecosystems Wherein Weedy Vegetation Has Resulted
in Increased Density and Activity of Natural Enemies

Crop	Pest(s)	Weed(s)	Natural enemies
alfalfa	alfalfa caterpillar	many	parasitic wasps
apple	caterpillars	many	parasitic wasps
	spider mites		predatory mites
citrus	spider mites	many	predatory mites
cole crops	aphids, diamondback moth, cabbageworm	pigweed, lamb's quarters, shepherd's purse	lady beetles, lacewings, parasitic wasps
corn (maize)	European corn borer	giant ragweed	tachinid fly
cotton	boll weevil	ragweed	parasitic wasps
	cotton bollworm	curly dock	stinkbugs
grape	grape leafhopper	blackberry	parasitic wasps
	spider mites	johnsongrass	predatory mites
peach	Oriental fruit moth	ragweed	parasitic wasp
sorghum	greenbug	sunflower	parasitic wasps
soybeans	Mexican bean beetle	grasses	predatory Hemiptera
sugarcane	sugarcane weevil	spurges	tachinid fly
sweet potato	tortoise beetle	morning glory	parasitic wasp

From Altieri and Letourneau, 1982; Andow, 1991; Bendixen et al., 1981.

in monocultures, where they tend to remain on individual plants (Bach, 1980; Wetzler and Risch, 1984). In my own research (Horn, 1981, 1988a), I found reduced populations of specialist herbivores (cabbage aphids, diamondback moth, and imported cabbageworm) on collards planted in weedy backgrounds versus numbers of these same herbivores on collards planted against bare soil or plastic mulch. This influence of weeds intensified once the weeds became as tall as the collards, effectively allowing the collards to "hide" amid the weeds. Table 1.2 lists other examples wherein presence of weeds enhanced biological control of crop pests by increased predation and parasitism. The same phenomenon can be seen over time in crop rotation; the frequent replacement of one crop by another keeps specialist herbivore populations below economic injury levels. Field crop producers in the midwestern U.S. can prevent the increase of corn rootworm populations by rotating from corn to soybeans every 2 to 3 years. (The western corn rootworm has recently developed resistance to annual corn-soybean rotation in parts of the Midwest.)

1.8   SCALE AND ECOLOGICAL MANAGEMENT

As mentioned earlier, an important consideration in assessing a pest problem and developing ecological management is the scale of the area involved. The perception of insect problems, economic injury levels, and approach to management can vary greatly depending on scale. One may view an ecosystem at the level of an individual plant, a research plot, a field, a whole farm, and/or the regional "agro-pastoral" (Altieri, 1994) landscape, the watershed, and so forth. At the level of the metapopulation, the global dynamics of a life system are very different from local dynamics, and although a pest population may appear to show a measurable equilibrium throughout a regional landscape, there is no equilibrium at the local level of interest to pest management. Thus, results from small plot research may not be applicable to a higher scale (Kemp et al., 1990). Local movement of pests may be less important at lower scales (i.e., the individual plant) but very influential in population dynamics at a regional landscape level, especially if this includes surrounding habitat. For instance, the Mexican bean beetle overwinters in hedgerows and along field edges, so that soybean fields nearest overwintering sites are likely to become infested earlier and bean beetle populations subsequently will be higher. Soybeans located near bush and pole beans (which are more suitable hosts for the bean beetle) are also likely to develop economically important infestations earlier (Stinner et al., 1983). Natural enemies often move from unmanaged edge habitat and nearby forests into adjacent farm fields; the nature of this movement may be very important to local suppression of pests.

Unfortunately, more intensive agriculture often leads to reduction in surrounding unmanaged communities with their rich store of associated biodiversity, including natural enemies. Many studies have shown that there are increased numbers and activity of natural enemies near field borders when there is sufficient natural habitat to provide cover and alternate prey and hosts, as well as food in the form of nectar and pollen. This function of wild border areas significantly enhances biological control (van Emden, 1965; Marino and Landis, 1996).

1.9   EXAMPLES OF PRACTICAL APPROACHES

Altieri (1994) and others (Andow, 1983, 1991; Collins and Qualset, 1999; Horn, 1988; Marino and Landis, 1996; Risch et al., 1983; and Vandermeer, 1981) have suggested that pest outbreaks can be mitigated by designing agroecosystems with attention to the following characteristics: (1) high diversity within the cropping system; (2) crop rotations and short-term, fast-maturing cultivars; (3) smaller field size and intervening areas of uncultivated, relatively closed ecosystems; (4) perennial crops such as orchard trees; (5) reduced tillage and tolerable weed backgrounds; and (6) high genetic diversity within the crop. Many outstanding examples of ecological control exist (Table 1.3). Flint and Roberts (1988), Altieri (1994), and Collins and Qualset (1999) cite many more. The examples given here serve to illustrate the range of possibilities for ecological (or biointensive) control in agroecosystems.

1.9.1   Multicropping

Growing several crops in the same space has been shown to reduce pest problems relative to monocultures of the same species. Andow (1991) (Table 1.1) cited numerous examples, and Altieri (1994) pointed out that over 90% of tropical legumes are produced in intercropped systems. Helenius (1989), Horn (1988a), and others have found increases in natural enemies in polycultures. A classical case of the adaptation of polyculture to pest management is that of planting blackberries among grapevines in central California to control grape leafhopper (Doutt and Nakata, 1973). The parasitoid *Anagrus epos* attacks the eggs of both the grape leafhopper and the

Table 1.3   Examples of Agroecosystems Wherein Increased Biodiversity Through Intercropping and Multicropping Reduced Outbreaks of Insect Pests

Primary Crop	Intercrop(s)	Pest(s) Controlled
beans	winter wheat	potato leafhopper, bean aphid
cassava	cowpeas	whiteflies
cole crops	beans	cabbage flea beetle, cabbage aphid
	clover	imported cabbageworm
corn (maize)	beans	fall armyworm, leafhoppers
	clover, soybeans	European corn borer
	squash	aphids, spider mites
cotton	alfalfa	*Lygus* bugs
	corn (maize)	cotton bollworm
	cowpeas	boll weevil
	sorghum	cotton bollworm
cucumbers and squash	cole crops and corn	cucumber beetles
melons	wheat	aphids, whiteflies
oats	beans	aphids
peaches	strawberries	Oriental fruit moth
peanuts	beans	aphids
tomato	cole crops	flea beetles

From Altieri and Letourneau, 1982; Andow, 1991.

leafhopper *Dikrella cruentata* on blackberry. By encouraging blackberries between alternate grape arbors, a constant supply of eggs of both leafhoppers is available to the parasitoid, which then persists in populations high enough to bring the grape leafhopper under biological control.

Relay cropping can be considered intercropping over time rather than space. Two (or more) different crops are grown on the same area in successive seasons; for instance, soybeans following winter wheat. The seasonal change from one crop to another, especially if they are distantly related (legume and grass) prevents the increase of specialist pests. Soybean pests are less abundant when soybeans are relay cropped with wheat.

1.9.2   Strip Harvesting

Francis (1990) noted that planting and harvesting corn and beans in alternate strips rather than solid monocultures reduced pest insects on both crops. In the Imperial Valley of California, alfalfa is grown throughout the year, and it is possible to harvest on a 3 to 4 week rotation when half the field is cut in strips. Natural enemies of the alfalfa weevil, alfalfa caterpillar, and aphids are conserved in the regrowth, so that there are alternative food sources and hiding places for these predators and parasitoids year round, in consequence of which they are always present and suppress pest populations to below damaging levels (Stern, 1981). This relationship can be applied to control of *Lygus* bugs in cotton by planting alfalfa adjacent to cotton, allowing the increase of natural enemies of *Lygus* in the alfalfa. These natural enemies move into the cotton and control *Lygus* bugs there (Stern, 1981).

1.9.3 Interplanting

Interplanting follows the same principles as multicropping, except that the inter-planted species are not a crop plant. Many studies have shown that weedy vegetation in or near crop fields support a diverse fauna, including natural enemies of pests on the crop plants (Table 1.3). Altieri and Whitcomb (1980) clearly demonstrated this beneficial aspect of weeds. In my research (Horn, 1981, 1988a), I found that this depended somewhat on the specific mix of weeds present; for instance, if the weeds were particularly attractive to aphids and their natural enemies, this would enhance aphid control on a commercial crop. Weeds such as pigweed (*Amaranthus*), lambs quarters (*Chenopodium*) and shepherd's purse (*Capsella*), when heavily infested with aphids, served as a nursery for production of aphid predators and parasitoids which then moved onto neighboring collard plants when surrounded with these weed species.

Several studies have shown that floral undergrowth in orchards provides resources to adult parasitic wasps and flies and therefore increases parasitism of phytophagous insects (particularly Lepidoptera) on the trees (Altieri and Schmidt, 1985, 1986; Leius, 1967). Andow and Risch (1985) noted that the presence of floral resources and alternate prey was particularly favorable to populations of generalist predators such as the lady beetle *Coleomegilla maculata.* (Altieri, 1994; Altieri and Nicholls, 1999) cite many additional examples in which biological control is enhanced by interplanting of non-crop plants.

1.10 CONCLUSIONS

All agroecosystems are complex, and the relative impact of alternate crops, weeds, and natural enemy competitors, and associated organisms on the life systems of pest species may be highly variable and difficult to predict. It is within this complexity that we need to develop ecological approaches to insect control and in the absence of predictability we often are forced to proceed on an ad hoc basis. As we proceed to devlop specific pest management options, ecological control needs to be considered in relation to insecticides, host-plant resistance, conventional control, and biological control. Ecological control is really nothing more (or less) than intelligent environmental management with due regard for the place of insect pest populations within a complex and interconnected ecosystem. Appreciation of this fact alone can lead to more sustainable pest management with reduced input costs.

REFERENCES

Altieri, M. A., *Agroecology: the Scientific Basis of Alternative Agriculture.* Westview Press, Boulder, CO, 433 pp., 1987.

Altieri, M. A., *Biodiversity and Pest Management in Agroecosystems.* Food Products Press, New York, 185 pp., 1994.

Altieri, M. A. and D. L. Letourneau, Vegetation management and biological control in agroecosystems. *Crop Prot.* 1, pp. 405-430, 1982.

Altieri, M. A. and C. I. Nicholls, Biodiversity, ecosystem function, and insect pest management in agricultural systems. In *Biodiversity in Agroecosystems*. CRC Press, Boca Raton, pp. 69-84, 1999.

Altieri, M. A. and L. L. Schmidt, Cover crop manipulation in northern California orchards and vineyards: effects on arthropod communities. *Biol. Agr. Hort.* 3, pp. 1-24, 1985.

Altieri, M. A. and L. L. Schmidt, The dynamics of colonizing arthropod communities at the interface of abandoned organic and commercial apple orchards and adjacent woodland habitats. *Agr. Ecosys. Envir.* 16, pp. 29-43, 1986.

Altieri, M. A. and W. H. Whitcomb, Weed manipulation for insect pest management in corn. *Environ. Management* 4, pp. 483-489, 1980.

Andow, D. A., The extent of monoculture and its effects on insect pest populations with particular reference to wheat and cotton. *Agr. Ecosys. Envir.* 9, pp. 25-35, 1983.

Andow, D. A., Vegetational diversity and arthropod population response. *Annu. Rev. Entomol.* 36, pp. 561-586, 1991.

Andow, D. A. and D. R. Prokrym, Plant structural complexity and host finding by a parasitoid. *Oecologia* 62, pp. 162-165, 1990.

Andow, D. A. and S. J. Risch, Predation in diversified ecosystems: relations between a coccinellid predator *Coleomegilla maculata* and its food. *J. Appl. Ecol.* 22, pp. 357-372, 1985.

Bach, C. A., Effects of plant diversity and time of colonization on an herbivore-plant interaction. *Oecologia* 44, pp. 319-326, 1980.

Bendixen, L. E., K. U. Kim, C. Kozak, and D. J. Horn, An annotated bibliography of weeds as reservoirs for organisms affecting crops. II. Arthropods. Ohio Agr. Res. Devel. Center Research Bull. #1125, 117 pp., 1981.

Cappuccino, N. and P. W. Price, eds., *Population Dynamics: New Approaches and Synthesis.* Academic Press, New York, 429 pp., 1995

Chesson, P. L., Models for spatially distributed populations: the effect of within-patch variability. *Theor. Pop. Biol.* 19, pp. 288-325, 1981.

Clark, L. R., R. D. Hughes, P. W. Geier, and R. F. Morris, *The ecology of insect populations in theory and practice*, Methuen, London, 282 pp., 1967.

Collins, W. W. and C. O. Qualset, eds., *Biodiversity in Agroecosystems*, CRC Press, Boca Raton, 334 pp., 1999.

Denno, R. F., Life history variation in planthoppers. pp. 163-215 In *Planthoppers, their Ecology and Management*, Chapman & Hall, New York, 1994.

Doutt, R. L. and J. Nakata, The *Rubus* leafhopper and its egg parasitoid: an endemic biotic system useful in grape-pest management. *Envir. Entomol.* 2, pp. 381-386, 1973.

Duelli, P., Biodiversity evaluation in agricultural landscapes: an approach at two different scales. *Agr. Ecosys. Envir.* 62, pp. 81-91, 1997.

Ehrlich, P. R., R. R. White, M. C. Singer, S. W. McKechnie, and L. I. Gilbert, Checkerspot butterflies: a historical perspective. *Science* 188, pp. 221-228, 1975.

Flint, M. L. and P. A. Roberts, Using crop diversity to manage pest problems: some California examples. *Amer. J. Altern. Agr.* 3, pp. 164-167, 1988.

Francis, C. A., *Sustainable agriculture in temperate zones*. John Wiley & Sons, New York, 487 pp., 1990.

Frisbie, K. E. and J. W. Smith, Jr., Biologically intensive IPM: the future. pp. 151-164 In *Progress and Perspectives for the 21st Century.* J. J. Menn and A. E. Steinhauer, eds. Entomological Society of America Centennial Symposium. Entomological Society of America, Lanham, MD, 1991.

Gilpin, M. and I. Hanski, *Metapopulation dynamics: empirical and theoretical investigations.* Academic Press, San Diego, CA, 336 pp., 1991.

Helenius, J., The influence of mixed intercropping of oats with field beans on the abundance and spatial distribution of cereal aphids (Homoptera: Aphididae), *Agr. Ecosys. Envir.* 25: 53-73, 1989.

Higley, L. G. and L. P. Pedigo, eds., *Economic Thresholds for Integrated Pest Management.* Univ. Nebraska Press, Lincoln, NE, 327 pp., 1996.

Higley, L. G. and W. K. Wintersteen, A novel approach to environmental risk assessment of pesticides as a basis for incorporating environmental costs into economic injury levels. *Amer. Entomol.* 38, pp. 34-39, 1992.

Horn, D. J., Effect of weedy backgrounds on colonization of collards by green peach aphid, *Myzus persicae*, and its major predators. *Envir. Entomol.* 10, pp. 285-289, 1981.

Horn, D. J., Parasitism of cabbage aphid (*Brevicoryne brassicae*) and green peach aphid (*Myzus persicae*) (Homoptera: Aphidae) on collards in relation to weed management. *Envir. Entomol.* 17, pp. 354-358, 1988a.

Horn, D. J., *Ecological Approach to Pest Management.* Guilford Press, New York, 285 pp., 1988b.

Horn, H. S., Regulation of animal numbers: a model counter-example. *Ecology* 49, pp. 776-778, 1968.

Hunter, A. F., Traits that distinguish outbreaking and non-outbreaking Macrolepidoptera feeding on northern hardwood trees. *Oikos* 60, pp. 275-282, 1991.

Kemp, W. P., S. J. Harvey, and K. M. O'Neill, Patterns of vegetation and grasshopper community composition. *Oecologia* 83, pp. 299-308, 1990.

Kogan, M. ed., *Ecological Theory and Integrated Pest Management Practice.* John Wiley & Sons, New York, 362 pp., 1986.

Kogan, M., IPM: Historical perspectives and contemporary development. *Annu. Rev. Entomol.* 43, pp. 243-270, 1995.

Leius, K., Influence of wild flowers on parasitism of tent caterpillar and codling moth. *Canad. Entomol.* 99, pp. 444-446, 1967.

Lotka, A. J., Analytical notes on certain rhythmic relations in organic systems. *Proc. Nat. Acad. Sci.* 7, pp. 410-415, 1920.

Lowrance, R., B. R. Stinner, and G. J. House, eds., *Agricultural Ecosystems: Unifying Concepts.* John Wiley & Sons, New York, 233 pp., 1984.

MacArthur, R. H. and E. O. Wilson, *The Theory of Island Biogeography.* Princeton Univ. Press, Princeton, NJ, 203 pp., 1967.

Marino, P. C. and D. A. Landis, Effects of landscape structure on parasitoid diversity in agroecosystems. *Ecol. Appl.* 6, pp. 276-284, 1996.

Murdoch, W. W., Population regulation in theory and practice, *Ecology* 75, pp. 271-287, 1994.

National Research Council, *Ecologically based pest management: New solutions for a new century.* Washington, D.C., 160 pp., 1996.

Pearl, D. K., R. Bartoszynski, and D. J. Horn, A stochastic model for simulation of interactions between phytophagous spider mites and their phytoseiid predators. *Exp. Appl. Acarol.* 7, pp. 143-151,1989.

Rabb, R. L., A sharp focus on insect populations and pest management from a wide area view. *Bull. Entomol. Soc. Amer.* 24, pp. 55-60, 1978.

Risch, S. J., D. Andow, and M. A. Altieri, Agroecosystem diversity and pest control: data, tentative conclusions and new research directions. *Environ. Entomol.* 12, pp. 625-629, 1983.

Root, R. B., Organization of a plant-arthropod association in simple and diverse habitats: the fauna of collards (*Brassica oleraceae*). *Ecol. Monogr.* 43, pp. 95-124, 1973.

Southwood, T. R. E. and M. J. Way, Ecological background to pest management. In *Concepts of Pest Management*, R. L. Rabb and F. E. Guthrie, eds., North Carolina State Univ. Press, Raleigh, NC, 242 pp., 1970.

Stern, V. M., Environmental control of insects using trap crops, sanitation, prevention and harvesting. pp. 199-207 In *CRC Handbook of Pest Management in Agriculture,* D. Pimentel, ed., CRC Press, Boca Raton, 1981.

Stern, V. M., R. F. Smith, R. van den Bosch, and K. S. Hagen, The integration of chemical and biological control of the spotted alfalfa aphid: the integrated control concept. *Hilgardia* 29, pp. 81-101, 1959.

Stiling, P., Density-dependent processes and key factors in insect populations. *J. Anim. Ecol.* 57, pp. 581-594, 1988.

Stinner, D. H., B. R. Stinner, and E. Martsolf, Biodiversity as an organizing principle in agroecosystem management: case studies of holistic resource management practitioners in the USA. *Agr. Ecosys. Envir.* 62, pp. 199-213, 1997.

Stinner, R. E., C. S. Barfield, J. L. Stimac, and L. Dohse. Dispersal and movement of insect pests. *Ann. Rev. Entomol.* 28, pp. 319-335, 1983.

Strong, D. R., Density-vague population change. *Trends in Ecology and Evolution* 1, pp. 39-42, 1986.

Tahvanainen, J. O. and R. B. Root. The influence of vegetational diversity on the population ecology of a specialized herbivore, *Phyllotreta cruciferae* (Coleoptera: Chrysomelidae). *Oecologia* 10, pp. 321-346, 1972.

Tilman, D., D. Wedin, and J. Knops, Productivity and sustainability influenced by biodiversity in grassland ecosystems. *Nature* 379, pp. 718-720, 1996.

Vandermeer, J., The interference production principle: an ecological theory for agriculture. *BioScience* 31, pp. 361-364, 1981.

Vandermeer, J., The ecological basis of alternative agriculture. *Annu. Rev. Ecol. Syst.* 26, pp. 201-224, 1995.

van Emden, H. F., The role of uncultivated land in the biology of crop pests and beneficial insects. *Sci. Hort.* 17, pp. 121-126, 1965.

Weires, R. W. and H. C. Chiang, *Integrated Control Prospects of Major Cabbage Insect Pests in Minnesota: Based on the Faunistic, Host Varietal and Trophic Relationships.* Univ. Minn. Agr. Exp. Sta. Tech Bull. 291, 1973.

Wetzler, R. E. and S. J. Risch, Experimental studies of beetle diffusion in simple and complex crop habitats. *J. Anim. Ecol.* 53, pp. 1-19, 1984.

Physical Control

CHAPTER **2**

Physical Control of Insects

Christian Y. Oseto

CONTENTS

1-56670-478-2/00/$0.00+$.50
© 2000 by CRC Press LLC

2.1   INTRODUCTION

Physical control of insects started when humans first picked insects off their bodies or crushed insects with available materials. Early physical and mechanical techniques emphasized control of agronomic and horticultural insect pests. Some of the techniques developed for commodity pests have been adapted for urban and stored-product pests. Modern physical and mechanical techniques involve direct or indirect human participation, and the degree of sophistication ranges from simple handpicking to the elaborate use of machines. In some cases, the simplest technique may be the most elegant and effective. Physical and mechanical measures may exclude insects or may reduce or eliminate existing pest populations, and many of these measures may have been in use since antiquity without encountering resistance problems commonly associated with insecticide use. Development of effective physical and mechanical control methods must be based on a detailed understanding of the pest's biology, behavior, and physiological requirements. Adoption of physical and mechanical controls depends on the level of effectiveness, convenience and ease of use, and economic considerations. Many of the physical and mechanical techniques have been refined over the years to increase effectiveness.

2.2 NON-RADIANT TRAPS

Traps, in general, serve to determine insect movement and establishment into new areas; to estimate temporal and spatial distribution of insects; and to evaluate need for control and effectiveness of control measures. In the past, traps provided the sole method of controlling pests. Early trapping recommendations indicate the nonsensical nature of the trapping techniques and the obvious need, in some cases, to understand fully the biology of the insect pest. An early popular treatment to control insects attacking cultivated plum trees involved orchardists building a 2.7-meter fence around trees with the hope that the fence would prevent access and oviposition by the plum curculio, *Conotrachelus nenuphar* (Herbst). Another recommendation suggested hanging dead mice from the trees so that weevils would oviposit on the decaying animal flesh and not on the fruit. Today, these remedies seem amusing because proponents of these measures failed to understand the biology and ecology of the pests (Waite et al., 1926).

In the mid-1800s, a simple control technique involved placing boards or other materials around a field or near plants to control the plum curculio. Growers cleared debris surrounding each tree and placed bark chips, stones, or other similar materials around each cleared tree. Growers then collected insects from beneath the trap materials and destroyed the insects (Chapman, 1938). A simple trap used in many places across the U.S. but no longer used in the numbers as they once were is the strip of sticky fly paper. These fly trap strips were placed on the outside of screen doors at the top to catch flies gathered at the door (Washburn, 1910). As flies land, they orient to narrow, vertical objects and adhere to the sticky material (National Academy of Sciences, 1969).

Traps to monitor and survey insect populations have gained popularity over the past years because of the development of effective food and visual attractants. A few studies have clearly demonstrated trap effectiveness in reducing pest insects below economic levels (Hardee et al., 1971; Lindgren and Fraser, 1994). Besides reducing pest populations, trap data can provide useful information on the spatial and temporal patterns of pest insects (Wagner et al., 1995) critical in making pest management decisions. For any trap to be effective, a systematic observation of the pest's behavior can provide important information why some traps of basically similar design catch more insects than other traps (Phillips and Wyatt, 1992). Traps have assumed a variety styles including flat traps, bucket traps, wing traps, delta or triangular traps, cylindrical traps, cone traps, and bag traps (Alm et al., 1994; Ali-Niazee et al., 1987; Dowd et al., 1992; Goodenough, 1979; Riedl et al., 1989; Finch,

1990; Anonymous, 1991; Reynolds et al., 1996; Barak, 1989; Goodenough and Snow, 1973; Byers, 1993; Uchida et al., 1996).

In recent years, much research has been reported on the use of various baits, especially pheromones and trap designs to maximize insect attraction to traps. The amount of literature dealing with attractants and traps is voluminous (Hartsack et al., 1979; Burkholder, 1985; Whitcomb and Marengo, 1986; Barak et al., 1990; Faustini et al., 1990; Mueller et al., 1990; Gauthier et al., 1991; Foster and Hancock, 1994; Heath et al., 1995; Hardee et al., 1996; James et al., 1996; Mason, 1997; Phillips, 1997; Pickett et al., 1997; Dowdy and Mullen, 1998) and an exhaustive treatment is beyond the scope of this chapter.

Mathematical models support the use of baits or lures in traps to enhance trap effectiveness. Baited electrical grid traps captured more tobacco budworms than did unbaited and baited light traps and sticky traps (Goodenough and Snow, 1973). The use of oil traps with pheromones successfully reduced populations of the pink bollworm, *Pectinophora gossypiella* (Saunders), in Sao Palo, Brazil. Oil traps employing a high dose of pheromone suppressed pink bollworm populations. A trap density of 20 traps per hectare was placed in the field at the first presence of bolls. The long lasting viscosity of the oil and the long life of the pheromone made oil traps an effective pink bollworm control technique (Mafra-Neto and Habib, 1996).

Mass trapping with pheromones has not been feasible in the U.S. because of the high cost and the labor-intensive activities associated with installing and maintaining traps. Compounding the non-use of pheromone traps for control, the initial pheromone trials proved to be ineffective in reducing pest numbers. The majority of pheromone traps function to monitor population levels as part of an integrated pest management system.

2.2.1 Bands

Several recommendations to control cankerworms appeared in the early popular press. A band of chestnut burrs tied around the tree excluded cankerworm larvae. Another technique involved scraping the bark and placing bands of hair rope around trees. Lead gutters filled with lamp oil were used to prevent cankerworm larvae and wingless females from moving over the trap into the trees (Howard, 1900).

In 1840, Joseph Burrelle advocated wrapping materials around the trunk of a tree or placing cloth in the crotch of a tree to collect codling moth, *Cydia pomonella* (L.), larvae. The materials, containing the trapped larvae, were placed in a hot oven and killed. A further refinement was made to this technique by scrapping bark off the trunk and clearing weeds beneath the trees to force larvae into the bands. Scraping in combination with banding effectively reduced populations of codling moth when compared with just banding or scraping. Various materials have been used as banding materials such as hay rope, wrapping paper, building paper, flannel cloth, canvas, and burlap. Regardless of the materials used, the traps had to be checked routinely and trapped larvae killed for the technique to be effective (Baker and Hienton, 1952). Overwintering larvae provided the most accurate estimate of banding effects. During the three years of the study, 35.2% of larvae in the untreated trees completed development while 20.7% completed development on scraped trees, and 13.9% of larvae developed into adults on scraped and banded trees (Baker, 1944; Baker and Hienton, 1952). The age of banded trees appeared to influence banding efficacy. In orchard trees of 13 to 50 years of age, 5.3% of the population was trapped. In trees less than 12 years old, the average trapped was 22.4%. The researcher did not state if 5.3 or 22.4% larval mortality was sufficient to control the codling moth (Barrett, 1935).

Benjamin Walsh, in his reply to a recommendation based on weak scientific evidence, decried the use of banding to control all tree-injuring pests. "The worm in fruit trees! As if fruit trees were not afflicted by hundreds of different worms, differing from each other in size, shape, color, and habits of life, time of coming to maturity, etc. as much as a horse differs from a hog. Yet the universal bandage system is warranted to kill them all. Does the apple worm bore your apples? Bandage the butt of the tree, and he perisheth forthwith. Does the web worm spin his web in the branches? Bandage the butt, and he dieth immediately. Does the caterpillar known as the red-humped prominent or the yellow-necked worm strip the leaves off? Bandage the butt of the tree, and hey! presto! he quitteth his evil ways. Does the Buprestis borer bore into the upper part of the trunk? Still you must bandage the butt with the same universal calico, and in a twinkling he vamoseth the ranch...Long live King Humbug! He still feeds on flapdoodle, and many of them have large and flourishing families, who will perpetuate the breed to the remotest generation." (Howard, 1900)

Sticky barrier bands and burlap bands provided a way to control gypsy moths (Raupp et al., 1992). In 1895, an infestation of several species of tree infesting insects appeared in many eastern cities. A broad, thick strip of raw cotton tied around the trees with a string was, at that time, the most economical and effective means of control (Howard, 1896). Through the Works Progress Program (WPA) in 1936, workers, as part of the program to control the gypsy moth in Connecticut, scouted for the insect. The WPA workers applied 80,942 bands to trees throughout the state and the bands killed 199,982 larvae (Britton et al., 1937).

Prior to the use of arsenicals to control cankerworms in trees, barriers of cotton, wool, or printer's ink placed around the trunk of the tree prevented wingless females from crawling into the tree canopy. These bands remained in place through late fall, the winter, and into spring until oviposition ceased. Tanglefoot, an adhesive, replaced the bands of cloth material or printer's ink. Most growers preferred to apply insecticides rather than banding trees because of the efficacy of the arsenicals (Pettit and Hutson, 1931). Sticky barriers around the bole of the seed orchard trees reduced injury by a weevil, *Lepesoma lecontei* from 25% in the controls to 6%. A metal baffle placed around the bole failed to prevent damage by the weevil, and a sticky barrier had the advantage of being inexpensive and needed only to be applied to those trees producing a crop in any given year (Sexton and Schowalter, 1991).

2.2.2 Livestock Insect Traps

The development of fly resistance to insecticide impregnated ear tags lead to a reevaluation of the walk-through fly traps developed nearly a century ago but unsuccessfully adopted (Haseman, 1927). Walk-through fly traps are passive control devices for capturing horn flies, face flies, and stable flies. The trapping elements, placed along the sides of the trap, function as inverted cones with wire window screening folded into a "Z" pattern. Small holes located along the apex of each fold allow flies to pass through to the outside of the trap. An exterior screen prevents flies from returning into the trap and back to the cattle. In one year of a study, the majority of trapped flies were horn flies accounting for 62 to 79% of the total catch, stable flies 13 to 27%, and face flies 2 to 13%. Given the number of stable flies caught in the study, the walk-through fly traps hold promise for control of stable flies in confined operations (Hall and Doisy, 1989). A prototype fly trap was modified to control horn flies on dry cattle and milkers in western Florida and Alabama. For dry cattle, the traps reduced 96.9% of horn flies and 90.2% of horn flies on milkers. Trapping reduced the need for insecticide treatments and offered a sustainable method of horn fly control (Tozer and Sutherst, 1996).

A modified Hodge-type trap with a single 40-W blacklight fluorescent bulb and a reflector economically reduced house fly populations in a caged-layer poultry facility. This ingenious fly trap was attached to the top of a garbage can, and the flies entered the trap and moved into the top of the trap (Figure 2.1) as a response to light (Washburn, 1910). During a 30-day test period, three traps placed in a poultry house captured over 1.1 million flies. The researchers failed to evaluate trap efficacy in controlling house fly populations, but the low cost of each trap might make trap use a feasible means of control. A 50,000 poultry operation with an associated

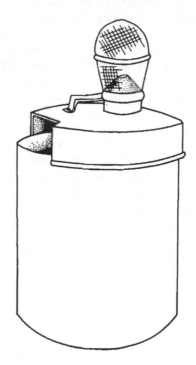

Figure 2.1   Hodge's fly trap, showing cut-away view of lid. (Redrawn from Washburn, 1910.)

one million house fly population would require 20 traps placed at 14-m intervals along the center aisle and five traps placed at 28-cm intervals along each side wall to capture enough flies to cause a steady or declining population level (Pickens et al., 1994).

2.2.3 Color and Traps

Color, as the only attractant, has been tested and used to attract insects (Table 2.1). How an insect responds to color depends on the trap position, ground composition, physiological state of the insect, and quality of the incident wavelengths hitting the traps (Prokopy and Owens, 1983). Numerous studies have tested color in combination with different trap types such as yellow water traps, (Heathcote, 1957; Capinera and Walmsley, 1978; Finch, 1990) and yellow sticky traps (Broadbent et al., 1948; Alderz, 1976; Samways, 1986; Zoebisch and Schuster, 1990; Sanderson and Roush, 1992), along with baits or pheromones.

The selection of different colors used in trap studies mirrors the host plant's spectral reflectance or wavelength. Typically, these colors are white, blue, green, and yellow. White, blue, or yellow traps caught higher numbers of the cabbage maggot, *Delia radicum* (L.); the seed corn maggot, *D. platura* (Meigen); the turnip maggot, *D. floralis* (Zetterstedt); and a radish maggot, *D. planipalpis* (Stein) than did green or uv-reflecting white traps (Vernon and Broatch, 1996). Painting different parts of fluorescent-yellow water traps black increased trap efficacy in capturing *D. radicum* (Finch, 1991). Color response by *Delia* spp. maggot complex varied, depending on the crop development stage and background color. Response differences were noted within and between sexes for the same color. In addition to these factors, the stage of plant development was considered when selecting or testing different trap colors (Vernon and Broatch, 1996). Unfortunately, visual attractants may lure pest insects along with beneficial insects, especially those traps with a sticky or an insecticidal material (Neuenschwander, 1991).

Green- and yellow-colored sticky traps in the laboratory and solutions used in McPhail traps in the field were the most attractive to male and female Mexican fruit flies, *Anastrepha ludens* (Loew). During the course of the study, attractiveness of red, orange, and yellow doubled from spring to autumn in the field. Trap placement around the tree influenced the number of flies caught with more flies recorded from traps placed on the north side of the trees (Robacker et al., 1990).

Colored spheres attracted several genera of tephritid fruit flies (Nakagawa et al., 1978; Cytrynowicz, 1982; Prokopy, 1975; Sivinski, 1990). Red spheres coated with a sticky substance and hung in apple trees in an orchard were effective at capturing female apple maggot flies, *Rhagoletis pomonella* (Walsh) and thus protected fruit from fly damage. No pheromones or other baits were used in the trap (Prokopy, 1975).

The height and position of traps may influence attractiveness to insects (Deay and Taylor, 1954). In studies with the apple blotch leafminer, *Phyllonorycter crataegella* (Clemens), horizontal red triangles collected more adults than any other color or orientation (Green and Prokopy, 1986). Color traps, such as yellow sticky traps, have been used to monitor species composition and population levels of

Table 2.1   Positive Response of Insects to Various Colored Traps Without The Use
of Baits or Pheromones

Insect	Trap Color(s)	Trap Type(s)	Reference(s)
aphids (*Aphis spiraecola, Anuraphis middletonii,* and *Myzus persicae*)	yellow	sticky (cylindrical)	Alderz (1976)
boll weevil (*Anthonomus grandis grandis*)	blue, green		Cross et al. (1976)
flower thrips (*Frankliniella tritici*)	white	water, sticky (cylindrical)	Lewis (1959) Southwood et al. (1961)
apple maggot (*Rhagoletis pomonella*)	red, yellow		Prokopy (1968, 1975) Reissig (1975)
palestriped flea beetle (*Systema blanda*)	yellow	water	Capinera and Walmsley (1978)
aster leafhopper (*Macrosteles fascifrons*)	orange	water, sticky	Capinera and Walmsley (1978)
leafhoppers (*Aceratagallia uhleri* and *Balclutha negelecta*)	orange	water, sticky	Capinera and Walmsley (1978)
sugarbeet root maggot (*Tetanops myopaeformis*)	yellow		Harper and Story (1962)
cabbage maggot (*Delia radicum*)	white, yellow, blue	sticky	Vernon and Broatch (1996)
turnip maggot (*Delia floralis*)	white, blue, yellow	sticky	Vernon and Broatch (1996)
radish maggot (*Delia planipalis*)	white, blue,	sticky	Vernon and Broatch (1996)
seed corn maggot (*Delia platura*)	white, blue, uv white	sticky	Vernon and Broatch (1996)
onion fly (*Delia antigua*)	white, blue		Judd, Borden, and Wynne (1988)
Mexican fruit fly (*Anastrepha ludens*)	green, yellow	sticky	Robacker, Moreno, and Wolfenbarger (1990)
Caribbean fruit fly, females only (*Delia suspensa*)	orange, green, white	sticky spheres	Sivinski (1990)
Mediterranean fruit fly (*Ceratitis capitata*)			Nakagawa, Prokopy, Wong, Ziegler, Mitchell, Unago, Harris (1978)
South American fruit fly (*Anastrepha fraterculus*)	yellow rectangles yellow spheres (females)	sticky	Cytrynowicz, Morgante, De Souza (1982)
Mediterranean fruit fly (*Ceratitis capitata*)	red and black sticky spheres (females)		Cytrynowicz, Morgante, De Souza (1982)
Apple blotch leafminer (*Phyllonorycter crataegella*)	red	sticky triangles	Green and Prokopy (1986)
thrips (*Frankliniella bispinosa*)	white	sticky	Childers and Brecht (1996)

beneficial insects such as the coccinelid, *Coleomegilla maculata*; the sevenspotted
lady beetle, *Coccinella septempunctata* L. (Udayagiri et al., 1997), *C. transversalis*,
and the twospotted lady beetle, *Adalia bipunctata* (Mensah, 1997).

Traps used in combinations with insecticides have reduced the amount of insecticide needed to control insects. To manage populations of the olive fruit fly, *Bactrocera oleae* (Gmelin), an effective combination of a fast knockdown insecticide, a strong phagostimulant, a male sex attractant, and a female aggregation pheromone were soaked into sticky boards. While insecticides were used in the trap boards, the volume of chemicals decreased from 1000 mg AI to 10 mg AI per tree (Haniotakis et al., 1991). In colored traps used with pheromones, male lilac borers were more attracted to brown or black traps over white traps. Dark colors attracted pheromone-stimulated males and knowledge of color preference among pest insects is important in maximizing trap catches (Timmon and Potter, 1981).

2.2.4   Plant Materials as Traps

A recommended control tactic in 1838 to control cutworms was to place compacted plant materials, such as elder sprouts, milkweed, clover, or other green plant material in every fifth row and sixth hill. These compacted plant materials were examined for cutworms and killed with a sharp instrument. To eliminate the need for regular examination of the plant materials, farmers later incorporated poison. Traps also caught wireworm adults in corn fields and squash bugs in home gardens (Howard, 1900).

One method of trapping insects used parts of the host plant such as banana pseudostems to control a banana weevil, *Cosmopolites sordidus* (Germar) (Coleoptera: Curculionidae). Banana pseudostems were split lengthwise and placed near banana suckers. The age of the banana pseudostems played a significant role in capturing weevils. Based on trap catch numbers, one-week-old traps collected 1.5- to 1.7-fold more adults than 2- to 3-week-old traps. Traps monitored for 11 months reduced weevil populations by 50%. Pseudostems, as traps, required extensive monitoring and worked where inexpensive labor was available (Koppenhofer et al., 1994). Adults and nymphs of a variegated grasshopper. *Zonocerus variegatus* L., feeding on cotton plants have been trapped using laos weed as bait (Gahukar, 1991).

2.2.5   Fermentation Traps

When pheromones were first developed and employed in traps, there was great promise for reducing lepidopterous pests (Roelofs et al., 1970). Pheromone baits used in traps have been directed largely to attract males, and food baits eliminate any sex bias. Moths are naturally attracted to molasses, fermenting fruit, tree sap, honeydew, and flower nectar (Norris, 1933). Sugar-based solutions have been used to attract and kill the oriental fruit moth, *Grapholita molesta* (Busck) (Frost, 1926, 1928, 1929) and the codling moth, *Cydia pomonella* (L.), in fruit orchards (Eyer, 1931). Corn earworm moths, *Helicoverpa zea* (Boddie), were attracted to and killed in a poisoned molasses and vinegar solution. Traps baited with molasses or unrefined palm sugar captured significant number of a noctuid, *Mocis latipes*, and the age of the bait and the ratio of the ingredients affected the efficacy of the solution. Research is needed to isolate and identify those odorants which serve to attract moths to bait stations (Landolt, 1995). Thus, sugar-solutions might provide useful attractants for monitoring pest populations or for developing attracticidal approaches to suppress pest populations.

Ephestia figulilella Gregson, commonly called the raisin moth, had been a serious pest of dried fruit in California. Attempts were made to control populations using various baits including tea; vinegar; diluted cranberry sauce; tea, milk, and sugar; malt, syrup, yeast, and nicotine sulfate; and cranberry sauce and yeast. A solution of malt syrup, water, and yeast placed in a pail covered with a screen to exclude large moths and butterflies was the most effective trap material. The water control trap collected an average of 30.5 moths and the syrup-yeast mixture collected 3007 moths over the study period (Donohoe and Barnes, 1934). Fermenting baits consisting of different combinations of sour milk, molasses, potatoes, yeast cake, grapes, peach juice, and crushed figures failed to capture sufficient numbers of fig beetles to make traps an effective control tactic (Nichol, 1935).

2.3   BARRIERS

Barriers deny insects access to feeding and oviposition sites. A variety of materials and techniques have been used as barriers, including screens, row covers, mulches, trenches, various particles, bags, shields, and packaging.

2.3.1   Screens

One of the most effective and inexpensive means of insect control is to prevent entry into dwellings. Properly fitted door and window screens can exclude nearly all insects from entering homes and other dwellings. Where drywood termites are commonly present, screening placed over vents, cavities, and windows serves to exclude termites (Bennett et al., 1997). A fine mesh of high-alloy with openings of 0.66 × 0.45 mm excluded termites over a wide size range including species of *Coptotermes*, *Reticulitermes*, and *Heterotermes*. The mesh effectively excluded the various termite populations found in Australia. Work is now required to determine the most effective means of incorporating the barrier mesh into building design and construction (Lenz and Runko, 1994).

A screen consisting of aluminum chains at 78 chains to the meter and hung from a rail deterred flies, wasps, and bees from passing through the screen. Because the screen was not attached to the floor, the screen served as a door. Small insects such as midges and mosquitoes were able to pass through the screen (Anonymous, 1990).

In the early 1940s, openings to tobacco warehouses, including doors, windows, and other openings, were recommended to be screened and small gaps between door and window frames caulked to prevent infestations by insects, such as the cigarette beetle, *Lasioderma serricorne* (Fabricius). An infestation in a tobacco warehouse would reinfest fumigated tobacco or infest manufacturing plants. To control invading insects, the floors of these tobacco warehouses were often constructed from creosote soaked boards (Reed and Vinzant, 1942).

Several mechanical control methods unsuccessfully controlled the tarnished plant bug. A wire mosquito screen was attached to a light wooden frame measuring 2.1 × 0.76 m, and a coating of tree tanglefoot was applied to the surface of the screen. The shields were carried along the plant rows and, to make the screens more effective, branches with leaves were attached to the screen to force the insects onto the sticky screens. The screens proved to be ineffective (Crosby and Leonard, 1914). In another study, populations of apple leafhopper and tarnished plant-bug were reduced and the screens were recommended for use (Haseman, 1913).

Exclusion of insects from greenhouses was realized by using insect-proof screens. Keeping insects out of greenhouses eliminates the need to apply insecticides and reduce the potential for insect-borne plant diseases (Van Steekelenburg, 1992). A set of barrier screens for use in greenhouses was evaluated for control of five common insect pests. The barriers tested consisted of a woven mesh of polyethylene strands, a filter of unwoven polyester, a woven brass strainer cloth, and a high-density polyethylene sheet perforated in the center. The thoracic width of the test insects could not be used to indicate which barrier would exclude the test insects. The two important determinants of the efficacy of the barrier were the holes' construction. Suprisingly, the barrier specifically designed to exclude insects from greenhouses failed to restrain any of the test insects. The size openings required to exclude insect pests varied for each species, and the optimum barrier was the one which reduced the greatest air flow (Bethke and Paine, 1991).

2.3.2 Row Covers

Row covers used in broccoli and Chinese cabbage production effectively reduced damage by *D. radicum*. Other row covers tested such as tarpaper collars and diatomaceous earth had no impact on cabbage maggots. Broccoli greenchop mulch, sand, and wood ash increased the number of maggots. Insecticide applications of diazinon reduced maggot numbers as did row covers, but yields were higher than with diazinon treatments (Matthews-Gehringer and Hough-Goldstein, 1988).

2.3.3 Trenches

When chinch bugs first appeared in outbreak numbers during the mid-1800s, many based control recommendations on the movement of the insect from one crop to another. The use of a creosote furrow barrier soon became standardized in the 1930s in the U.S. Unfortunately, a strong breeze blew many chinch bugs over the creosote barrier, rendering ineffective the furrow. Other drawbacks to this technique included the soil type, which made digging an appropriate furrow difficult. To overcome these problems, a paper barrier soaked in creosote eliminated many of the problems associated with the creosote furrow barrier. The paper strip, about 11.4 cm high, was placed on the ground and covered with soil until about 5.1 cm remained exposed (Harris and Decker, 1934).

Plastic-lined trenches proved to bar movement of Colorado potato beetles into or out of potato fields. Adult beetles were able to walk on plastic but the beetles had difficulty in moving over plastic covered with fine soil particles. The trench slope of 46 degrees or greater retained an average of 84% of the beetles in the trench, and the efficacy of the traps diminished when it rained and resumed when the plastic dried. Surrounding a potato field with plastic-lined trenches, growers reduced populations of overwintering adults by 47 to 49% and the summer population by 40 to 90% when compared with non-trenched fields. Trenching with plastic was shown to reduce egg populations. Data analysis indicated that other control measures should be used along with plastic-lined trenches to manage populations of the Colorado potato beetle. (Boiteau et al., 1994) To dig effective "V" shaped trenches and to lay plastic, a machine was developed to perform both tasks. The "V" shaped trenches lined with plastic averaged 95% effectiveness at keeping adult beetles in the trenches (Misener et al., 1993).

Another chinch bug barrier method used boards set on edge and soaked with kerosene. Coal tar barriers were used in 1871 and the first use of creosote barriers was in 1913 in Illinois (Flint, 1935). Physical barriers of earth, metal, tar paper, or corrugated paper all required some type of repellent to be effective. The most promising low-cost barrier was tar paper (Flint et al., 1935). Strawberry root weevils were prevented from entering strawberry fields by placing a barrier of tarred boards around the periphery of new fields (Metcalf and Metcalf, 1993).

A field study, to determine the efficacy of barrier materials used by or recommended to growers to control chinch bugs, tested four materials: creosote, Tarvia M. T., and two water gas tar formulations of different specific gravities. The population of moving chinch bugs averaged five insects per linear foot of wheat. Data analysis indicated no significant differences between water gas tar and creosote and between water gas tar and Tarvia M. T. Unfortunately, the data presented for water coal tar showed great variability because of the inherent variations in the water coal tar derived as by-products of artificial gas plants. (Huber and Houser, 1935). A long narrow line of coal tar was in place in the late 1880s and was supplanted with the development of a creosote barrier in Illinois in 1913 to 1914 (Flint, 1935). Regardless of the barrier, proper timing of barrier placement and maintenance of the barriers were critical to success (Sorenson, 1995).

In 1770, a widespread outbreak of the armyworm caused extensive damage to wheat and corn as larvae moved unimpeded through fields. Desperate farmers threw ropes over plants to dislodge larvae, which only resulted in delaying the inevitable damage. Trenches, dug in front of the advancing larvae, soon filled with larvae. Subsequent larvae crossed the ditches over the backs of trapped larvae. A slight modification was made to the ditches to make them more effective. Holes, spaced 2 or 3 feet apart were dug into the ditches. As larvae fell into the ditches and into the holes, sticks were used to crush the larvae (Webster, 1914).

2.3.4 Particle Barriers

Materials such as sand, granite, glass splinters or globules, and fossilized coral of a specific size served as termite barriers. These materials were large enough to prevent termites from moving them, and the spaces between the materials were too small for the termites to move through (Ebeling and Pence, 1957). A 20-cm thick layer of sand particles between 2.0 to 2.8 mm effectively excluded species of *Reticulitermes* and *Coptotermes*, but these physical barriers had to be monitored and maintained to be effective (Pearce, 1997). Other studies indicated that particle sizes of 2.00 to 2.36 mm and 2.36 to 2.80 mm are best where there are active colonies of the Formosan subterranean termites, *C. formosanus* Shiraki, whereas mixed particle sizes of 1.70 to 2.36 mm allowed slight penetration (Su and Scheffrahn, 1992).

Worker castes of *C. formosanus*, and the eastern subterranean termites, *R. flavipes* (Kollar), were unable to penetrate a 5-cm thick layer of ground coral particles. Coral particle size ranged from 0.5 to 4.0 mm in diameter. The recommendation was to mix uniformly the particles to obtain a mixture of 1.18 to 2.80 mm particles to prevent termite penetration (Su et al., 1991). In Hawaii, a refined sandblast sand of 1.7 to 2.8 mm size range was used to effectively exclude *C. formosanus* especially when applied before the foundation was poured (Tamashiro et al., 1991).

Sand, as a barrier to exclude the western subterranean termite, *R. hespersus* Banks, failed to perform better than a 1% permethrin chemical barrier. The sand barrier might be more effective in new construction rather than attempting to modify barriers to fit existing structures (Lewis et al., 1996). Another form of barrier used polystyrene beads. In Dar es Salaam, a floating layer of expanded polystrene beads on the surface of latrines and septic tanks controlled populations of the southern house mosquito, *Culex quinquefasciatus* (Chavasse et al., 1995).

2.3.5 Inert Dusts

Inert dusts have been used as a physical control measure against stored-product insects (Golob, 1997). Early use of grain protectants is attributed to the Aztecs who mixed maize with lime. Grain protectants have been placed into five categories (Golob, 1997): group 1, non-silica dusts; group 2, sands, kaolin, paddy haskash, wood ash, and clays; group 3, diatomaceous earth; group 4, synthetic and precipitated silicas; and group 5, silica aerogels. Because inert dusts work mechanically to remove the protective waxy layer of the exoskeleton, their modes of action are much slower than that of chemical insecticides (Ebeling, 1971, 1978). Inert dusts have controlled several stored-product pests (Strong and Sbur, 1963; White and Loschiavo, 1989; Permual and Le Patourei, 1990; Subramanyam et al., 1994).

Benefits of inert dusts are the low mammalian toxicity and the nearly negligible resistance development. Concerns are the particle sizes of inert dusts which pose potential respiratory hazards to handlers and the large doses required to control pests which may alter the physical properties of the grain (Gobol, 1997). While inert dusts have been used in stored-product environments, not much is known about the interaction of inert dusts and temperatures (Nickson et al., 1994). The granary weevil, *Sitophilus granarius* (L.), and the lesser grain beetle, *Rhyzopertha dominica* (Fabricius), were more susceptible at 30°C than 20°C but the confused flour beetle, *Tribolium confusum* Jacquelin du Val, was susceptible at lower temperatures (Aldryhim, 1990, 1993).

An inexpensive and effective means of controlling the cowpea weevil, *Callosobruchus maculatus* (Fabricius), in northern Cameron involved the use of ash mixed with cowpeas. Wood ash was sieved and the large pieces of ash removed and discarded. The sieved ash was thoroughly mixed with 35 to 45 kg of cowpeas in a container and then placed into the storage container. The top of the ash-cowpea mixture was pressed by hand to compact the mixture, which was then topped by a 3-cm layer of ashes. The ash-cowpea mixture prevented cowpea weevils from emerging from the stirred seeds (Kitch and Ntoukam, 1991a).

2.3.6   Bags

In a preliminary study, bags constructed from different materials were placed over branches in an attempt to exclude tarnished plant-bugs on peach trees. Cheesecloth bags, paper bags, and mosquito-netting bags used in the study did not appreciably alter the growth of the peach foliage, and the study did not discuss the efficacy of each bag type to exclude tarnished plant bugs (Crosby and Leonard, 1914).

In Hawaii, an early recommendation was to cover young fruit to prevent oviposition by the Mediterranean fruit fly. One method was to line a bag with cheesecloth and the entire bag slipped over trees and tied to trunks. The difficulty in knowing whether female Mediterranean fruit flies were trapped in the bags and the problem of plant breakage caused by the weight of the bag and subsequent winds made this recommendation impractical. Another method was to cover individual fruit with paper bags but this method required much labor and a high degree of patience in covering each fruit (Back and Pemberton, 1918).

Exclusion bags consisting of spunbound polyester, polyethylene row covers, and nonwoven polyethylene pollinating bags were taped to flowers of red ginger and evaluated for yield and insect damage. Bagging flowers reduced the number of bigheaded ants, *Pheidole megacephala* (Fabricius) and *Technomyrmex albipes* and banana aphids, *Pentalonia nigronervosa* Coquerel. A single foliar application of chlorpyrifos reduced only the damage caused by banana aphids. When chlorpyrifos was applied before bagging, a significant number of insects were controlled compared with only spraying or bagging. Bagging, however, did cause damage to the flowers with the level of damage dependent on the type of bag used (Hata et al., 1995).

2.3.7   Shields

Metal shields and caps placed around the foundations of homes were once recommended as a way to exclude termites, but because termites can easily cross over these barriers, this practice is no longer recommended (Su and Scheffrahn, 1990; Bennett et al., 1997). Stainless steel mesh (a marine grade 316 steel mesh) placed around buried wood in Hawaii prevented attack by *Coptotermes formosanus* except where slight gaps existed because of improper installation. As with any termite barrier, proper installation and maintenance are crucial to prevent termites from circumventing the barrier (Grace et al., 1996).

Mechanical barriers to control oviposition by the roundheaded appletree borer, *Saperda candida* Fabricius, were not effective, but forced the beetles to oviposit above the guards which made detection of borers easy. A mouse guard placed at the base of the apple tree served to trap ovipositing females when trap tops were plugged with burlap or other similar material. The trapped female beetles died in the mouse guards and failed to oviposit (Hess, 1940).

2.3.8 Packaging

Insects are the main cause of losses to dry, packaged goods. As food travels across the country by various means and stored in different environments, insect infestations remain a constant threat. Packaging as a barrier to prevent insect infestations has seen major advancements with the development of new packaging materials (Newton, 1988). Polymer films, laminations, and extrusions can protect packages from insect infestations, whereas polyester, polyurethane, or polypropylene films resist insect penetration. The integrity of the package must remain intact for the protective covering to be effective. Rough handling during transit, storage, and shelving must be avoided. In large storage warehouses and manufacturing operations, the dictum of "first-in-first-out" is an excellent way to lessen insect infestations (Highland, 1991). Packaging and temperature manipulation provide consumers with the best combination of food safety and reliability (Mason, 1997).

Insects can enter packages through seams or directly through packing materials. Several stored-product pests that can bore into packaging are *Rhyzopertha dominica*; *Lasidoerma serricorne*; the cadelle, *Tenebroides mauritanicus* L.; the warehouse beetle, *Trogoderma variable* Ballion; the rice moth, *Corcyra cephalonica* (Stainton), and the almond moth, *Cadra cautella* (Walker). Because the female Indian meal moth, *Plodia interpunctella* (Hubner), is able to detect food in sealed packages, the insect has served as a rapid method of determining suitability of packages to deter insect infestations (Mullen, 1994).

Second and last instars of 11 stored product insects were tested for their ability to penetrate packages made of paper, polyester, cellophane, polyethylene, polyvinylchloride, aluminum foil, and polypropylene. Larvae of three species: the merchant grain beetle, *Oryzaephilus mercator* (Fauvel); the squarenecked grain beetle, *Cathartus quadricollis* (Guerin-Meneville), and the flat grain beetle, *Cyptolestes pusillus* (Schonherr), were unable to penetrate any of the packaging materials. *Tenebroides mauritanicus* and *Trogoderma variable* were the only two species that penetrated all seven packaging materials. The hide beetle, *Dermestes maculatus* De Geer; *Lasioderma serricorne*; *Ephestia cantella*; *Corcyra cephalonica*, and *P. interpunctella* larvae were able to penetrate only five of the seven packaging materials. *T. variabile* was the only insect to penetrate polypropylene packaging. Of the locations penetrated by small and large larvae, the majority of test insects penetrated package folds compared with the top, bottom, or middle portions (Cline, 1978).

In another test, 14 different packaging films were tested against penetration by 11 species of stored product pests. Aluminum foil proved to be the most resistant to insect penetration but was not insect-proof (Gerhardt and Lindgren, 1954). Resistance of polymer films to penetration by *Rhyzopertha dominica* was related to the type of resin and the manufacturing process. Of the materials tested, polyurethane and polyester films were most resistant to penetration (Highland and Wilson, 1981).

To prevent infestation by clothes moths, an effective control technique was to deny the insect access to clothing. Before the advent of airtight plastic containers, furs and other garments were stored in boxes or trunks lined with heavy tar paper rather than in cedar chests which lose their effectiveness during the course of a few years with a resultant loss of protection. Other storage techniques involved large pasteboard boxes which, after the items have been placed, were sealed with strips of gummed wrapping paper. A steel comb was run through the fur, and the furs stored in tar paper-lined boxes or in closets lined with tar paper (Marlatt, 1908).

Various plastic materials have been tested for penetration by *Callosbruchus maculatus*. Small cowpea packets made from saran, polyethylene, ethylene vinyl acetate, cellophane, polyvinyl chloride, polyester, polypropylene, polyurethane, polybutylene along with paper and aluminum were tested. After 39 days, polyurethane, saran, and a combination of polyester, aluminum foil, and polyethylene packets showed no penetration by weevils nor did any weevils emerge from the packets (Highland, 1986). Weevils stored in sealed plastic films died because of the low oxygen transmission rates of the films.

Food pouches sealed with a vacuum of 48.8 mm Hg quickly killed larvae and adults of *Cadra cautella*; the red flour beetle, *Tribolium castaneum* (Herbst), and *Trogoderma variable. Lasioderma serricorne*, however, survived for one week (Cline and Highland, 1987). In another study, unvacuumized polyester film bags and vacuumized polyester film bags resisted penetration by red flour beetles, cigarette beetles, almond moths, merchant grain beetles, and Indian meal moths. All insects penetrated vacuumized polyethylene bags, and the rigidity of the vacuumized bags may have made the bags vulnerable to penetration (Highland, 1988). Some stored product insects survived a 24-h exposure to a vacuum of 160 mm Hg. (Calderon and Navarro, 1968).

Triple-bagging of cowpeas is an inexpensive method of controlling cowpeas in developing countries and takes advantage of the low oxygen transmission of plastic films. In the Cameroon, harvested cowpeas are dried and placed into three 50 kg clear plastic bags, commonly available. The bags must be free of any holes or tears. The top of the first bag is folded, placed into the second bag, and the top of the second bag folded. Both bags are then placed into a third bag, and the top folded. The first bag is gently rocked to eliminate air pockets and the top sealed securely, folded over, and tied a second time. The tying procedure is repeated for the other two bags. The tied bags are kept sealed for a minimum of two months to kill all cowpea weevils (Kitch and Ntoukam, 1991b).

E. figulilella infests figs kept in temporary boxes before shipping. A tobacco shade cloth, a loosely woven fabric, was tested to exclude the raisin moth from apricots, nectarines, peaches, and raisins. In all commodities, the covered fruit sustained lower infestations of the raisin moth than did uncovered fruit (Donohoe et al., 1934).

2.4 PHYSICAL DISTURBANCES

2.4.1 Shaking

The plum curculio and the pecan weevil were dislodged from infested trees by jarring the trunk or shaking the larger branches. The dislodged adults or "June drops" (Chapman, 1938) were collected on sheets placed beneath the trees and the adults killed. Shaking trees to remove pecan weevils reduced populations by 50% (Baker and Hienton, 1952). These techniques may appear to be ludicrous by today's control standards, but were once held to be a potential control method.

Adult pentatomids, *Dolycoris baccarum*, and meloid beetles, *Cneorrhinus globatus*, in Japan in the early 1930s were shaken off plants, collected into containers, and killed (Clausen, 1931). Agitation or shaking rugs, furs, clothing, and other materials attacked by clothes moths was recommended as one of the best control methods. Furriers, who store furs for their customers, would thoroughly and vigorously beat the fur with small sticks to dislodge loosened fur and to remove larvae or moths. In addition to shaking, potential targets of the clothes moth were exposed as long as possible to sunlight in early spring (Marlatt, 1908).

2.4.2 Jarring

In the process of cutting down trees infested with roundheaded appletree borers, workers observed large numbers of dead pupae in felled trees. Pupal death was thought to have been attributed to the jarring during tree removal. To test this hypothesis, researchers struck ten young saplings ten times each with a large padded mallet, trees felled, and examined for insects. A total of five dead adults and eight dead pupae were collected. Based on this experiment, jarring was offered as a potential technique to control the roundheaded apple tree borer (Hess, 1940).

In small plots of asparagus, beating plants infested with the asparagus beetle would knock slow-moving larvae to the ground, where most beetles died. Jarring of asparagus would be effective only in small plots because of the high labor required and considerable time involved (Drake and Harris, 1932). Limb jarring has been used as a method to determine pest population levels as part of a decision-making process. Jarring limbs to dislodge pear psylla resulted, in part, to the development of an action threshold of 1.0 to 1.2 pear psyllas per limb jar (Adams and Los, 1989).

2.4.3 Mechanical Disturbances

Moving or turning grain has been studied as a method of reducing insect infestations in stored grain (Bailey, 1962; Joffe, 1963; Bryan and Elvidge, 1977; Loschiavo, 1978). During grain movement, insects infesting grain are subject to shaking, jarring, vibrations, and centrifugal forces which can be fatal to insects, and grain turning can reduce grain temperatures to unfavorable levels for insect development (Muir et al., 1977). The type of infested grain and method of movement can influence mortality levels (Muir et al., 1977).

Frequent impacts or disturbances during the development of *Sitophilus granarius* caused substantial mortality to the immature stages (Banks, 1987). In addition, the rusty grain beetle, *Cryptolestes ferrugineus*, sustained 96% mortality when small wheat-filled bags containing insects were dropped several times (Loschiavo, 1978). Physically disturbing wheat at least two or more times a week might prevent immature stages of *S. granarius* from completing development (Bailey, 1969).

Movement of infested grain from cell to cell by screw conveyor, bucket elevator, and two pneumatic conveyors resulted in high mortalities (80 to 90%) of *Cryptolestes* spp., *Rhyzopertha dominica*, and the rice weevil, *S. oryzae* (Cogburn et al., 1972). By handling grain with a pneumatic conveyor during grain movement, 80% of the adults and 60% of the larvae of *C. ferrugineus* were killed (Banks, 1987).

In another study, infested wheat moved by auger caused 89% mortality to adult *Tribolium castaneum* and 94% mortality of *C. ferruigenus*. All larvae of both species were killed based on sampling with a Berlese funnel. Moving infested wheat by pneumatic conveyor from a bin into a trunk and into another bin resulted in total mortality of *C. ferrugineus* adults and *T. castaneum* adults and larvae. Pneumatic movement of infested corn killed 97% of *T. castaneum* adults, 72% of *C. ferrugineous* adults, and 100% of the plaster beetle, *Cartodere constricta* (Gyllenhal). Moving wheat with a pneumatic conveyor offered an effective way to physically control stored product pests (White et al., 1997).

An Entoleter, a machine containing a spinning disk with several steel pegs at the edge of the disk, was constructed to kill stored-product insects. The infested commodity was placed in the Entoleter, which flung materials against the pegs and the machine casing. An Entoleter running at half speed killed 99% of free living insects without damaging the grain, and the full potential of this machine has not been realized (Banks, 1987).

2.4.4 Hand-destruction

Noctuids, *Spodoptera littoralis* and *Pectinophora gossypiella* are major cotton pests in Egypt. An effective integrated pest control system for these pests include handpicking egg masses of *S. littoralis* during the first part of the growing season and ceasing irrigation of clover fields after 10 May. Handpicking and burning of infested and dry bolls reduced the population of next season's pink bollworm populations (Brader, 1979). Early practices to control cotton boll weevil involved handpicking eggs and larvae and infested plant parts (Bottrell and Adkisson, 1977).

Rice stem borers in Asia consist of eight species, which are widely distributed throughout the rice growing regions of temperate and tropical Asia. While biological control agents, cultural practices, host plant resistance, and insecticide applications have been used to control rice stem borers, the earliest use of hand destruction of eggs was reported in the late 1880s. The success of these cultural practices, including hand destruction, must be performed over several years over a large area to effect any meaningful control (Kiritani, 1979).

Cassava provides a major source of energy for 300 to 500 million people, and farmers with limited access to technology grow cassava throughout the tropical regions of the world. In West Africa, cassava production provides the most economical means with the lowest risk for subsistence farmers. Numerous pests attack cassava, but non-chemical control measures are limited to hand-picking a cassava hornworm, *Erinnyis ello*, removing and burning infested plant parts to control larvae of the *Lagochirus* spp., a cerambycid, and cutting and burning plants infested with various species of scales (Bellotti and van Schoonhoven, 1978). Hand removal of insects in stored grain by peasant farmers, small retailers, and women assisted by children involves great patience. Sifting contained grain with a sieve having openings smaller than the grain is an improvement on hand sifting (Appert, 1987).

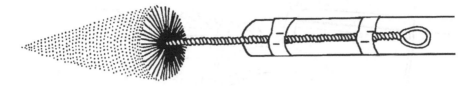

Figure 2.2 Bottle brush used to remove nests of tent caterpillars. (Redrawn from Britton, 1913.)

The browntail moth, *Euproctis chrysorrhoea* (L.), was once considered a pest in the New England states along with the gypsy moth, *Lymantria dispar* (L.). The browntail moth, because of its strong flight ability, could not be managed with quarantines and inspections, and control was targeted at hand removing the nests from infested trees. In 1913, the children of Newfields, New Hampshire collected egg clusters of the apple-tree tent-caterpillar for which they were paid ten cents per hundred clusters. Reportedly, 1,237,500 eggs were destroyed at a modest cost of $8.25. In another offer, one student collected 4000 tent-caterpillar nests which were then burned (Britton, 1913). Bottle brushes, (Figure 2.2) attached to a long pole, were thrust into nests of tent caterpillars and twisted to remove the tents (Howard, 1900).

Grasshopper outbreaks in the Upper Great Plains of the U.S., especially damage by the Rocky Mountain locust, in the late 1880s were met with techniques which had changed very little from ancient times. These techniques involved hand destroying eggs, paying a bounty for eggs, crushing with mechanical rollers, trapping in ditches, or by burning the grasshoppers (Sorenson, 1995).

One of the contributing factors causing the decline of hand removal of pests was the mechanization of American agriculture. Machinery increased the number of cultivated acreage per person, making hand removal impractical. Where farms were small and labor plentiful, hand removal was practical. In fact, the terms "abraupen" in German and "decheniller" in French reflect the common practice of hand removal in these countries (Sorenson, 1995).

2.4.5 Pruning

Selective pruning is used to remove pests from landscape plantings (Raupp et al., 1992), and pruning has been used in Illinois in an attempt to save Dutch elm diseased trees. Pruning was restricted to trees with 5% or less of the crown exhibiting wilt symptoms, and early detection was important for pruning to be effective. While pruning did not impact population numbers of the elm bark beetles, the procedure was effective in certain situations (Himelick and Ceplecha, 1976). While pruning served to reduce Dutch elm disease, pruning of willow, *Salix lasiolepis*, resulted in increased densities and total numbers of a sawfly, *Euura lasiolepis*, feeding on willow (Hjalten and Price, 1996).

To control the green peach aphid, *Myzus persicae* (Sulzer), pruning can remove most of the eggs oviposited on *Prunus* spp., the aphid's primary host. Pruning has a limited value in reducing damage caused by the buffalo treehopper, *Stictocephala*

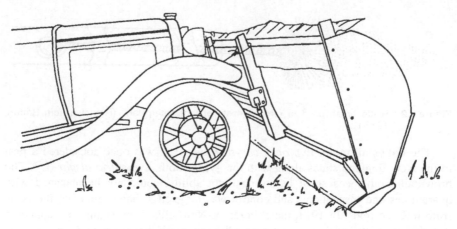

Figure 2.3 Hopperdozer attached to vehicle to control grasshoppers (Original drawing.)

bisonia Kopp and Yonke. Pruning infested twigs was practiced in the 1930s but was practical only when occasional twigs were infested. Cutting off all infested twigs in heavy infestations might destroy one- or two-year-old trees (Yothers, 1931).

One of the oldest control techniques for *Saperda candida* was to remove borers from the trees with a knife and a piece of wire. This procedure, termed "worming," involved scraping away materials from the base of the tree, which might hamper a search for borer castings. At the site of the castings, the bark was carefully removed to trace the path of the borer. The wire was then pushed into the tunnel to hook and remove larvae from the tree. An unrealistic requirement was to be certain that all borers be killed in the orchard (Brooks, 1915).

2.4.6   Hopperdozer

An innovative physical technique used to control grasshoppers was the hopperdozer (Figure 2.3) based on the grasshoppers' leaping behavior. The hopperdozer consisted of a long narrow trough, filled partly with water, creosote, or coal oil, which was placed on a wooden skid pulled by humans or horses (Washburn, 1910, 1912). Hopperdozers were later attached to the front of cars or trucks. As the skid moved slowly through a grasshopper-infested field, grasshoppers would jump into the air, strike the backing, and fall into the trough (Sorenson, 1995). In the Red River Valley of North Dakota and Minnesota in the early 1910s, the hopperdozer was recommended as an inexpensive and effective method of controlling grasshoppers. The Minnesota hopperdozer was constructed of a wooden frame to which cloth sides and a back were attached. Soaking the side wings with kerosene added to the mortality of the grasshoppers clustered on the cloth. The device could be used only when the grasshoppers were able to jump high enough to fall into the pan or trough (Washburn, 1912).

A variant of the hopperdozer was used in 1907 to control populations of the rose chafer, *Macrodactylus subspinosus* (Fabricius), feeding on grapes (Figure 2.4). A pan about 2.1 m long, 56 cm wide, and 2.5 cm deep was constructed of galvanized iron. To the frame was attached a 0.9-m high cloth backing. Pieces of cloth soaked

Figure 2.4 Rose chafer catcher used in a vineyard. (Redrawn from Pettit, 1908.)

in water were placed into the pan and covered with kerosene. The unit was then placed alongside a vine and the vine beaten with broom corn switches to dislodge beetles into the pan. The unit appeared to work well when beetle populations were high. In situations of low beetle density, one person with a milk pan containing oil rags could effectively collect beetles. No mention was made about the efficacy of this method in reducing rose chafer damage (Pettit, 1908).

2.5 SANITATION

Sanitation, a key component in many early pest management programs (Yothers, 1934; Larson and Fisher, 1938), played a major role in reducing or eliminating insect pests from many agricultural and urban situations. The first official sanitation effort may have been related to the spread of typhus, which occurred during the winter of 1795 (Service, 1996) in England. A voluntary board of health sought to deal with the problem, and the group would disband when the crises ended. The relationship between insects and disease transmission was not understood fully until the 1890s. For example, the relationship of mosquitoes and malaria became evident in 1892; this provided control opportunities based on the mosquito's preferences for breeding in standing water.

2.5.1 Structures

Sanitation is an appropriate and often the most effective way to control insect pests in zoos, museums, homes, and hospitals where pesticide applications may not be appropriate. Sanitation along with caulking, screening, and other exclusion techniques can help in long-term pest management programs. Eliminating or reducing pest habitats outdoors can effectively reduce indoor pests (Bennett et al., 1997).

Pest-proofing or denying ants access into structures by sealing entry points is an effective non-chemical management technique based on the ants' foraging behavior. Ants follow existing edges, such as wires, pipes, and conduits, to and from their nests (Klotz and Reid, 1992). Caulking entry points of these utility lines effectively excluded ants from dwellings. Removal of wood debris, such as stumps and tree roots, form boards used during construction, etc., around the buildings will eliminate termite breeding areas (Mallis, 1982).

Sanitation to control cockroaches involves cleaning premises to remove food and water. This simple procedure is enough to reduce German cockroach populations (Bennett et al., 1997) without the use of insecticides. In food plants, sanitation provides an excellent nonchemical management program for stored insect pests. Where insecticides were applied, sanitation helped to decrease cockroach populations (Schal, 1988). Sanitation in food storage and food handling facilities consists of removing spilt grain and flour (Stern, 1981; Fields and Muir, 1996), but using sanitation techniques in the tropics may be difficult (Young and Teetes, 1977). Sanitation, including cleaning of warehouses prior to storage and rapid removal of infested materials, played an important role in preventing infestations by approximately 38 species of insects, mostly Coleoptera (Belloti and van Schoonhoven, 1978). Boxes and sacks placed on pallets away from the wall permitted cleaning on all sides, and delivery vans were regularly inspected and cleaned to prevent infestation of the cargo. The use of drop ceilings, boxed-in pipe runs, cable ducts, etc. create hiding places for insects (Bateman, 1992). Wind blows many insects into structures and opening of doors and windows should be minimized. Well-planned and managed landscaping can assist in reducing insect breeding and harborage areas (Thorpe, 1992).

2.5.2   Animal Habitats

Several sanitation practices have reduced populations of insects affecting livestock. Cleanliness of the confinement area and good clean dust baths are part of an effective sanitation program against poultry lice, as are removal of water-soaked rotten straw and disposal of manure. Removal of other fermenting organic matter helped reduce stable flies along with the removal of carcasses denied fleece worms or wool maggots' breeding site (Metcalf and Metcalf, 1993; Mallis, 1982). Stable flies spend relatively short time on hosts so that management techniques must be directed away from the host. Immatures develop in manure, spilled feed, and decaying vegetation, and the elimination of larval breeding sites can aid in stable fly control if combined with other control techniques (Foil and Hogsette, 1994).

Sanitation can effectively reduce flea populations in the home. Standard vacuuming may remove more than 90% of flea eggs in the carpet but only 15 to 27% of the larvae are extracted. A more effective method is to clean pet bedding to eliminate off-host stages of fleas (Hinkle et al., 1997). Weed and grass removal in greenhouses eliminated carry-over sites for aphids and other insects (Hanan et al.,

1978). Sanitation along with stock rotation and trapping could serve as a non-chemical control technique in grocery stores, but a survey of 322 grocery stores revealed that most grocers lacked knowledge of IPM practices (Platt, 1998).

2.5.3 Field Crops

A scarab, *Oryctes rhinoceros*, caused severe damage to coconuts by boring through the unopened leaves, causing leaves to drop or distorting emerging leaves. Management options include field sanitation, which consisted of destroying breeding sites and removing dead trees. These relatively simple methods were effective and less costly than insecticide applications (Brader, 1979).

A millet stem borer, *Coniesta ignefusalis*, caused economic damage throughout the Sahel of Africa. The low commercial value of millet requires farmers to use the most economical control method available. Removing or destroying millet residue reduced larval populations by 61 to 84% and pupal populations by 98 to 100%. Residue removal, however, exposes unprotected soil to wind erosion (Nwanze, 1991).

The alfalfa seed chalcid, *Bruchophagus roddi* (Gussakovsky), attacks alfalfa seed production in the western U.S. Chemicals cannot reach developing larvae feeding within seeds, and insecticide applications would be detrimental to pollinators. Because seed chalcids develop on volunteer alfalfa and bur clover growing adjacent to production fields, sanitation achieved through burning straw and chaff after harvest and covering trucks hauling seeds have restricted off-target development sites (Stern, 1981).

Strawberry root weevil, *Otiorhynchus ovatus* (L.), feeding stunts strawberries and other plants. The insect occurs in field rubbish, and the destruction or burning of plant residues immediately following cultivation was of value in managing this pest (Pettit, 1906). Cleanup of fallen, rotting papayas effectively reduced populations of the Oriental fruit fly, *Bactrocera dorsalis* Hendel, and field sanitation has been recommended as an integral component of a pest management program (Liquido, 1993).

A pyrrhocoridae, *Dysdercus voekleri* Schmidt, and *Pectinophora gossypiella* are common pests of cotton in Central and West Africa and crop residue destruction was recommend as a control tactic (Gahukar, 1991). Spiny bollworm larvae, *Earias insulata* Boisduval and *E. biplaga* Walker, bore into shoots of growing cotton, causing death of the growing point. Removal and destruction of these infested points reduce surviving numbers, but only until after damage had been caused. Thus, most of the mechanical measures such as hand destruction and removal of infested materials reduced next season's pest populations (Gahukar, 1991). Stalk destruction before the cotton boll weevil entered diapause removed food and breeding sites of the insect (Bottrell and Adkisson, 1977).

Removal and destruction of infested corn stalks was recommended as a means of controlling the sugarcane borer in Louisiana. In 1926, workers were taught how to recognize borer-infested stalks for removal and destruction, and workers were able to reduce by 90% the borers in a field by disposing 5 to 20% of the stalks in the field. Based on this study, a recommendation was made that if more than 25% of the stalks were infested then all stalks in the field must be removed. To be effective, the hard butt portion of the stalk was destroyed by being run through a stalk-chopping machine, buried in a deep furrow, or submerged in water to prevent moth emergence (Hinds and Spencer, 1927). Another method was to collect the dead plants in bags and to burn the bags. Collecting plant material was feasible only when there was sufficient number of unpaid labor such as convicts used in Louisiana (Holloway et al., 1928).

By 1929, the spread of the European corn borer caused great concern in newly infested areas of Connecticut. To control the European corn borer, Connecticut issued a European corn borer cleanup order. The order required all cornstalks be disposed on or before April 10 by feeding to livestock, burning, or plowing. As might be expected, the spotty execution of these orders and the failure of some growers to meet stated deadlines required inspectors to check fields for compliance (Britton and Zappe, 1931).

Dutch elm disease and efforts to control its spread was studied in Syracuse, New York for 20 years. Sanitation played an essential role in managing the insect and the disease. Destroying dead and dying elms was one of the three recommended methods for controlling Dutch elm disease. Areas not practicing sanitation suffered 5.8 to 15% loss of elm trees to Dutch elm disease, whereas maximum sanitation efforts showed 0.84 to 1.63% loss (Miller et al., 1969). Management of *Sphaeropis* die-back disease transmitted by the eastern pine weevil, *Pissodes nemorensis* Germar and a scolytid, *Orthotomicus erosus* to trees was managed by removing diseased trees (Wingfield and Swart, 1994).

2.6 EXTRACTION

2.6.1 Digging

Zomocerus variegatus is a widespread pest in West and Central Africa. The grasshopper commonly congregates in dense groups, which contributes to its pest status and affords local people a way of managing the pest without chemicals. In West Africa, the variegated grasshopper attacks subsistence crops, and alternative chemical controls were recommended, such as knocking early instars off resting sites into large nets and then placing the nets into a water and gasoline mixture. Grasshoppers remain inactive in the early morning hours and can be easily caught (Page, 1978).

Another effective control measure, based on the congregating habit of the grasshopper, was to dig eggs from the soil. Grasshoppers oviposit in conspicuous and concentrated areas. While this has been criticized by some, a study to determine the effectiveness of digging up the eggs was conducted at two sites. Analysis of the nymphal emergence data indicated that 83% and 91% of the eggs were destroyed at the two sites, respectively (Page, 1978; Chapman et al., 1986).

2.6.2 Vacuuming

Vacuuming plays a critical role in sanitation programs in homes and other structures. Large numbers of the Asian lady beetle, *Harmonia axyridis* (Pallas), invade homes and other dwellings during the fall, and vacuuming is recommended to remove the Asian lady beetle. After vacuuming the beetles into the bag, the bags are removed from the house or the bag destroyed to prevent reinfestation (Bennett et al., 1997).

Vacuuming and trapping with sticky traps led to significant reductions of German cockroach populations. Flushing agents applied before vacuuming resulted in a higher population reduction and removal of hard to reach gravid females (Kaakeh and Bennett, 1997). German cockroaches were repelled by an air flow at 4 m/s, which is the velocity of air coming from forced heating and air conditioning systems. Studies indicated that airflow can be redirected to force cockroaches into areas treated with insecticides or into closer proximity to poisoned baits (Appel, 1997).

A tractor-mounted vacuum collector was constructed to control populations of the Colorado potato beetle. The collector removed 40% of small larvae and 48% of the adults, but only 27% of the large larvae were removed. Improvements will be needed to increase the catch of larvae and adults. Regardless of the anticipated improvements to the vacuum collector, 13% of the adults, 3% of the small larvae, and 23% of the large larvae fell to the ground as the vacuum approached the insect. Thus, the maximum collection efficiency would be no greater than 97% for small larvae, 87% for adults, and 77% for large larvae. Compounding the problem was the number of dislodged larvae and adults that returned to the plant. One positive side benefit of the vacuum collector was the removal of a certain number of aphids on potato plants (Boiteau et al., 1992). A similar device was developed to control insects in celery and potato crops by dislodging insects with air projected from side vents and simultaneously vacuumed. Populations of whiteflies, leafhoppers, and aphids were reduced 50-75%, and the removal effects lasted for several weeks (Weintraub et al., 1996).

2.7 IRRIGATION

Water has been used to control termites in the Sudan by overirrigating crops or trees. Some termites, however, can survive in air pockets created between interstitial spaces of soil particles. Overirrigation reduced termite damage in irrigated cotton (Pearce, 1997). Early and heavy rains increased infestations of the cotton whitefly, *Bemesia tabaci* (Gennadius), on selected cultivars grown in Central Africa. The next season's whitefly populations were controlled by stopping irrigation earlier than normal, which restricted the late development of the whitefly (Gahukar, 1991).

The diamondback moth, *Plutella xylostella* (L.), a pest of crucifers has developed resistance to registered insecticides. An alternative to chemical control of the dia-mondback moth was to use overhead irrigation to prevent oviposition and to reduce damage to watercress (Tabashnik and Mau, 1986) and cabbage (McHugh and Foster, 1995). Overhead irrigation of watercress, throughout the day and night, reduced by seven-fold diamondback moth oviposition. The irrigation study in cabbages, which compared the efficacy of overhead with drip irrigation, found that overhead irrigation applied intermittently during the evening hours had the greatest potential of disrupt-ing oviposition (McHugh and Foster, 1995).

Attempts to control the soybean looper, *Pseudoplusia includens* (Walker), with irrigation showed no significant difference in eggs oviposited, number of larvae present, and defoliation when compared with non-irrigated soybeans. (Lambert and Heatherly, 1995). A whitefringed beetle, *Graphognathus leucoloma* (Boheman), can increase its population if perennials such as alfalfa are irrigated, especially in areas where alfalfa is not normally irrigated (Matthiessen and Learmonth, 1992).

Azalea lace bug, *Stephanitis pyrioides* (Scott), infestations on azaleas were minimized by ensuring that plants were sufficiently irrigated (Trumble and Denno, 1995). Intermittent irrigation of rice fields in Japan decreased abundance of larval mosquitoes, *Culex tritaeniorhynchus* Giles and *Anopheles sinensis* Wiedemann (Mogi, 1993).

An economical method suggested to control the sugarcane borer, *Diatraea saccharalis* (Fabricius), was to soak seed sugarcanes in water of "ordinary temperatures" for 72 h to kill 27 to 100% of the borers in the cane. For sugarcane not used for seed, higher water temperatures of 55°C for 20 minutes destroyed over 90% of the borers. The recommended water temperatures for seed sugarcane protected the sprouted and unsprouted eyes. Based on the success of immersing sugarcanes in water, stubble cane fields were flooded 3 or 4 days after harvest to kill borers in the cane trash and stubble. Because of the costs involved, flooding becomes feasible only where economically practical (Ingram and Bynum, 1941).

Fig beetle, *Cotinis texana* Casey, eggs and first instars in the soil cannot survive in water-saturated soil for 48 h or longer. Flooding Fig beetle oviposition sites in Texas during late August and in September reduced beetle populations (Nichol, 1935). Stable flies, breeding in trickling filters of sewage disposal plants, were controlled by flooding the maggots for 12 h twice a week (Nettles, 1934).

A reverse of irrigation is to remove water to control insects. The rice water weevil is a semi-aquatic insect feeding and mating above or below the surface of the water. Early instars feed in or among the roots of the rice plant and later instars live in the mud around the roots in which they feed. Rice fields drained at the proper time averaged 17.8% more rice than the controls (Isley and Schwardt, 1934). Mosquito control has received more attention than any other group of arthropods affecting vertebrates (Harwood and James, 1979). Source reduction of breeding sites, such as removing standing water, often leads to permanent control of mosquitoes and horse flies (Mallis, 1982). Construction of drainage systems or filling in breeding areas have initial high costs, but reduces the need for seasonal control measures (Lyon and Steele, 1998; Service, 1996). Periodic flushing of small isolated pools of mosquito-infested water has been used effectively in India and Malaysia to control mosquitoes (Service, 1996).

2.8 MULCHES

Mulches can impact insect populations (Wilcox et al., 1932; Zehnder and Hough-Goldstein 1990; Brust, 1994) and can have positive effects on soil temperatures and plant growth. In a study of tomatoes, color mulches were examined for impact on tomato growth, damage caused by *Lycopersicon esculentum*, and the impact of mulch on tomato mottle virus transmitted by the silverleaf whitefly, *Bemisia argentifolii* (Bellows and Perring). Mulch colors tested were blue, orange, red, yellow, white, black, and aluminum. In general, fewer aphids were trapped on aluminum and yellow mulches than the other colors and were most numerous on blue mulch. Larger fruit size was obtained on blue than on white mulches, and red mulch showed higher tomato yields than on black mulch. Season yields of extra-large fruits were better on orange than white mulches in the presence of the tomato mottle (Csizinsky et al., 1995).

Cabbage grown in rye mulch sustained lower insect pest populations of the diamondback moth, imported cabbage worm, cabbage looper, and the green peach aphid than cabbage grown in the absence of mulch. The bacterium, *Bacillus thuringiensis* Berliner, (Bt), applied to plants grown with a rye mulch crop, was more effective in reducing diamondback moth damage than Bt applied to non-mulched plants. As in previous studies, plants grown with mulch failed to yield as well as plants grown under conventional tillage (Bottenberg et al., 1997a, 1997b).

Straw mulch was applied to rotated and nonrotated potato fields in Virginia to reduce Colorado potato beetle populations. Mulched potatoes had greater negative effects on overwintered adult and egg mass density than did the standard insecticide treatment. Potato feeding damage was more severe in nonmulched plots than in mulched plots (Zehnder and Hough-Goldstein, 1985). Mulches of black, white, and clear were tested against aphids in watermelons grown in Mexico. The clear mulch had the lowest aphid populations in contrast with bare soil, which had the highest aphid populations. Highest watermelon yields and weights were obtained with clear mulch. By delaying aphid population buildup, mulches may decrease the need for insecticides to control aphid populations (Farais-Larios and Orozco-Santos, 1997a, 1997b). Aluminum-painted plastic mulch performed as well as aluminum plastic film and was superior to non-painted plastic on reducing populations of aphids and thrips on pepper and tomato (Kring and Schuster, 1992).

Populations of *Delia radicum* were significantly lower on sprayed-on wood fibers plus adhesives than on unmulched broccoli plots. Early-season populations of *D. radicum* with the correct combination of color and hydromulch were used to reduce damage (Liburd et al., 1998).

Weed strips planted in a commercial apple orchard in Switzerland showed no significant difference in insect species diversity compared with unweeded areas. However, significantly higher numbers of predators and alternative prey were collected from the weed strips (Wyss, 1996). Sessids or clearwing moth borers attack tree boles, which have been damaged during grass mowing and weeds around the trees. Mulch placed around the base of trees protects the boles from mower injury, thereby reducing the risk of attack by the clearwing borers (Raupp et al., 1992).

In the semi-arid tropics of India, groundnuts are dried directly on the soil which provided an opportunity for termites. *Microtermes obsei* and *Odontotermes* spp. Mulches of dried neem cake or *Ipomoea fistulosa* showed 80 to 90% lower termite damage than groundnuts dried without neem mulch (Gold et al., 1991).

2.8.1   Cover Crops

Cover crops and mulches have been used to manage arthropod pests. (Muma, 1961; Stern et al., 1969; Koptur, 1979; Price et al., 1980; Ryan et al., 1980; Lawton, 1982; Rogers, 1985; Andow et al., 1986; Fleisher and Gaylor, 1987; Smith, 1988; Bugg and Dutcher, 1989; House and Alzugaray, 1989; Russell, 1989; Bugg and Ellis, 1990; Bugg et al., 1990 Phatak, 1992; Liang and Huang, 1994). Cover crops can benefit predators (Bugg et al., 1991; Bugg, 1992) in the absence of prey or under low prey densities. Integration of cool- and warm-season cover crops may provide sufficient secondary prey to predators. Species numbers of lady beetles showed variable response to the 18 cover crop regimes tested, as did aphid populations. Cover crops in orchards may enhance the control of one insect pest species and exacerbate the damage caused by another species (Bugg and Waddington, 1994).

Different cover crops support complexes of beneficial and pest arthropods. Cover crops, however, impact the orchard microclimate and nutritional status of the crop which influence pest dynamics (Bugg and Waddington, 1994). Cover crops hold promise in enhancing populations of predators but further research on cover crops and insects is needed (Bugg et al., 1990).

In California almond orchards, cover crops of soft chess, strawberry clover, or resident vegetation supported populations of the pavement ant, *Tetramorium caespitum* (L.) and the southern fire ant, *Solenopsis xyloni* McCook, which feed on fallen nuts. The very same cover crops accelerated the decomposition of fallen almonds, which removed the overwintering niche of the navel orangeworm, *Amyelois transitella* (Walker) (Bugg and Waddington, 1994).

Mixed results have been obtained using living and dead cover crops to manage insect pests and to maximize crop yields. Living mulches of cereal rye cover crop and perennial rye grass as living mulch did little to reduce insect damage in snap beans. None of the mulches effectively reduced damage by cabbage pests. Mulches require more management to be effective than conventional fall-tillage with trifluralin. Other researchers reported success in managing pests when the cover crops were mowed and used as dead mulch (Masiunas et al., 1997). In a study combining rye and hairy vetch as dead mulch or cover crops, the dead mulch gave better control of *Delia radicum* and *Plutella xylostella* larvae than the incorporation of cover crops. Dead mulch, however, did result in an increased number of slugs (Mangan et al., 1995).

2.8.2   Floating Row Covers

Floating row cover (FRC) along with transparent polyethylene mulch (TPM) were evaluated along with petroleum and sunflower oil sprays to reduce populations of aphids and whiteflies attacking cantaloupe in a semi-arid tropic region of Colima, Mexico. TPM reduced aphid and whitefly populations and the resultant incidence of viral diseases that contributed to the difference of 322 carton/ha between untreated and treated plots. The FRC excluded insects which increased yields, in part, over the control and delayed the onset of viral diseases by two weeks (Orozco et al., 1994).

FRC on cantaloupe grown in Colima, Mexico excluded aphids, *Bemisia tabaci*, and the vegetable leafminer, *Liriomyza sativae* Blanchard, as long as the covers remained in place over the plants. Transparent plastic mulch reduced aphid and whitefly populations but not those of *L. sativae*. Yields from plants grown with mulch alone and with covered plants when the cover was removed during vegetative growth was higher than plants grown on bare soil. Transparent mulch reduced whitefly populations and viral incidence when compared against bare soil (Orozco et al., 1995).

2.9 LIGHT TRAPS

Blacklight, another name for ultraviolet radiant energy, encompasses wave-lengths from 320 to 380 nm. Traps with blacklight lamps are either unidirectional with a cover making the traps visible from only one direction or omnidirectional with light visible from all directions (Deay et al., 1965).

Light traps to control insect pests are based on the assumption that light exerts a powerful influence on insect activity. One of the first experiments to test the efficacy of light traps as a control technique was designed to disrupt oviposition of the codling moth (Herms, 1929). Although some reduction in oviposition was detected in apples on trees near the light traps, the results could not support the use of light traps as a control technique. In other studies, light traps placed in apple orchards reduced the amount of codling moth damage when compared with areas where light traps were not used but cover sprays of insecticides were required (Parrot and Collins, 1935).

Light traps were tested to reduce damage by *Cydia pomonella*; the artichoke plume moth, *Platyptilia carduidactyla* (Riley), and the eastern grape leafhopper, *Erythroneura comes* (Say). After reviewing the light trap catch data, researchers could recommended light traps as a secondary source of insect control and not as a primary control measure (Herms and Ellsworth, 1934). To reduce house fly popu-lations by 50%, the number of traps would be impractical (Pickens and Thimijan, 1986). Light trap efficiency can be improved by placing the traps in dark or dimly lit areas or in areas where pests congregate. In addition to trap placement, bulb orientation and background illumination may influence trap catch (Pickens and Thimijan, 1986).

Lamp selection played an important role in attracting insects to the standard blacklight (Harding et al., 1966). A study involving the standard cool white lamp and three other lamps with different phosphors: conventional (barium silicate), Philips (barium strontium magnesium silicate), and phosphors (strontium borate) were tested for attractancy. The phosphors were significantly better 37% of the time at attracting a particular group of insects. The next effective lamp was Philips at 25.9%, followed by cool white at 14.8%, and the conventional lamp failed to perform better than any of the lamps. In developing light traps or analyzing trap catch data, the type of lamp used in the trap must be considered. For example, while cool white lamps collected less insects than the other lamps, the cool white lamps were more effective in attracting geometrids, scarabs, leafhoppers, and lacewings than the other traps. Over the 39 nights of the lamp test, conventional and new phosphors lamps had the highest mosquito catch (3470 and 3154, respectively), the cool white lamps collected 1211, and the Philips collected 2399 mosquitoes. Some lights are not as attractive to insects and these lamps can be used as part of management programs to minimize insect infestations where lights must remain on (Barrett and Broersma, 1982). Ultraviolet lamps, used in insect electrocutor trays, flickering at twice the main's AC frequency, captured 75% more flies than traps with non-flickering lamps. Lamp flicker should be considered when comparing trap efficacy data (Syms and Goodman, 1987).

2.9.1 Electrocuting Traps

Insect-electrocuting light traps have been used in food and pharmaceutical facilities, with an estimated 200,000 industrial light traps in use (Gilbert, 1984). To be effective, light traps must be used along with appropriate sanitation measures. To maximize trap efficiency, trap placement and trap height are important factors, as are lamp type and reflector surface (Gilbert, 1984).

Electrocutor traps have been effectively used to reduce house fly populations in two dining halls in San Antonio, Texas. Electrocutor traps reduced house fly populations in the dining room by 72% compared with the traps, which were not operational (Lillie and Goddard, 1987). The placement of electrocuting traps to control house flies in restaurants must consider where the dead flies will land. To answer the concern of users and health inspectors as to the correct placement of these traps, a study was conducted in food preparation or handling facilities. Electrocuted house fly scatter patterns are influenced by trap design, air velocity, and height of traps above the floor. A formula, developed from the study, was be used to determine the fraction of total scatter for a wall-mounted trap placed 1 and 2 m above the floor. While an explanation of both formulae is beyond the scope of this review, a few basic points bear mentioning. A suggested placement of an insect containment tray, to collect at 89% of the electrocuted flies in an unbaffled trap, should be no further than 40 cm from the top of the electrocuting grid. The containment trap should measure 20 cm wide and 5 cm deep (Pickens, 1989).

A visual house fly and stable fly trap was constructed of white and yellow pyramids situated on top of a white base with large openings on each side. When flies entered the trap, they were killed by solar-powered electrocuting grids. When three traps were placed near a manure dump, a total of 1360 house flies and 1190 stable flies were caught daily in cool (<23°C) or warm (>30°C) weather. More house flies were collected on the yellow portion of the trap on cool mornings and on the white in warm afternoons. When temperatures reached over 30°C, house flies and stable flies congregated in the shaded part of the trap. (Pickens and Mills, 1993). Trap efficacy can be increased if placed in areas having the highest fly activity and incorporating muscalure. Muscalure (Z-9-tricosene) placed in electrocuting blacklight traps increased trap efficacy 76% in poultry houses (Rutz et al., 1987). Modified traps or solar-powered electrocution grid traps effectively reduced stable fly populations (Foil and Hogsette, 1994) and were more effective than sticky traps (Pickens, 1991).

Commercial blacklight traps have been advertised to control mosquitoes and other biting Diptera in backyards or other areas occupied by humans. Trap catches from blacklight traps revealed that only 31 (or 0.22%) biting flies were caught out of a total of 13,789 insects. Of the non-target insects, 194 insect families representing 12 orders were collected. Of this group, there were 1868 (or 13.5%) predators and parasites and 6670 (or 48.4%) non-biting aquatic insects. The difference between the overwhelming number of beneficial and non-pest status insects caught and the extremely low number of biting flies caught makes the use of blacklight traps worthless and counterproductive (Frick and Tallamy, 1996).

Other researchers experienced better pest control with light traps. In a two-year study, substantial reductions in codling moth damage to apples were noted in an orchard in southern Indiana (Hamilton and Steiner, 1939). Light trap efficacy was compared against a series of different insecticide combinations to control codling moth populations. Light traps used in conjunction with insecticide treatments did influence codling moth populations over insecticide treatments without light traps. The cost of light traps was ignored in the study, but would be a factor in adopting this practice (Parrott and Collins, 1935).

Different aspects of European corn borer adult control with light traps were studied by several workers (Ficht and Hienton, 1939, 1941; Ficht et al., 1940). Complete control was never obtained in the studies, although gravid females were captured by the light traps. The violet-blue band proved the most attractive to the European corn borer. Traps placed in the high spots within the fields caught the highest number of moths. Based on these results, the use of light traps to control insects could not be recommended. In other studies, light traps equipped with fluorescent blacklight bulbs reduced damage to cucumbers from striped and spotted cucumber beetles (Deay et al., 1959; 1963). Some success was obtained in using light traps to control tomato and tobacco hornworms.

A problem of using light traps to reduce pest populations is the efficiency of the traps regardless of the light source to capture the majority of the insects in the study area. Only a few studies have concentrated on the trap efficiency to establish the most effective trap spacing (Hartsack et al., 1968). Trap spacing along with field size and shape affect the probability of catching moths. Populations of cabbage loopers and tobacco budworms may be reduced substantially through the use of light traps if placed according to theoretical calculations. The predicted degree of control can be obtained if certain assumptions are met, such as the insects removed have not oviposited; no migration into the area occurs; and each trap functions independently and performs the same as other traps (Hartsack et al., 1971). Light traps installed at 3 traps per 2.6 km² reduced female tobacco hornworm populations by 55% and 89% for male tomato hornworm (Stanley et al., 1964) and were supported by egg and larval counts inside and outside the study area. In a follow-up study, hornworm populations were reduced 80% compared with control fields.

In another study, pans filled with oil and water were used to determine trapping efficiency. A total of 144 pans were placed within 6 concentric circles surrounding a central light trap. Using a series of formulas, the researchers calculated trap efficiencies for catching bollworms and cabbage loopers. Trap efficiency was 10.7 to

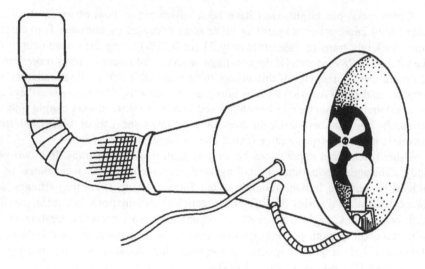

Figure 2.5 Suction light trap with collecting bottle. (Redrawn from Reed et al., 1934.)

50.0% for the bollworms and 8.21 to 38.4% for cabbage loopers indicating that a large number of insects were not attracted to the light (Hartsack et al., 1968).

2.9.2 Suction Light Traps

The congregating behavior of large numbers of cigarette beetles to windows and skylights during sunset or cloudy days led to the development of a light trap attached to a suction device (Figure 2.5). While numerous approaches were taken to control this beetle in warehouses, none proved satisfactory in reducing populations. Sticky fly paper was suspended beneath electric lights, which remain on during the night. The largest reported catch on a 20 × 36 cm fly paper was 1865 beetles. Other attempts were to place sticky materials such as tanglefoot, caster oil, cylinder oil, etc., on windows to capture beetles. Based on the high number of beetles caught in the suction trap, researchers concluded that this apparatus would be an effective control device. Additional information on the accurate infestation levels in warehouses would be needed before the true efficacy of such a device could be determined (Reed et al., 1934).

2.10 IRRADIATION

Electromagnetic energy ranked in order of increasing quantum energy consists of radio waves, infrared, visible, ultraviolet, X-rays, and gamma rays; differ from one another in wavelength; and travel through space as electromagnetic waves with the speed of light. The electromagnetic spectrum can be divided into two basic types: ionizing radiation and nonionizing irradiation. Ionizing radiation consists of gamma rays and electron beam irradiation (Nelson, 1967).

Nonionizing irradiation lacks sufficient energy to dislodge electrons from molecules and includes radio waves, infrared waves, visible light, and microwaves. The use of high-frequency electrical energy in agriculture has been reviewed by several researchers (Nelson, 1973, 1987, 1996; Highland, 1991; Moy, 1993; Fields and Muir, 1996).

2.10.1 Microradiation

The dielectric heating of microwave radiation involved frequencies starting above 500 MHz with application frequencies of 869, 915, and 2450 MHz (Nelson, 1996). *Tribolium confusum* and *Plodia interpunctella* exposed to 2450 MHz of microradiation died as a result of their bodies reaching the lethal temperature of 80°C for the most resistant life stage. Intermittent exposures of 1 or 5 min intervals more effectively killed both species than continuous exposure. The number of insects surviving increased as the moisture levels increased from 6 to 12%. No temperature hot spots were detected on the medium, which had been a concern in other studies (Shayesteh and Barthakur, 1996).

Delia radicum eggs and pupae were treated in a microwave oven at 2450 MHz with 0 to 6kW. Eggs treated at 10 s at 2100 W, 20 s at 1600 W, 20 s at 2100 W, 25 s at 1600 W, and 25 s at 2100 W inhibited egg hatch along with lethal effects on the cabbage. Treatments of pupae after harvest at 10 s at 3000 W and 10 s at 4000 W inhibited adult emergence with no impact on the plant. Post-harvest treatment of the cabbage would be preferable to spring treatment because of the high energy levels required to kill pupae in the soil (Biron et al., 1996).

Wheat and flour infested with *T. confusum* and *Sitophilus granarius* were microwaved at an output of 12.25 cm for 21 s. At this rate, the temperature reached 70.4°C which was lethal to all insects after one week. The number of eggs experiencing mortality varied according to the length of exposure and temperature. A temperature of 80.6°C for 18 s was lethal to 100% of flour beetle larvae (Baker et al., 1956).

2.10.2 Gamma Radiation

Gamma rays damage organisms by breaking chemical bonds or by causing the production of ions or free radicals, which are highly reactive. Numerous commodities have been treated with gamma radiation to disinfest stored product insects. A review of the pests and life stages, commodities, level of treatment, and locality are given in Table 2.2.

The eradication of screwworms, *Cochliomyia hominivorax* (Coquerel), from the U.S. and in most parts of Mexico serves as a cornerstone of the use of gamma radiation to control insects. In 1937, Knipling observed the aggressive sexual behavior of male screwworms and reasoned that large numbers of sterilized males released into the environment would outcompete feral males for females. The hoped-for result would be reduced populations or complete eradication of the screwworm. Male screwworms were sterilized with gamma-irradiation from 60 Co source (Metcalf and Metcalf, 1993). A pilot test was conducted in Florida which involved 2 million sterilized flies per week released at a rate of 1000 flies per 400/km²) over a 5000 km²

Table 2.2   Effect of Irradiation on Survival of Selected Stored-Product Insects

Commodity	Country	Insect	Stage Tested	Radiation Dose (kGy) and Mortality	Citation
Dates	Iraq	Cadra cantella (= Ephestia cautella Oryzaephilus Surinamensis	larvae	0.75 (effective only when used with heat at 40°C.)	Ahmed 1991
	Iran	Tribolium confusum, Oryzaephilus surinamensis, Ephestia cautella	All stages	0.75	Zare et al. 1993
Pulses: Lentil Mashkalai Mung	Bangladesh	Callosobruchus chinensis	eggs	0.04 (prevented hatching)	Bhuiya et al. 1991
			4th instar	0.28–0.32 (prevented molt to adult)	
Oilseeds: Mustard		Angoumois grain moth	eggs, larvae, pupae	0.4 (complete inhibition of metamorphosis)	
		Sawtoothed grain beetle	eggs, larvae, pupae	0.4	
		Cigarette beetle	eggs, larvae, pupae	0.4	
Tobacco leaves		Cigarette beetle	eggs	0.05 (killed eggs)	
			pupae	0.5 (adults emerged, but all killed in 7 days)	
Stored broad beans	Egypt	Bruchus rufimanus Bruchidius incarnatus	adults	0.8 (reinfestation after 150 days post-treatment)	El Kahdy 1991
Dried mushrooms	Hungary	Nemapogon granellus	eggs, larvae	0.2 (both controlled)	Kovacs 1991
		Plodia interpunctella	eggs	0.4 (hatch prevented)	
			adults	0.4 (sterilized)	
Wheat germ		Flour beetles	adults	0.2–0.4 (all dead 20 days post-treatment)	
			adults	0.8 (all dead 10 days post-treatment)	
Copra	Philippines	Copra beetle (Necrobia rufipes)	eggs, early instars	0.05 (no adult emergence)	Manoto 1991
			late instars, pupae	0.10–0.25 (no adult emergence)	

Table 2.2 (continued)   Effect of Irradiation on Survival of Selected Stored-Product Insects

Commodity	Country	Insect	Stage Tested	Radiation Dose (kGy) and Mortality	Citation
Wheat	China	Sitophilus zeamais	adults	0.6–0.8 (all died 20 days post-treatment)	Cheng 1991
		Bruchus pisorum	adults	0.8–1.0 (all died 15–30 days post-treatment)	
Cocoa beans	Malaysia	Tribolium castaneum	adults	0.05–0.1 (all dead in 4–7 weeks post-treatment)	Rahim bin Muda et al. 1991
		Oryzaephilus surinamensis	adults	0.2–1.0 (all dead in 1.5–3 weeks)	
		Lasioderma serricorne	adults	2.0 (all dead in 1 week)	
Coffee beans	Indonesia	Araecerus fasciculatus	all stages	0.10 (all insects dead 20 weeks post-treatment)	Soemantaputra 1991
Leaf tobacco		Lasioderma serricorne	all stages	0.30–0.60 (increase in numbers after 4 months post-treatment)	

area. The natural population was reduced 90%. In 1962, a program to eliminate the screwworm from the southwestern U.S. was started and after two years, no screwworms were found in Texas or New Mexico. Because flies migrated into the U.S. from Mexico, a cooperative project with Mexico pushed the screwworm-free area north of the Isthmus of Tehuantepec. Screwworms have been eliminated from most of Mexico, and the last reported sighting of screwworm in the U.S. was 1982 (Klassen, 1988).

A group of 100,000 Mediterranean fruit fly larvae exposed to gamma radiation at 40 Gy resulted in complete larval mortality with similar results obtained for third instars. The low-dose radiation level makes gamma radiation a suitable quarantine technique (Mansour and Franz, 1996; Heather et al., 1991). Gamma radiation has been used to conserve archival materials such as books, textiles, historical relics, etc. No harmful effects were detected when objects were irradiated below 870 Gy, but these objects sustained damage above 870 Gy. Rates below 870 Gy were sufficient to kill insects even at 43.5 to 130.5 Gy. The treated insects suffered damaged reproductive function, sterilization, and even mortality (Huifen, et al., 1993).

Gamma radiation at a range of 25 to 125 Gy (at five dose levels) against 7- to 10-day-old adults of the sawtoothed grain beetle, Oryzaephilus surinamensis (L.), responded differently depending on the dose level. All died at 125 Gy with LD 50 = 55.6 Gy and LD99.9 = 216 Gy (Tuncbilek, 1997). The effect of irradiation to control Tribolium confusum can be enhanced if used in combination with exposure to nitrogen. Irradiation only at 120 to 1000 Gy caused complete mortality to T. confusum adults in 12 to 15 days. The time to mortality of adult beetles decreased to

17 h when 10 day post-treated irradiated adults were exposed to nitrogen (Buscarlet et al., 1987). In another study involving insect-infested wheat germ and wheat bran, a dose range of 0 to 0.8 kGy, eggs of *T. confusum* were the most sensitive to radiation, followed by larvae and pupae (Kovacs et al., 1986).

Male and female tobacco budworm adults were sterilized with 45 krd of gamma radiation and reduced longevity in both sexes 0 to 10%. The mating ability of sterilized males was approximately the same as that of nonsterilized males, and sterilized females mated as frequently as nonsterilized females. The study attempted to determine if gamma radiation might be used as it was in the eradication of screwworms. The critical aspect was the mating of sterile males to untreated females, which resulted in 25% fewer eggs per female over controls. A combined program of sterile male release augmented with biological control might serve to eradicate the tobacco budworm (Flint and Kressin, 1968).

Dates sealed in polyethylene bags and placed in standard carton boxes were irradiated with low doses of gamma-radiation at 0.46 ± 0.20 kGy. All insects within the bags were killed and no reinfestation occurred. The combination of the standard carton boxes and polyethylene bags lead to the effectiveness of the low gamma radiation in controlling the insects (Ahmed et al., 1994). Although irradiation of food can effectively disinfest various foodstuffs, consumer acceptance of irradiated food as safe and wholesome remains a major obstacle (Moy, 1993). As of 1996, irradiation of food to prevent spoilage and sprouting along with insect disinfestation has been approved by over 30 countries.

2.10.3 Infrared Radiation

Between the radiofrequencies and the visible spectrum exists the infrared region categorized into near-infrared, intermediate-infrared, and far-infrared. Infrared radiation is associated with heat radiated by hot objects and the absorption of infrared energy produces heating in the absorbing material. Any object above absolute zero radiates energy, and the amount of energy radiated is proportional to the fourth power of the absolute temperature, according to the Stefan-Boltzman law (Nelson, 1967).

Surface characteristics of the object impact the radiation and absorption of energy, termed the emissivity of the surface. Infrared radiation to control pests can be categorized into direct application to the insect or infested material or to alter the insect's ability to detect or to be lured to infrared radiation (Evans, 1964; Callahan 1965b).

A few studies exist on the attraction of insects to infrared radiation (Blazer, 1942; Peterson and Brown, 1951; Callahan 1965a). The majority of this effort has been directed toward managing mosquitoes via infrared radiation. When temperatures reached 15°C or below, mosquitoes were attracted to heated objects (Brown 1951). When temperatures reached or rose above 15°C, moisture on clothing had a greater attractancy than simply increasing the temperature of the test object. People with warm hands attracted more mosquitoes than cool skin hands (Smart and Brown, 1956).

Blazer (1942) was the first to suggest the use of infrared radiation to control insect pests. Infrared lamps of 3.18 amperes and 114 volts were placed within 12.7 cm of *T. confusum*; *T. molitor*; the bean weevil, *Acanthoscelides obtectus* (Say); *Attagenus piceus* (Oliver), and the large milkweed bug, *Oncopeltus fasciatus* (Dallas) and exposed to infrared radiation between 5 to 240 s. A 30-s minimum exposure time killed all larvae and adults of *T. molitor* and the other beetles (Frost et al., 1944).

An elaborate wheat treatment study using infrared energy was developed, consisting of a conveyor belt lined with banks of infrared lamps mounted above and below the belt. With the lamps, uniform temperatures were obtained in a 9.8-mm layer of wheat. After the wheat was treated with infrared, the wheat was held for 10 min in a closed tunnel with a temperature of 60°C to kill all stored-grain insects. No impact on germination, baking quality, or thiamin content was detected. The high cost and poor penetrating characteristic of infrared made the conveyor belt treatment too costly and ineffective for commercial use. Infrared treatments of the confused flour beetle, yellow mealworm, and a few other insects raised the insects' temperatures and killed the insects. Only slight mortality differences were detected between larvae and adults of the confused flour beetle treated with infrared. When adult beetles were restrained and prevented from turning on their backs, adults required a shorter exposure time to die than unrestrained larvae. These mortality differences may be attributed to the darker body of the adults which resulted in higher absorption of the radiation and longer retention by the heavily sclerotized body of the adults (Nelson, 1967).

Infrared treatments of the pea weevil, *Bruchus pisorum* (L.), in seed peas at 3 min with lamps held 25 to 20 cm from a single layer of seeds killed the adults, but seed pea germination remained untouched. A gas-fired infrasource ceramic heater was effective in killing all immature rice weevil larvae in rice with 12% moisture with infrared raising rice temperature to 56°C. Larvae of immature lesser grain borers required infrared treatments to obtain a lethal temperature of 68°C for complete mortality. In later experiments with the Angoumois grain moth, *Sitotroga cerealella* (Olivier), rice weevils, and lesser grain borers, complete mortality was obtained with low temperatures, low intensity, and long exposures. The three species could be controlled in rice with 14% moisture with infrared treatments raising the temperature to 65 to 70°C (Nelson, 1967).

While some positive responses have been reported using infrared to control insects, the full potential of this technique has not been realized. In a review of the use of infrared nearly 30 years ago, a statement was made that "It is still too early to evaluate the possibilities of infrared radiation for insect control in all of its ramifications." In recent years, infrared technology has seen major uses in military applications, but not so in pest control (Nelson, 1966).

2.10.4 Pulsed Electric Fields

Pulsed electric fields were used to inactive microorganisms in liquid prepared foods as a way to prolong shelf life and to prevent food poisoning. This technique was applied to *Anastrepha ludens*, eggs and third instars. Pulsed electric fields at 9.2 kV/cm2 delivered in 50 mu s pulses reduced percentage egg hatch to 2.9%. Third instars treated with > 2.0 kV failed to survive to the adult stage. More entomological and engineering research are needed to make pulsed electrical fields practical (Hallman and Zhang, 1997).

Drosophila melanogaster Meigen exposed to a power frequency field above 220 kv/m lost control of body movement and experienced bouncing which lead to reduced life expectancy. Based on this observation, a high-voltage power line might present a barrier to *Drosophila* spp. movement and an electric fence might prove to be a barrier against certain insects (Watson et al., 1986). Reduced mobility of *D. melanogaster* occurred when they were subjected to a 50 Hz electric field. A prototype electrical barrier suppressed flying penetration at 400 kv/m, and crawling penetration was reduced to 6% after 10 min at 640 kv/m (Watson and Neale, 1987).

2.11 TEMPERATURE

Insects develop within a limited temperature range and lowering or raising this range will cause insects to die faster. Insect control has taken advantage of the poikilothermic nature of insects. During pre-neolithic period in the Middle East (9000 to 7000 B.C.) and neolithic Europe (from 4500 B.C. and on), foodstuffs were stored in underground stores to take advantage of cool temperatures (Sigaut, 1980). Hermetic storage of commodities underground continues to be used in India (Girish, 1980), Egypt (Kamel, 1980), Cyprus, and Kenya (DeLima, 1980).

Temperature tolerance of stored product insects can be divided into those that require 30°C for optimum population increase and those which require 26°C. The hot spot temperature in stored grain reaches about 42°C and insects move away from that temperature. For most stored-product insects, 25 to 33°C is optimal for development and at 13 to 25°C or at 33 to 35°C, insects can complete development and produce young. At less than 13°C or higher than 35°C, stored product insects will die. The lethal temperatures will vary depending on the species, life stage of the insect, prior temperature acclimation, relative humidity, moisture content of grain, and the rate of cooling and heating of the grain (Evans, 1987; Fields, 1992; Ali et al., 1997).

For the majority of stored grain pests, an exposure of 14°C will cause complete elimination of the pest. In addition to species differences, life stages respond differently to cold temperatures. For most species, the egg is the most tolerant stage (Banks and Fields, 1995). Extreme temperatures to disinfest stored product pests have been recently reviewed (Mason and Strait, 1998). Heat and cold treatments have successfully disinfested stored grain of insect pests, and complex biological and physical interactions must be considered when using temperatures to control stored product pests (Evans, 1987; Longstaff, 1987). If the moisture content is not high, a temperature of –20°C or lower provides quality pest control and 60°C or above can be used to disinfest grain.

2.11.1   Heat

Stored-product insects exposed to high temperatures require less exposure time to be lethal than cold temperatures (Fields, 1992). One of the first reports of effectively using heat to control stored product pests was reported in 1911 by researchers in Kansas, and the equipment used to heat the grain was installed in several mills throughout the state (Wilson, 1911).

Hot air, especially air found in the head spaces of grain bins or ambient temperatures during the summer, might have potential for managing insect pests in stored grain. Because of the greater volume found in horizontal grain storage versus vertical storage, using heat would have greater potential in horizontal storage (Heather, 1994).

Fluidize and sprouted beds can heat grain from 60 to 120°C to control stored-product insects (Mason and Strait, 1998). Fluidized beds can heat a continuous flow of grain or a single batch. Sprouted beds use air to circulate grain at temperatures of 80 to 180°C. The mechanical disturbance of the grain along with the insect causes negligible damages to the grain and secondarily contributes to insect mortality (Mason and Strait, 1998).

Current hot-air treatments for fruit involve temperatures between 44 and 47°C ranging from exposure times of a few minutes to several hours (Armstrong et al., 1989; Corcoran, 1993). The relationship between temperature and treatment time is probably not linear because more than one physiological system is impacted by heat (Baker, 1939). The upper limit of normal development temperatures of 35 to 40°C or the rapidly lethal temperatures of 60 to 80°C have been investigated to control insects (Dermott and Evans, 1978; White, 1988). Commodities with the potential to harbor Mediterranean fruit flies were treated with heat to kill fruit flies (Baker, 1939). To control *D. melanogaster*, temperatures of 46.5 or 47°C were needed for citrus, 44°C for tomatoes, and 45°C for zucchini (Baker, 1939; Heather, 1994).

Forced air at 48°C killed all third instar Caribbean fruit flies on Florida-grown navel oranges (Sharp and McGuire, 1996), and eggs and larvae infesting carambola were killed when exposed to high temperature of 45.5°C for 90 to 120 min of exposure. (Sharp and Hallman, 1992). Hot air treatments have been used successfully to control a wide range of different fruit fly species in grapefruit (Mangan and Ingle 1994; Sharp and Gould, 1994) and in mangoes (Mangan and Ingle, 1992). While forced-air effectively reduced pest populations on selected commodities, avocado treated at 43°C rendered the fruit unacceptable because of the increased rate of weight loss, susceptibility to vibration injury, and loss of fresh avocado flavor (Kerbel et al., 1987).

An electric sterilizer was advocated in 1930 to kill insects in sealed packages. The machine used "high voltage, high frequency corona" to destroy all stages of insects infesting packaged cereals. Cartons on a conveyor passed beneath electrodes which emitted a corona and the current selected the path of least resistance, which was through the insect acting as a dielectric flux. Boxes containing metal foil or metal paint rendered the machine ineffective. The insects tested, *Plodia interpunctella*,

Tribolium confusum, Silvanus surinamensis, and *Lasioderma serricorne,* were all killed with no damage to the packages (Garman, 1931).

The temperatures of cowpeas, exposed only to the sun's rays, can be to over 60°C which is above the 57°C needed to kill cowpea weevils kept in an oven for one hour. Cowpeas were placed in a simple solar heater constructed of a 3 m × 3 m black and clear plastic sheets. Prior to disinfestation, a 3 m × 3 m ground surface was overlaid with a mattress of dried weeds to insulate the heater. A black plastic sheet was placed on top of the weed mattress on which infested cowpeas were spread out in a single layer. The edges of the clear and black plastic sheets were folded and tucked under to prevent air circulation around the seeds. Small rocks, placed around the sheet, helped secure the heater. Seeds remained in the solar heater for a minimum of two hours (Murdock and Shade, 1991; Ntoukam and Kitch, 1991; Kitch and Ntoukam et al., 1992).

Powderpost beetles cause damage to wood, especially in wood used for flooring. These small beetles tunnel through the wood and create small piles of fine sawdust. The scarcity of good wood for building in 1906 lead to recommendations of using liberal applications of pure kerosene, benzene, gasoline, formalin, and other similar materials to infested wood. Another method was to steam the wood in a tight room or to subject the wood to the highest practical dry heat in a drying kiln. These techniques were practical only if the wood were portable and not finished (Pettit, 1906).

2.11.1.1 Steaming

Fruits and vegetables may be heated to disinfest insects by exposure to hot water, to vapor-heat (water-saturated hot air), to hot dry air, to infrared radiation, and to microwave radiation (Covey, 1989). The first reported use of vapor-heat treatment was applied to eggs and larvae of the Mediterranean fruit fly in Florida in 1929. The treatment process was refined in subsequent years. An unforeseen benefit of the vapor-heat process was the detection of bruised fruit as a result of the exposure to high temperatures. These damaged fruits could be culled before shipment (Baker, 1939).

Scale insects in California were treated with steam in the same manner as gas fumigation. The generated steam was introduced into tents covering infested fruit trees, and the temperature inside the tent was raised to 48.9°C with care taken not to raise the temperature to 51.7°C which would kill buds, fruit, and blossoms. Major difficulties associated with recharging the steam generator and the cost of the steam plant gave way to the standard insecticide application (Marlatt, 1897).

Vapor-heat treatment temperatures of 43.3 to 43.9°C applied for 1.5 hours was effective in controlling narcissus fly and the lesser bulb fly damage to narcissus bulbs (Latta, 1939). A steam and air mixture generated by a prototype machine killed over 56% of adult Colorado potato beetles with acceptable injury levels to potato plants (Pelletier et al., 1998). Hot air treatments of mangoes infested with Caribbean fruit fly, *Anastrepha suspensa* (Loew), larvae sustained 99.9% mortality when treated for 101 min at 51.7°C (Sharp et al., 1991). Heated water, as a quarantine treatment, effectively controlled infestations of several species of fruit flies by immersing fruit for 10 to 70 min at 41.6°C (Sharp and Picho-Martinez, 1990).

2.11.1.2 Burning

Flaming and burning are cultural control methods which use heat to kill insects and to eliminate breeding sites. Chinch bugs, overwintering in bunch grasses in Kansas (Wilson, 1911) and in Illinois (Flint, 1935), were killed by burning the sites, which resulted in reduced damage the following year. Burning wheat stubble was recommended in the 1880s to reduce populations of the Hessian fly and was recommended only when all-natural parasites had emerged from the Hessian flies (Osborn, 1898). Burning webbed hay was recommended to reduce populations of the clover-hay worm prior to planting a new cover crop (Pettit, 1908).

In 1910, grasshoppers remained a major concern in Minnesota and burning a tract of land containing swarming grasshoppers indicated growers' frustrations in dealing with this pest. Growers had to balance the benefits of burning with the damage caused to the hay crop (Washburn, 1910). Burning the short woody segments of straw destroyed the overwintering immature larvae and pupae of the wheat jointworm. In most years, the infestation of wheat jointworm arose from the failure to burn or bury the previous season's wheat stubble (Pettit, 1906; Metcalf and Metcalf, 1993).

In southern Alberta, five burning regimes of seed alfalfa showed control of insects and an increase in carbohydrates, total N, NO_3-N, and extractable K during the eight years of the study (Dormaar and Schaber, 1992). Burning of alfalfa and chemical applications to control a variety of insect pests had variable effects, depending on the insect species (Schaber and Entz, 1991; 1994). Larval alfalfa weevil populations were reduced by burning alfalfa every autumn versus burning every spring or alternate year burning (Schaber and Entz, 1991). A combination of six burning regimes and two insecticide treatments on alfalfa reduced *Lygus* spp. populations and those of the alfalfa plant bug (Schaber and Entz, 1994).

2.11.1.3 Flaming

Flaming, with kerosene as the fuel, started in the western U.S. to control chinch bugs and the greenbug (National Academy Sciences, 1969). An accidental burn in a Georgia alfalfa field resulted in control of the alfalfa weevil, and the burn created an interest in using this technique against the weevil. (Titus, 1910) While flaming was extensively tested against the alfalfa weevil, the cost of the LP-gas was prohibitive and the procedure has not been adopted (Schaber and Entz, 1988).

The impact of burning alfalfa stubble in the spring to control insect pests was dependent of the insect species, and the height at which alfalfa was burned did affect both pests and beneficial insects. The alfalfa plant bug, *Adelophocoris lineolatus* (Goeze), populations were reduced when plants were burned before growth and at the 20 to 25 cm plant height stage. In contrast, Lygus bug, *Lygus* spp., populations were reduced only in the before-growth burn and at the 20 to 25 cm plant stage. No impact was detected on beneficial insects at pregrowth burn but the burn at the 20 to 25 cm growth stage did have a significant negative effect on the number of spiders and ladybird beetles (Schaber and Entz, 1988).

Propane flaming of cotton had major impact on caged tarnished plant bugs and convergent lady beetles. At 10- and 20-cm weed heights, a 100% mortality occurred to both insect species at ground level. Insect mortality was lower at 10- and 20-cm heights at 100 kPa LP gas pressure. At 175 kPa, lady beetles showed lower mortality at a height of 20 cm above the ground than at the 10-cm height and soil surface. While lady beetles suffered mortality, the level of mortality was not as high as that for the tarnished plant bugs (Seifert and Snipes, 1996).

Adult Colorado potato beetles disperse from senescing plants in late summer to overwintering habitats. Young potato foliage served to concentrate the highest number of adult beetles. By concentrating adults in a defined area, propane flamers or vacuuming can be used to reduce populations (Hoy et al., 1996).

2.11.1.4 Radio-frequency Energy

The radio-frequency portion of the electromagnetic spectrum, generally accepted to be between 1 and 100 MHz, causes heating of biological materials, especially wood, stored grain, and foodstuffs, and this dielectric heating characteristic has been studied as a potential for insect control (Headlee and Burdette, 1929; Frings, 1952; Baker et al., 1956; Nelson, 1962, 1965, 1966; Ahmed, 1991). To be effective as a control measure, the thermal tolerance of the host must be sufficiently lower than that of the pest to create differential or dielectric heating. The host medium, therefore, plays a critical role in determining the success of radio-frequency in controlling insect pests (Nelson, 1967). For example, frequencies between 3 and 27 MHz applied to fruits and vegetables caused undesirable heating of the plant materials.

Death caused by dielectric heating can be altered by the frequency and average voltage gradient of the radio-frequency field; the vertical fraction of the field occupied by the treated object if an air space is present; the physiological state of the insect; and the shape of the insect (Frings, 1952). A test of dielectric heating was performed on cotton seed to destroy *Pectinophora gossypiella* larvae. Seeds sterilized at 65.6°C held for 30 s or longer *P. gossypiella* larvae, but a more efficient and uniform control method was required. Results from a dielectric heating test indicated that larvae were killed by the generated heat. Survival of the *P. gossypiella* depended on the final temperature and the exposure time used. Intensities of 512 to 803 volts per cm at 14 to 29 s of exposure raised the temperature to 71.7 to 78.3°C, sufficient for complete mortality. Because the rate of heating increased as voltage increased, the exposure times were reduced (Lowry et al., 1954).

Various species of *Tribolium*, which infest packaged flour and dry cereals, were controlled at various field intensities. A field intensity of 1350 volts per cm at 30 s created a maximum temperature of 68.1°C and caused 99% adult and 100% larval mortalities. When the field intensity was increased to 1560 volts per cm at 22 s, the insect's body temperature was raised to 70.5°C and caused 98% adult and 100% larval mortalities. With an increase of the field intensity to 1790 volts per cm, the body temperature was raised to 75°C and caused a comparable mortality to adults and larvae recorded with 1560 volts per cm. In most cases, arcing within the packages occurred before the lethal temperatures were obtained. The study indicated that the heat loss from the irradiated surface required higher temperatures in the center of the mass to accomplish complete mortality. An external source of heat might be necessary along with the irradiation to maintain the desired temperature at the center of the treated object (Webber et al., 1946).

An important component of radio-frequency use is to determine the conductivity characteristics of the host materials in which the host's dielectric loss factor must be lower than that of the insect to create differential heating. Limited success has been noted in radio-frequency control of borers in wood. Treatments of 1 min at radio-frequency fields of 37 and 76 MHz to blocks infested with powderpost beetles, *Lyctus brunneus* (Stephens), had the same efficacy as heat used in a normal drying kiln. With the radio-frequency fields used and the exposure time, heat from the insect was conducted into the wood, reducing the differential temperature (Nelson, 1966).

In general, the greatest mortality caused by radio-frequency occurred in adults rather than to the immature stages. Among adults, older adults were more susceptible to radio-frequency than younger adults. All adult rice weevils infesting wheat were killed when radio-frequency treatments applied for a few seconds raised the grain temperature to 38°C. This apparent discrepancy in the mortality caused by low temperature may be explained by the rice weevil absorbing radio-frequency energy at a faster rate than the grain. Complete mortality of immature rice was obtained when radio-frequencies raised the grain temperature to 60°C (Nelson, 1967).

Larvae of *Tenebroides mauritanics* experienced greater mortality than adults when exposed to less than 100% lethal dosage, but the immature and adult stages experienced the same mortality at lethal temperatures. These mortality differences were explained on differences in the physiological responses of the nervous systems in the two life stages (Headlee and Burdette, 1929). The greater mortality experienced by adults may be related to the high current concentrations in the insect's legs, perhaps as a result of the damage to the histoblasts (Nelson, 1966). Because of their small size, larvae may have reached the minimum size to affect differential heating. Larvae feeding within seeds are partially protected and may account, in part, for mortality differences between larvae and adults. Seeds are poor conductors of electricity; this may account for the survival of larvae in seeds and may be related to geometric factors (Headlee and Burdette, 1929). An analysis of other insect pests killed by radio frequencies besides tissue heating showed lethal effects to the thoracic nerve ganglia. In the Oriental cockroach, possible synaptic blocking in the ganglia caused death as a result of energy absorption and heating in nerve tissue. Caution must be exercised when reviewing the physiological changes related to radio-frequency treatments, because of the difficulty in attributing mortality effects to thermal and nonthermal events (Nelson, 1967).

Species responded differently to the same level of radio-frequencies. In one study, insects ranked according to increasing resistance to radio-frequencies were rice and granary weevils, sawtoothed grain beetle, confused and red flour beetles, a dermestid (*Trogoderma parable* Beal), cadelle, and lesser grain borer. Associated with the level of variable response to the same level of treatment was the variable time to mortality after exposure. Rice weevil mortality occurred one day after treatment and peaked at one week with no subsequent mortality. Of the insects exposed to radio-frequency, the lesser grain borer exhibited the greatest delayed mortality of the insect pests studied, and this mortality increased substantially during the second week post-treatment than during the first week of post-treatment (Nelson, 1966).

Confused flour beetles in wheat required a radio-frequency to raise the temperature to 49°C and 60°C in wheat shorts where total mortality occurred. Granary weevils found in corn and wheat with similar moisture contents survived better in wheat than in corn when radio-frequency levels caused less than 80% mortality. For radio-frequencies causing higher than 80% mortality, the host medium failed to show mortality differences (Nelson and Kantack, 1966).

A series of insect species including *Sitophilius oryzae*, *S. granarius*, *Trogoderma parable*, *T. glabrum*, *Oryzaephilus surinamensis*, *Tenebroides mauritanicus*, *Plodia interpunctella*, and *Lasioderma sericorne*, were exposed to 39-megacycle radio-frequency. No differential heating was obtained for flour beetles and granary weevils treated at a high of 2450 MHz. To control flour beetles, a temperature exceeding 82°C is required and to control granary weevils temperatures exceeding 77°C are required to control the immature stages. Thus, the lower frequency range of 1 to 50 MHz showed differential heating, and this range had more control possibilities. Moisture content of stored grain showed variable impact on the efficacy of radio-frequencies. Rice weevil adults held in wheat with a moisture content of 11.4 and 12.8% showed no differences in mortality when subjected to radio-frequencies. When moisture levels increased from 12 to 16%, the efficacy of the radio-frequency increased over the weevils held at the lower moisture levels (Nelson and Kantack, 1966).

While much work has been done on radio-frequencies to manage insect pests, no practical large scale operations have been developed. One reason for the failure to develop radio-frequencies is the cost associated with the development or large-scale installations for practical use. Cost estimates, made 25 to 30 years ago, for using radio-frequencies to control stored insect pests revealed a cost of 3 to 5 times more cost than chemical fumigation (Nelson, 1973; 1987; 1996). In the absence of new discoveries of lethal mechanisms of a nonthermal nature, radio-frequencies and microwaves will not replace current control methods (Nelson, 1996).

2.11.2   Cold

The first use of cold temperatures to manage apple maggot was in 1889 (Harvey, 1890). Eggs and larvae of the apple maggot have been controlled by continuous exposure to temperatures between –0.56 to 0.56°C, the range normally used to hold fruit in cold storage. All eggs and larvae of the apple maggot were killed within 35 days at 0°C (Chapman, 1933, 1941; Evans, 1983, 1987).

Grain stored at 15°C was tested to determine the efficacy of chilled grain in reducing infestations. Data from the three-year study showed mixed results in three different storage volumes ranging from 580, 1700, and 5400 tonnes and a two-year study in a 15,000-tonne storage facility. Storage time ranged from 5 to 10 months. Final grain temperatures varied from 5 to 12°C. In 4 out of the 10 trials, insects were found during outloading. Cold treatment was as effective as a chemical treatment and the cost appeared feasible (Elder et al., 1983). Wheat cooled from 34°C to 10°C in 7 weeks and held for 7 months provided a practical and effective physical control of stored-product pests in bulk grain without altering grain quality. Unlike thermal disinfestation, grain can be held under refrigeration for over 3 months (Hunter and Taylor, 1980).

Cold hardiness of stored-product insects varies between- and within-species. A proportion of *Cryptolestes ferrugineus, Oryzaephilus surinamensis,* and *Sitophilus granarius* survived exposure to 13.5°C for 6 months but all died when exposed to 9°C. Populations of *S. zeamais* and *S. oryzae* failed to eclose or metamorphose when subjected to grain held at 10°C. At 15°C, a small number of *S. zeamais* emerged but populations of *S. oryzae* were completely suppressed (Nakakita and Ikenaga, 1997). Chilled aeration lowered populations of *Plodia interpunctella* and the hairy fungus beetle, *Typhaea stercorea* (L.). Cooler grain temperatures prevented establishment of infestations rather than controlling populations. Popcorn stored under cold temperatures does not emit a "popcorn" odor, which might be attractive to the two insect species (Mason, et al., 1997).

Controlled ambient temperatures applied in the fall coupled with chilled aeration in the summer showed significant promise as a non-chemical management technique of *S. zeamais* (Maier et al., 1996). Cold storage of fruits to control eggs, larvae, and pupae of the melon fly, *Bactrocera cucurbitae* Coquillett, was successful when storage temperatures were 4.4 to 7.2°C for 7 weeks, 0.5 to 4.4°C for 3 weeks, or from 0 to 0.5°C for 2 weeks with humidity ranges from 80 to 91 (Back and Pemberton, 1917). Postharvest treatments of three apple cultivars, Anna, Golden Delicious, and Jonathan, subjected to 50°C for 5 or 10 hours and 46°C for 10 hours controlled all developmental stages of the San Jose scale. However, apples were damaged at 50°C, but could withstand 36°C for 12 h (Lurie et al., 1998).

Supercooling of the webbing clothes moth, *Tineola bisselliella* (Hummel), can kill the larvae at −13°C but the recommendation is to treat infested materials at −29°C for prompt disinfestation of all life stages (Chauvin and Vannier, 1997). Clothes, especially furs, were placed in cold storage subjected to a temperature of 4.4°C. Experiments conducted at 4.4°C and at 7.8°C on the survivability of the clothes moth indicated that larvae are able to survive in storage held uniformly at 7.8°C. When the storage temperature was raised to 4.4 or 10°C and returned to 7.8°C, the larvae died when retransferred to the warmer temperature. Based on this experiment, fur storage facilities were advised to subject goods to two to three temperature changes before holding them at 4.4 to 5.5°C (Marlatt, 1908).

While temperatures have been used extensively to manage store product insects, temperatures have not been used in dwellings. The effectiveness of cold temperatures was tested in a mock-up wall void to control wood destroying insects such as the western drywood termite, *Incisitermes minor* (Hagen), *I. snyderi; Coptotermes formosanus*, and the southern lyctus beetle, *Lyctus planicollis* LeConte. The critical thermal minimum temperature, the threshold low temperature at which an insect succumbs from exposure, varies from species to species. To obtain this lethal temperature, every void would need to reach the critical temperature. This would require thermocouples strategically placed through the treatment area. A concern of using liquid N_2 to achieve the low lethal temperatures reduces the level of ambient O_2 in the work area endangering applicators (Rust et al., 1997).

Where insecticide applications are impractical, cold temperatures at 0°C for 60 min or 0°C for several hours have been used to control German cockroaches. Cockroach-infested objects were sealed in plastic bags and placed into a freezer (Bennett et al., 1997).

The integration of heat and cold was investigated to determine impact on Packham's Triumph pear and light brown apple moth, *Epiphyas postvittana* (Walker). All developmental stages of the light brown apple moth were held at either 30°C for 16 h or at 0°C for 1 h, or for 16 h under hypoxia. In one test, all treatments were combined. In the combined treatments, all stages were killed. A combined treatment of 30°C at 30 h under low O_2 plus a month's cold storage killed all light brown apple moth pupae. Companies with controlled atmosphere capabilities could implement the various treatment regimes (Chervin et al., 1997).

2.12   SOUND

Insects produce sounds — which may serve as a warning device — to communicate, find mates, and isolate species. Intraspecific sounds function to aggregate species. Insects produce sound by using a variety of mechanisms, including beating wings during flight. A few studies have focused on detecting sound generated by stored-product insects feeding in stored grain (Hickling and Wei, 1995; Hickling et al., 1997; Shuman et al., 1997; Pittendrigh et al., 1997; Mankin et al., 1997) or termites feeding in wood (Fujii et al., 1990; Erdman, 1994; Lemaster et al., 1997). There exists only limited research on the use of sound to control insects (Frings and Frings, 1958). The flight sound created by female mosquitoes attracts conspecific males and initiates copulatory behavior. A tuning fork vibrating between 300 and 350 Hz elicited the same type of behavior from male mosquitoes when exposed to flying female mosquitoes (Belton, 1994). Traps employing sound to trap mosquitoes have not attracted significant numbers of mosquitoes (Kahn and Offenhauser, 1949). Recent advances in trap design coupled with efficient fans and sound-storing integrated circuits should make traps smaller and more efficient. The primary use of mosquito sound traps is to collect males for sterilization and release (Belton, 1994). In a symposium addressing the topic of attractants for mosquito surveillance and control, nearly all participants agreed that the greatest priority was to develop more efficient and economical traps and attractants (Kline, 1994).

Sound, at 20Hz or greater, caused 90% of the larvae and adults of the Colorado potato beetle to drop from plants. More adults on the leaf surface dropped than those holding on to the leaf edge. Highest response occurred during the initial 2 s of exposure. While vibrations have been demonstrated in removing larvae and adults from plants, no studies have been reported on the level of reinfestation (Boiteau and Misener, 1996). Low-frequency sound waves adversely impacted the development of *Plodia interpunctella* (Mullen, 1973). A 5-minute exposure to 1MHz sound at 14.5 W/cm2 at 26°C in wheat killed all stages of *Sitophilus granarius* (Banks and Fields, 1995).

Sound has been used to collect adults of *Eupasiopterx depleta*, a tachinid predator of a mole cricket *Scapteriscus* spp. (Orthoptera: Gryllotalpidae). Stridulating sounds produced by *S. acletus* were synthesized and used to attract the tachinid. While sticky traps collected more adult tachinids, the sound trap was useful when healthy adults, especially females, were needed (Fowler, 1988).

The lesser wax moth, *Achroia grisella* (Fabricius) and the greater wax moth, *Galleria mellonella* (Linneaus), are pests of bee products. Males of both species produce sound through tymbals located on the tegulae, and females respond to males by fanning their wings. Males can be trapped by detectors, and females can be collected by acoustically-baited traps (Spangler, 1988).

2.13   CONTROLLED ATMOSPHERES

In recent years, there has been growing interest in controlled atmospheres (CA) to manage insect pests in stored grain. Much of the renewed interest in CA is driven by the public's concern of insecticide residues in foodstuffs; health and environmental agencies are targeting the use of insecticides and fumigants in stored grain and have been placing severe limitations on their use. Several excellent conferences have devoted a section of their programs to CA and to the use of non-chemical control of stored product pests (Shejbal, 1980; Ripp, 1984; Donahaye and Navarro, 1987; Highley et al., 1994).

2.13.1   Carbon Dioxide

Carbon dioxide has been known to be toxic to insects (Busvine, 1942) and has received considerable attention in disinfesting stored products. The problems with CO_2 are its slow action and the costs associated with sealing the storage bins. *Sitophilus oryzae* and diapausing *Trogoderma* spp. are some of the most CO_2 tolerant stored-product pests. Under high CO_2 concentrations, pupae (LT99 = 6.9 days in 65% CO_2) were the most tolerant followed by adults (LT99 = 1.5 days in 65% CO_2). At lower CO_2 concentrations, eggs were the most susceptible (LT99 = 8.5 days in 20% CO_2) (Annis and Morton, 1997).

Acanthoscelides obtectus infests grain legumes in many countries. All life stages were killed in 5 days at 25°C and in 3 days at 32°C in 88% CO_2 in air atmosphere. Other temperatures and percent CO_2 concentrations effectively killed the insects. Adults and eggs were the least tolerant and mature larvae and pupae were the most tolerant (Ofuya and Reichmuth, 1993). Eggs of the tobacco moth, *Ephestia elutella* (Hubner), *Plodia interpunctella*, adults of *Oryzaephilis surinamensis* and *Tribolium confusum* and five separate breeding stages of *Sitophilus granarius* were tested at 15°C and in an oxygen content of less than 3% (Reichmuth, 1986). All insect populations, except *S. granarius*, were controlled within 10 days. At a higher temperature of 20°C, the time for mortality was 6 days. At higher CO_2, mortality was achieved in 8 days at 15°C and 6 days at 20°C. Warehouses using this technique should be gas tight to make controlled atmospheres effective. A constant flushing of the system with a pressure of 10 Pascals is required to overcome the entry of external oxygen (Reichmuth, 1986).

Developmental stages of *Epiphyas postvittana* were treated at 30°C for 16 h under hypoxia, or at 0°C for one month. Both treatments killed all insect stages. A combined treatment of 30°C for 30 h under low O_2 plus one month cold storage killed all light brown apple moth pupae and all fifth instars of codling moth and Oriental fruit moths (Chervin et al., 1997). Various concentrations of CO_2, ranging from 10 to 90% were tested on two species of Psocoptera, *Liposcelis bostrychophila* and *L. entomophia*. Of the two species tested, *L. bostrychophila* was the most tolerant species under 45 or 60% CO_2. As expected, increasing exposure times resulted in corresponding increases in mortality for both species. The egg stage was the most tolerant of treatment of both species (Leong and Ho, 1994).

Eggs and adults of two species of bruchids, *Callosobruchus subinnotatus* and *C. maculatus*, died when exposed to 100% CO_2 at 32°C and relative humidity of 70%. Early instars died with an exposure of 48 h, and late instars required 72 h of treatment for complete mortality. The pupal stage was the most tolerant and required 5 to 6 days for complete mortality (Mbata et al., 1994).

The granary weevil appears to have different physiological requirements from those of other common stored-product insects in response to controlled atmospheres. With 98% N_2 and 2% O_2 at 75% relative humidity, all life stages of the granary weevil had shorter lethal exposure times as the temperature increased from 5°C to 30°C. Eggs held at 10°C took 42 days for complete mortality and 3 days for mortality at 30°C. High CO_2 atmospheres (60 to 90%) and N_2 levels (8 to 32%) and relative humidity at 75% showed similar mortality trends, as did the high N_2 treatments (Adler, 1994).

In a large-scale test of *Cydia pomonella* (L.) infested apples held at 0°C, 95 to 100% relative humidity, atmospheric components of 1.5 to 2.0% O_2, <1% CO_2, and the remainder consisting of N_2, no adults emerged from 142,021 immature larvae and 40,389 mature larvae. Thus, controlled atmospheres can be used as part of a potential quarantine treatment for codling moth eggs and larvae (Toba and Moffitt, 1991).

2.13.2 Carbon Dioxide and Nitrogen

Treatment of stored grain with CO_2 or N_2 leaves no insecticide residues and possesses fewer questions than ionizing radiation. *Tribolium confusum* pupae and adults exposed to pure nitrogen, CO_2, or air at temperatures between 38 and 46°C, CO_2 was the most effective against the pupae (Buscarlet 1993). A 99% mortality to *T. castaneum* pupae occurred when placed in an atmosphere of 85% N_2, 14% CO_2, and 1% O_2, and a temperature of 35°C for 44 h. At 26°C under the same conditions, all stages, especially eggs and adults, suffered lower mortality than exposure at 35°C (Donoahaye et al., 1996).

CO_2 served as an effective fumigant of stored wheat and barley. The insects, *Cryptolestes ferrugineus* and *Tribolium castaneum*, were controlled with 34% CO_2 and 15% O_2 with decreased temperatures from 18 to 10°C for 2 weeks (White and Jayas, 1993). In another trial, populations of *Cryptolestes ferrungineus* and the foreign grain beetle, *Ahasverus advena* (Waltl) and four species of mites were completely absent from airtight bins treated with CO_2 at a concentration of 15 to 50% by volume for 42 days. For incompletely sealed bins, frequent addition of CO_2 can compensate for gas loss (White and Jayas, 1991).

The use of CO_2 need not be restricted to grain bins. Based on the success of dry ice disinfestation of four hopper-type rail cars filled with flour and infested with *T. confusum*, CO_2 was useful in static and in-transit situations such as barges, truck-type containers, and ocean bound vessels (Jay and D'Orazio, 1984).

2.13.3 Carbon Dioxide and Pressure

A concern in using CO_2 in stored grain facilities was the 10-day or longer exposure periods needed to be effective, and these times may be unsuitable for quarantines (Annis, 1986). Exposure periods have been reduced by increasing temperatures, holding insects under vacuum, or a combination of reduced pressure and high temperatures. A study of CO_2 under high pressure was performed to determine efficacy against *Sitophilus oryzae*, *Rhyzopertha dominica*, *Tribolium castaneum*, and *Corcyra cephalonica*. Eggs, larvae, and pupae exposed to CO_2 at a high pressure of 30 kg/cm^2 for 5 min were effectively controlled (Aliboso et al., 1994).

Eggs of *Plodia interpunctella* have been reported to be relatively tolerant to CO_2 treatments under pressure. The mortality of eggs to treatment increases with age so knowing the age of the eggs within 2 h is important because if very young eggs are omitted from studies to determine lethal exposure times, a too-short exposure time would be recommended. Eggs less than one day old can survive CO_2 exposure for 40 min at 20 bar. All eggs 1 to 4 days old died in 20 min at 20 bar. The lethal effect was determined by the length of exposure and not on the rate of depressurization, because slowly increased or decreased pressure had toxic effects (Reichmuth and Wohlgemuth, 1994).

The effectiveness of CO_2 under reduced pressure was tested against insects infesting dried fruit. CO_2 was applied at three different temperatures (20, 25, and 30°C) and exposure times of 6, 8, 9, 24, 36, 48, and 60 h in a vacuum autoclave set at 34.6-44 kPa. The eggs, instars 1, instars 2, and mature larvae of *Ephestia cautella*, and *Plodia interpunctella*, and all life stages of *Tribolium castaneum* were tested. Mortalities of 95 and 99% were obtained after 55 and 64 h at 20°C, 41 and 48 h at 25°C, and 31 and 37 h at 30°C. Pupae were the most resistant stage, followed by mature larvae. CO_2 under reduced pressure was less effective than at higher pressures because at temperatures below the optimum for insect development, metabolism was slower, requiring less oxygen, reducing the impact of the controlled atmosphere (Suss and Locatelli, 1994).

Controlling stored product pests with high gas pressure was altered by testing the impact of sudden pressure loss. CO_2 at 20kg/cm^2 and a 5 min exposure time was sufficient to kill all adults of *Sitophilus zeamais*, *Rhyzopertha dominica*, *Tribolium castaneum*, and *Lasioderma serricorne*. When the body pressure rapidly equilibrated with the atmospheric pressure, the rapidly expanding CO_2 in the body damaged the integument. Thus, gas solubility appears to a major cause of mortality related to rapid decrease in CO_2 pressure (Nakakita and Kawashima, 1994).

The sudden or slow increase and decrease of CO_2 pressure to control pests were tested on the adult *L. serricorne*. A rapid increase and decrease of CO_2 with a pressure with 20 bar within 2 s resulted in higher mortalities than a slow increase and decrease within 2 min. Other iterations of this procedure indicated that a slow increase in pressure, followed by a rapid pressure decrease, resulting in higher beetle mortality was higher than with a rapid increase and slow decrease. Thus, decreasing pressure is more efficient than increasing pressure to obtain rapid mortality. (Ulrichs, 1994). Four species of insects: *Rhyzopertha dominica, Sitophilus oryzae, Oryzaephilus surinamensis*, and *Plodia interpunctella* were treated with CO_2 at 5 temperatures and 8 exposure times in a vacuum autoclave. Adults were the most susceptible at 20°C for 24-h exposure (Locatelli and Daolio, 1993).

2.13.4 Atmosphere Generators

Adults of *Tribolium castaneum* and *Sitophilus oryzae* were controlled with fumigation by on-site generation of CO_2. The generation of 80% pure CO_2 was possible by mixing naturally occurring limestone (25 kg) and dilute acid (15 liters of 6N HCl) or baking soda (200 g) mixed with citric acid (100g) along with 200 ml of water. The time required to generate the CO_2 was 90 min with 100% mortality in all tests for all commodities. This unique method may be applicable in developing countries where farm storage does not exceed 10 tonnes and in rural areas lacking bulk storage facilities (Krishnamurthy et al., 1993).

Another method of generating CO_2 used biogenerators in place of costly bottled gas. Wheat bran, sawdust, coffee husks, and a combination of wheat bran and sawdust were used as biogenerators of CO_2. Sawdust alone produced no more than 2% CO_2 while wheat bran generated 19% CO_2, wheat bran and sawdust generated nearly 20% CO_2, and coffee husks recorded 26% CO_2. Most CO_2 was generated when the moisture content was 40%. Biogenerators, using coffee husks, were attached to containers of sorghum infested with *Tribolium castaneum*. After 14 days, all insects in the test bins died while those in the control bin were still alive. Additional development is needed to determine the effectiveness of this plant waste biogenerator in rural parts of the developing world (Patkar et al., 1994).

An exothermic gas generator was used to produce inert atmospheres containing less than 2% oxygen from gases such as butane, propane, or natural gas. A practical test of an exothermic generator was conducted on a 2500-tonne vertical bin containing wheat. A 15-hour purge of the interstitial atmosphere at a flow rate of 220 m³/h of inert gas was followed by exothermic treatment for 4 days at the same flow rate. After 3 to 5 weeks of treatment, all *Sitophilus granarius* adults died, depending on the location of the sample. Between 87-94% of the hidden infestations died after 47 days of treatment at 12.5 to 15.5°C and 1.5% O_2. One of the few negatives associated with the use of exothermic gas generators was the higher cost of propane compared with CO_2 treatments. This procedure would be most useful in developed countries for special purposes where a need to create an inert atmosphere exists, such as in the storage of explosive dusts (Lessard and Le Torc'h, 1987). A commercial test of an exothermic inert atmosphere was conducted in a 39.6-m-high concrete elevator tank filled with 20,000 bu of wheat. Each of the two exothermic generators produced 424.7 m^3 of atmosphere consisting of <0.1% 02, 8.5-11.5% CO_2, and the remainder was N_2. Within 24 h, all adult confused flour beetles were killed, but the immature stages of the rice weevil were resistant when treated between 72 to 96 h of post-treatment (Storey, 1973).

The application of CO_2 in outdoor treatments was conducted in the Philippines against insects infesting maize. Storage cubes containing maize were constructed from polyvinyl chloride or reinforced chlorinated polyethylene sheets. The floor-walls and upper roof-wall were covered by a gas-proof zipper, forming a hermetic seal. The containers were stacked on wooden pallets and covered with tarpaulin sheets. A 35% CO_2 level was maintained for 11 weeks. In the tests, light infestations (0.33 to 0.67 live insects/kg) were observed after storage, but no significant increase in damaged kernels was noted. Hermetic storage of maize can be used for seed storage without significantly altering the qualities of maize (Gras and Bason, 1989; Alvinda et al., 1994).

A novel method of CO_2 delivery involved the use of foam as the carrier. CO_2 gas and air were applied in a non-toxic foam and was designed to be spread over crops or other objects infested with insects and formulated to suffocate the pests. Laboratory studies involved the German cockroach that had higher mortality with the CO_2 foam than with air foam. This technique could be applied to control urban and agricultural pets without pesticides (Choi et al., 1997).

One factor contributing to the lack of acceptance of controlled atmospheres for stored grain insect control was the absence of data relating insect mortality and controlled atmosphere dosage. This information would help to minimize treatment costs and maximize reliability. A review of available control atmosphere data sets was conducted in 1986 and a set of a priori assumptions were made to provide a convenient standard to describe the relationship between gas concentration and exposure required for mortality (Annis, 1986). Based on the perusal of the literature as defined by the assumptions, several conclusions were developed. Both oxygen-deficient and carbon dioxide-enriched atmospheres disinfested stored grain of insect pests. The majority of the 24 species reviewed died in less than 10 days. A few species did survive beyond 10 days with the exact time dependent on the composition of the atmosphere. Dosage regimes for the control of all stored product pests in oxygen deficient atmospheres should be based on the response of *Sitophilus oryzae* pupae; for carbon dioxide atmospheres, the pupae of *S. oryzae* and *Tribolium castaneum* adults. If *T. granarium* is the pest, then carbon dioxide exposures should be set for that species. For grain temperatures between 20 and 29°C, a series of recommendations are made on the per cent of oxygen and CO_2 (Annis, 1986).

2.13.5 Nitrogen Treatments

In addition to controlling insects in stored grain facilities with modified or controlled atmospheres, museums and archives must also contend with controlling insect pests in an economically and environmentally friendly manner. *Drosophila melanogaster*, because of its high reproductive rate, was used to test nitrogen efficacy in a museum environment. Each stage of *D. melanogaster* was placed into separate containers and flushed with nitrogen for 30 min at a rate of 200 ml/min at various temperatures and at 75% relative humidity. A nitrogen atmosphere effectively controlled all stages of *D. melanogaster*. The combination of high temperature and low relative humidity played an important role in decreasing the incubation time required for 100% mortality. Using nitrogen atmospheres as a model, researchers tested the efficacy of nitrogen against materials infested with termites, *Cryptotermes brevis*. The wood was treated with nitrogen at 200 ml/min at 40% relative humidity and 25°C for 8 h. Nitrogen killed all termites, and the method used in the study was safe, inexpensive, and easy to use (Valentin and Preusser, 1990).

Tineola bisselliella, along with other common museum pests, was controlled in a nitrogen atmosphere at 25°C at 65% relative humidity. Infested materials were wrapped in a commercially available plastic tent, and complete mortality was obtained after 8 days. Wrapping worked well when fumigation chambers were unavailable, or the materials were too brittle or delicate to move into chambers (Wudtke and Reichmuth, 1994).

In a test of pure nitrogen and pure CO_2 to control *Sitophilus granarius*, results indicated higher mortality in *S. granarius* pupae exposed to pure CO_2 than to pure nitrogen. Under anoxic conditions, granary weevil pupae produced lactate and the contents of lactate increased most strongly during the first 24 h of treatment. Lactate levels in pupae decreased to nearly zero, suggesting inhibition of the metabolic pathway. When compared with nitrogen, the CO_2-treated pupae had one-third level of lactate. The inhibition of glycolysis at lower lactate levels could explain the more rapid lethal action of CO_2 than that of nitrogen (Adler, 1994).

A 14-day treatment of 95% N_2 caused 100% mortality of the apple twig beetle, *Hypothenemus obscurus* (F.) (Delate and Armstrong, 1994). Nitrogen, argon, and CO_2 tested against several Coleopteran pests in different environments, such as plastic bags of low permeability, a fumigation vacuum chamber, and a fumigation bubble, showed that argon-modified atmosphere achieved the best control with the shortest exposure time (Valentin, 1993).

REFERENCES

Adams, R. G. and L. M. Los, Use of Sticky Traps and Limb Jarring to Aid in Pest Management Decisions for Summer Populations of the Pear Psylla (Homoptera: Psyllidae) in Connecticut. *Journal of Economic Entomology.* 82, pp. 1448-1454, 1989.

Adler, C. S., A Comparison of the Efficacy of CO_2-rich and N_2-rich Atmospheres Against the Granary Weevil *Sitophilus granarius* (L.) (Coleoptera: Curculionidae), in *Stored Product Protection Proceedings of the 6th International Working Conference on Stored-product Protection 17-23 April 1994 Canberra, Australia.* Highley, E., E.J. Wright, H.J. Banks, and B.R. Champ, Eds. CAB International, Wallingford, U.K., pp. 11-15, 1994.

Ahmed, M. S. H., A. A. Altaweel, and A. A. Hameed, Dose Distribution of Gamma Radiation in a New Geometric Configuration of a Standard Carton Date Package and Its Experimental Application for Disinfestation of Packed Dates. *Radiation Physics and Chemistry.* 43, pp. 589-593, 1994.

Ahmed, M. S. H., Irradiation Disinfestation and Packaging of Dates, in *Insect Disinfestation of Food and Agricultural Products by Irradiation. Proceedings of the Final Research Co-ordination Meeting, Beijing, China.* International Atomic Energy Agency, Vienna, pp. 7-26, 1991.

Alderz, W. C., Comparison of Aphids Trapped on Vertical Sticky Board and Cylindrical Aphid Traps and Correlation with Watermelon Mosaic Virus 2 Incidence. *Journal of Economic Entomology.* 69, pp. 495-498, 1976.

Aldryhim Y. N., Efficacy of the Amorphous Silica Dust, Dryacide, Against *Tribolium castaneum* Duval and *Sitophilus granarius* (L.). *Journal of Stored Products Research.* 26, pp. 207-210, 1990.

Aldryhim, Y. N., Combination of Classes of Wheat and Environmental Factors Affecting the Efficacy of Amorphous Silica Dust, Dryacide Against *Rhyzopertha dominica* (F.) *Journal of Stored Products Research.* 29, pp. 271-275, 1993.

Ali, M. F., E. F. M. Abdel-Reheem, and H. A. Abdel-Rahman, Effect of Temperature Extremes on the Survival and Biology of the Carpet Beetle, *Attagenus fasciatus* (Thunberg) (Coleoptera: Dermestidae). *Journal of Stored Products Research.* 33, pp. 147-156, 1997.

Aliboso, F. M., H. Hakakita, and K. Kawashima, A Preliminary Evaluation of Carbon Dioxide under High Pressure for Rapid Fumigation, in *Stored Product Protection Proceedings of the 6th International Working Conference on Stored-product Protection, 17-23 April 1994, Canberra, Australia,* Highley, E., E.J. Wright, H.J. Banks, and B.R. Champ, Eds. CAB International, Wallingford, U.K., pp. 45-47, 1994.

AliNiazee, M. T., A. B. Mohannad, and S. R. Booth, Apple Maggot (Diptera: Tephritidae) Response to Traps in an Unsprayed Orchard in Oregon. *Journal of Economic Entomology.* 80, pp. 1143-1148, 1987.

Alm, S. R., T. Yeh, M. L. Campo, C. G. Dawson, E. B. Jenkins, and A. E. Simeoni, Modified Trap Designs and Heights for Increased Capture of Japanese Beetle Adults (Coleoptera: Scarabaeidae). *Journal of Economic Entomology.* 87, pp. 775-780, 1994.

Alvinda, D. G., F. M. Calibosco, G. C. Sabio, and A. R. Regpala, Modified Atmosphere Storage of Bagged Maize Outdoors Using Flexible Liners: A Preliminary Report, in *Stored Product Protection Proceedings of the 6th International Working Conference on Stored-product Protection, 17-23 April 1994, Canberra, Australia.* Highley, E., E.J. Wright, H.J. Banks, and B.R. Champ, Eds. CAB International, Wallingford, U.K., pp. 22-26, 1994.

Andow, D. A., A. G. Nicholson, H. C. Wien, and H. R. Willson, Insect Populations on Cabbage Grown with Living Mulches. *Environmental Entomology.* 15, pp. 293-299, 1986.

Annis, P. C. and R. Morton, The Acute Mortality Effects of Carbon Dioxide on Various Life Stages of *Sitophilus oryzae. Journal of Stored Product Research.* 33, pp. 115-124, 1997.

Annis, P. C., Towards Rational Controlled Atmospheres Dosage Schedules: A Review of Current Knowledge, in *Stored Product Protection Proceedings of the 6th International Working Conference on Stored-product Protection, 17-23 April 1994, Canberra, Australia.* Highley, E., E.J. Wright, H.J. Banks, and B.R. Champ, Eds. CAB International, Wallingford, U.K., pp. 128-148, 1994.

Anonymous, Metal Screen Curtains Keep Insects and Birds Out of Food Premises. *International Pest Control.* 32, p. 103, 1990.

Anonymous, Pesticide Free Cockroach Traps. *International Pest Control.* July/August. 32, pp. 99-100, 1991.

Appel, A., Nonchemical Approaches to Cockroach Control. *Journal of Agricultural Entomology.* 14, pp. 271-280, 1997.

Appert, J., The Storage of Food Grains and Seeds. *The Tropical Agriculturist.* The Technical Centre for Agricultural and Rural Co-operation. MacMillian Press, London, U.K., 1987.

Armstrong, J. W., J. D. Hansen, B. K. S. Hu, and S. Brown, High-temperature Forced-air Quarantine Treatment for Papayas Infested with Tephritid Fruit Flies. *Journal of Economic Entomology.* 82, pp. 1667-1674, 1989.

Back, E. A. and C. E. Pemberton, The Mediterranean Fruit Fly. *United States Department of Agriculture Bulletin No. 640,* 1918.

Back, E. A. and C. E. Pemberton, The Melon Fly in Hawaii. *United States Department of Agriculture Bulletin No. 491,* 1917.

Bailey, S. W., The Effects of Percussion on Insect Pests of Grain. *Journal of Economic Entomology.* 55, pp. 301-310, 1962.

Bailey, S. W., The Effects of Physical Stress in the Grain Weevil, *Sitophilus granarius. Journal of Stored Products Research.* 5, pp. 311-324, 1969.

Baker, A. C., The Basis for Treatment of Products Where Fruit Flies are Involved as a Condition for Entry into the United States. *United States Department of Agriculture Circular No. 551,* 1939.

Baker, H. and T. E. Hienton, Traps Have Some Value. in *Insects. The Yearbook of Agriculture.* United States Department of Agriculture, Washington, D.C. pp. 406-411, 1952.

Baker, H., Effect of Scraping and Banding Trees Upon the Numbers of Transforming and Hibernating Codling Moth Larvae. *Journal of Economic Entomology.* 37, pp. 624-628, 1944.

Baker, V. H., D. E. Wiant, and O. Taboada, Some Effects of Microwaves on Certain Insects Which Infest Wheat and Flour. *Journal of Economic Entomology.* 49, pp. 33-37, 1956.

Banks, H. J., Impact, Physical Removal and Exclusion for Insect Control, in *Proceedings of the 4th International Working Conference on Stored-product Protection (formerly Stored-product Entomology), Tel Aviv, Israel, September 21-26 1986,* Donahaye, E. and S. Novarro, Eds. Agricultural Research Organization, Volcani Center, Bet Dagan, Israel, pp. 165-184, 1987.

Banks, J. and P. Fields, Physical Methods for Insect Control in Stored-Grain Ecosystems, in *Stored-Grain Ecosystems.* Jayas, D. S., N. D. G. White, and W. E. Muir, Eds. Marcel Dekker, Inc., New York, New York, 1995.

Barak, A. A., Development of a New Trap to Detect and Monitor Khapra Beetle (Coleoptera: Dermestidae). *Journal of Economic Entomology.* 82, pp. 1470-1477, 1989.

Barak, A. V., W. E. Burkholder, and D. L. Faustini, Factors Affecting the Design of Traps for Stored-product Insects. *Journal of the Kansas Entomological Society.* 63, pp. 466-485, 1990.

Barrett, J. R. Jr. and D. B. Broersma, Attractiveness of Three 15-W Blacklight Lamps and a Cool White Lamp to Insects. *Transactions of the American Society of Agricultural Engineers.* 25, pp. 450-455. 1982.

Barrett, R. E., A Statistical Method of Determining the Efficiency of Banding for Codling Moth, with Eight Years' Results. *Journal of Economic Entomology.* 28, pp. 701-704, 1935.

Bateman, P., Good Pest Control in Food Processing is Possible with Minimum Pesticide Usage. *International Pest Control.* 34, pp. 68-79, 1992.

Bellotti, A. and A. van Schoonhoven, Mite and Insect Pests of Cassava. *Annual Review of Entomology.* Annual Reviews, Inc., Palo Alto, California. 23, pp. 39-67, 1978.

Belton, P., Attraction of Male Mosquitoes to Sound. *Journal of the American Mosquito Control Association.* 10, pp. 297-301, 1994

Bennett, G., J. Owens, and R. Corrigan, *Truman's Guide to Pest Control Operations,* 5th ed. Advanstar Communications, Cleveland, OH, 1997.

Bethke, J. A. and T. D. Paine, Screen Hole Size and Barriers for Exclusion of Insect Pests of Glasshouse Crops. *Journal of Entomological Science.* 26, pp. 169-177, 1991.

Bhuiya, A. D., M. Ahmed, R. Rezaur, G. Nahar, S. M. S. Huda, and S. A. K. M. Hossain, Radiation Disinfestation of Pulses, Oilseeds and Tobacco Leaves, in *Insect Disinfestation of Food and Agricultural Products by Irradiation. Proceedings of the Final Research Co-ordination Meeting, Beijing, China.* International Atomic Energy Agency, Vienna. pp. 27-50, 1991.

Biron, D., C. Vincent, M. Giroux, and A. Maire, Lethal Effects of Microwave Exposures on Eggs and Pupae of the Cabbage Maggot and Cabbage Plants. *Journal of Microwave Power and Electromagnetic Energy.* 31, pp. 228-237, 1996.

Blazer, A. I., Heating and Drying, in *Insect Pests of Stored Rice and Their Control, United States Department of Agriculture Farmer's Bulletin No. 1906.* pp. 10-11, 1942.

Boiteau, G. and G. C. Misener, Response of Colorado Potato Beetles on Potato Leaves to Mechanical Vibrations. *Canadian Agricultural Engineering,* 38, pp. 223-227, 1996.

Boiteau, G., C. Misener, R. P. Singh, and G. Bernard, Evaluation of a Vacuum Collector for Insect Pest Control in Potato. *American Potato Journal.* 69, pp. 157-166, 1992.

Boiteau, G., Y. Pelletier, G. C. Misener, and G. Bernard, Development and Evaluation of a Plastic Trench Barrier for Protection of Potato from Walking Adult Colorado Potato Beetles (Coleoptera: Chrysomelidae). *Journal of Economic Entomology.* 87, pp. 1325-1331. 1994.

Bottrell, D. G. and P. L. Adkisson, Cotton Insect Pest Management. *Annual Review of Entomology.* Annual Reviews Inc., Palo Alto, California. 22, pp. 451-481, 1977.

Bottenberg, H., J. Masiunas, C. Eastman, and D. Eastburn, Yield and Quality Constraints of Cabbage Planted in Rye. *Biological Agriculture and Horticulture.* 14, pp. 323-342, 1997b.

Bottenberg, H., J. Masiunas, C. Eastman, and D. M. Eastburn, The Impact of Rye Cover Crops on Weeds, Insects, and Diseases in Snap Bean Cropping Systems. *Journal of Sustainable Agriculture.* 9, pp. 131-155, 1997.

Brader, L., Integrated Pest Control in the Developing World. *Annual Review of Entomology.* Annual Reviews, Inc., Palo Alto, California. 24, pp 225-254, 1979.

Britton, W. E. and M. P. Zappe, Measures for the Control of the European Corn Borer. *Connecticut State Entomologist Thirtieth Report 1930, Connecticut Agricultural Experiment Station Bulletin 327.* 1931.

Britton, W. E., J. T. Ashworth, and O. B. Cooke, Report of the Gypsy Moth, 1936. *Connecticut State Entomologist Thirtieth Report 1936, Connecticut Agricultural Experiment Station Bulletin 327.* 1937.

Britton, W. E., The Apple-Tree Tent-Caterpillar. *Connecticut Agricultural Experiment Station Bulletin 177.* 1913.

Broadbent, L., J. P. Doncaster, R. Hall, and M. A. Watson, Equipment Used for Trapping and Identifying Alate Aphids. *Proceedings of Royal Entomological Society of London (A).* 23, pp. 57-58, 1948.

Brooks, F. E., The Roundheaded Apple-tree Borer. *United States Department of Agriculture Farmer's Bulletin 675,* 1915.

Brown, W. A., Studies of the Responses of the Female *Aedes* Mosquito. Part IV. Field Experiments on Canadian Species. *Bulletin of Entomological Research.* 42, pp. 575-583, 1951.

Brust, G. E. Natural Enemies in Straw-mulch Reduce Colorado Potato Beetle Populations and Damage in Potato. *Biological Control.* 4, pp. 163-169, 1994.

Bryan, J. M. and J. Elvidge, Mortality of Adult Grain Beetles in Sample Delivery Systems Used in Terminal Grain Elevators. *Canadian Entomologist.* 109, pp. 209-213, 1977.

Bugg, L., Using Cover Crops to Manage Arthropods on Truck Farms. *HortScience.* 27, pp. 741-744, 1992.

Bugg, R. L. and C. Waddington, Using Cover Crops to Manage Arthropod Pests of Orchards —
A Review. *Agriculture Ecosystems and Environment*. 50, pp. 11-28, 1994.

Bugg, R. L. and J. D. Dutcher, Warm-Season Cover Crops for Pecan Orchards: Horticultural
and Entomological Implications. *Biological Agriculture and Horticulture*. 6, pp. 123-148,
1989.

Bugg, R. L. and R. T. Ellis, Insects Associated with Cover Crops in Massachusetts. *Biological
Agriculture and Horticulture*. 7, pp. 47-68, 1990.

Bugg, R. L., F. L. Wackers, K. E. Brunson, J. D. Dutcher, and S. C. Phatak, Cool-Season
Cover Crops Relay Intercropped with Cantaloupe: Influence on a Generalist Predator,
Geocoris punctipes (Hemiptera: Lygaeidae). *Journal of Economic Entomology*. 84,
pp. 408-416, 1991.

Bugg, R. L., S. C. Phatak, and J. D. Dutcher, Insects Associated with Cool-Season Cover
Crops in Southern Georgia: Implications for Pest Control in Truck-Farm and Pecan
Agroecosystems. *Biological Agriculture and Horticulture*. 7, pp. 17-45, 1990.

Burkholder, W. E., Pheromones for Monitoring and Control of Stored-product Insects. *Annual
Review of Entomology*. Annual Reviews, Palo Alto, CA. pp. 257-272, 1985.

Buscarlet, L. A., B. Aminian, and C. Bali, Effect of Irradiation and Exposure to Nitrogen on
Mortality of Adults of *Tribolium confusum* Jacquelin du Val, in *Proceedings of the 4th
International Working Conference on Stored-product Protection (formerly Stored-product
Entomology), Tel Aviv, Israel, September 21-26 1986*, Donahaye, E. and S. Novarro, Eds.
Agricultural Research Organization, Volcani Center, Bet Dagan, Israel, pp. 186-193, 1987.

Buscarlet, L. A., Study on the Influence of Temperature on the Mortality of Tribolium
Confusum Duval Exposed to Carbon Dioxide or Nitrogen. *Zeitschrift fur Naturforschung*.
48, pp. 590-594, 1993.

Busvine, J. R., Relative Toxicity of Insecticides. *Nature*. 150, pp. 208-209, 1942.

Byers, J. A., Orientation of Bark Beetles *Pityogenes chalcographus* and *Ips typographus* to
Pheromone Baited Puddle Traps Placed in Grids. A New Trap for Control of Scolytids.
Journal of Chemical Ecology. 19, pp. 2297-2326, 1993.

Calderon, M, and S. Navarro, Sensitivity of Three Stored-Product Insect Species Exposed to
Different Low Pressures. *Nature*. 218, pp. 190, 1968.

Callahan, P. S., Far Infra-red Emission and Detection by Night-flying Moths. *Nature*. 206,
pp. 1172-1173, 1965b.

Callahan, P. S., Intermediate and Far Infrared Sensing of Nocturnal Insects. Part 1. Evidences
for a Far Infrared (FIR) Electromagnetic Theory of Communication and Sensing in Moths
and Its Relationship to the Limiting Biosphere of the Corn Earworm. *Annals of the
Entomological Society of America*. 58, pp. 727-745, 1965a.

Capinera, J. L. and M. R. Walmsley, Visual Responses of Some Sugarbeet Insects to Sticky
Traps and Water Pan Traps of Various Colors. *Journal of Economic Entomology*. 71,
pp. 926-927, 1978.

Chapman, P. J., Mortality of the Apple Maggot in Fruit Held in Cold Storage. *United States
Department of Agriculture Circular No. 600*. 1941.

Chapman, P. J., The Plum Curculio as an Apple Pest. *New York State Agricultural Experiment
Station Bulletin No. 684*. 1938.

Chapman, P. J., Viability of Eggs and Larvae of the Apple Maggot *(Rhagoletis pomonella*
Walsh) at 32°F. *New York State Agricultural Experiment Station Technical Bulletin No.
206*. 1933.

Chapman, R. F., W. W. Page, and A. R. McCaffery, Bionomics of the Variegated Grasshopper
(Zonocerus variegatus) in West and Central Africa. *Annual Review of Entomology*.
Annual Reviews, Inc., Palo Alto, California. 31, pp. 479-505, 1986.

Chauvin, G. and G. Vannier, Supercooling Capacity of *Tineola bisselliella* (Hummel) (Lepidoptera: Tineidae): Its Implication for Disinfestation. *Journal of Stored Products Research*. 33, pp. 283-287, 1997.

Chavasse, D. C., J. D. Lines, K. Ichimori, A. R. Majala, M. N. Minjas, and J. Marijani, Mosquito Control in Dar Es-Salaam. 2. Impact of Expanded Polystyrene Beads and Pyriproxyfen Treatment of Breeding Sites on *Culex quinquefasciatus* Densities. *Medical and Veterinary Entomology*. 9, pp. 147-154, 1995.

Cheng, L. S., W. C. Yao, Z. S. Fen, and J. M. Yue, Optimum Dose Requirements for Irradiation Induced Mortality of *Sitophilus zeamais* and *Bruchus pisorum*, and the Effect of Irradiation on the Nutritional Factors in Corn, Wheat and Rice, in *Insect Disinfestation of Food and Agricultural Products by Irradiation. Proceedings of the Final Research Coordination Meeting, Beijing, China*. International Atomic Energy Agency, Vienna. pp. 127-135, 1991.

Chervin, C., S. Kulkarni, S. Kreidl, F. Birrell, and D. Glenn, A High Temperature Low Oxygen Pulse Improves Cold Storage Disinfestation. *Postharvest Biology and Technology*. 10, pp. 239-245, 1997.

Chervin, C., S. Kulkarni, S. Kreidl, F. Birrell, and D. Glenn, A High Temperature Low Oxygen Pulse Improves Cold Storage Disinfestation. *Postharvest Biology and Technology*. 10, pp. 239-245, 1997.

Childers, C. C. and J. K. Brecht, Colored Sticky Traps for Monitoring *Frankliniella bisponsa* (Morgan) (Thysanoptera: Thripidae) During Flowering Cycles in Citrus. *Journal of Economic Entomology*. 89, pp. 1240-1249, 1996.

Choi, C. Y., P. M. Waller, and T. M. Dennehy, Insect Control with Carbon Dioxide Foam, *Transactions of the American Society of Agricultural Engineers*. 40, pp. 1475-1480, 1997.

Claussen, C. P., Insects Injurious to Agriculture in Japan. *United States Department of Agriculture Circular No. 168*. 1931.

Cline, L. D. and H. A. Highland, Survival of Four Species of Stored-Product Insects Confined with Food in Vacuumized and Unvacuumized Film Pouches. *Journal of Economic Entomology*. 80, pp. 73-76, 1987.

Cline, L. D., Penetration of Seven Common Flexible Packaging Materials by Larvae and Adults of Eleven Species of Stored-product Insects. *Journal of Economic Entomology*. 71, pp. 726-729, 1978.

Cogburn, R. R., E. W. Tilton, and H. J. Brower, Bulk Grain Gamma Irradiation for Control of Insects Infesting Wheat. *Journal of Economic Entomology*. 65, pp. 818-821, 1972.

Corcoran, R. J., N. W. Heather, and T. A. Heard, Vapor Heat Treatment for Zucchini Infested with *Bactrocera cucumis* (Diptera: Tephritidae). *Journal of Economic Entomology*. 88, pp. 66-69, 1993.

Covey, H. M., Heat Treatment for Control of Postharvest Diseases and Insect Pests of Fruit. *HortScience*. 24, pp. 198-202, 1989.

Crosby, C. R. and M. D. Leonard, The Tarnished Plant-bug. *Cornell University Agricultural Experiment Station Bulletin 346*. 1914.

Cross, W. H., H. C. Mitchell, and D. D. Hardee, Boll Weevils: Response to Light Sources and Colors on Traps. *Environmental Entomology*. 5, pp. 565-571, 1976.

Csizinsky, A. A., D. J. Schuster, and J. B. Kring, Color Mulches Influence Yield and Insect Populations in Tomatoes. *Journal of the American Society for Horticultural Science*. 120, pp. 778-784, 1995.

Cytrynowicz, M., J. S. Morgante, and H. M. L. De Souza, Visual Responses of South American Fruit Flies, *Anastrepha fraterculus*, and Mediterranean Fruit Flies, *Ceratitis capitata*, to Colored Rectangles and Spheres. *Environmental Entomology*. 11, pp. 1202-1210, 1982.

Deay, H. O., J. R. Barrett, Jr., and J. G. Hartsock, Field Studies of Flight Response of *Heliothis zea* to Electric Light Traps, Including Radiation Characteristics of Lamps Used. *Proceedings of North Central Branch, Entomological Society of America.* 20, pp. 109-116, 1965.

Deay, H. O. and J. G. Taylor, Preliminary Report on the Relative Attractiveness of Different Heights of Light Traps to Moths. *Proceedings of the Indiana Academy of Science.* 63, pp. 180-184, 1954.

Deay, H. O., J. G. Hartock, and J. R. Barrett, Results on the Use of Light Traps to Control Cucumber Beetles. *Proceedings of the North Central Branch, Entomological Society of America.* 18, p. 37, 1963.

Deay, H. O., J. G. Taylor, and E. A. Johnson, Preliminary Results on the Use of Electric Light Traps to Control Insects in the Home Vegetable Garden. *Proceedings of the North Central Branch, Entomological Society of America.* 14, pp. 21-22, 1959.

Delate, K. M., J. W. Armstrong, and V. P. Jones, Postharvest Control Treatments for *Hypothenemus obscurus* (F.) (Coleoptera, Scolytidae) in Macadamia Nuts. *Journal of Economic Entomology.* 87, pp. 120-126, 1994.

DeLima, C. P. F., Field Experience with Hermetic Storage of Grain in Eastern Africa with Emphasis on Structures Intended for Famine Resources. *Controlled Atmosphere Storage of Grains: An International Symposium Held From 12-15 May, 1980 at Castelgandolfo (Rome) Italy.* Elsevier Scientific Publications, Amsterdam, Holland, 1980.

Dermott, T. and D. E. Evans, An Evaluation of Fluidized-bed Heating as a Means of Disinfesting Wheat. *Journal of Stored Products Research.* 14, pp. 1-12, 1978.

Donahaye, E. J. and S. Navarro, *Proceedings of the 4th International Working Conference on Stored-product Protection (formerly Stored-product Entomology), Tel Aviv, Israel, September 21-26 1986.* Agricultural Research Organization, Volcani Center, Bet Dagan, Israel,. 1987.

Donahaye, E. J., S. Novarro, M. Rindner, and A. Azrieli, The Combined Influence of Temperature and Modified Atmospheres on *Tribolium castaneum* (Herbst) (Coleoptera: Tenebrionidae). *Journal of Stored Products Research.* 32, pp. 225-232, 1996.

Donohoe, H. C. and D. F. Barnes, Notes on Field Trapping of Lepidoptera Attacking Dried Fruits. *Journal of Economic Entomology.* 28, pp. 1967-1072, 1934.

Donohoe, H. C., D. F. Barnes, C. K. Fisher, and P. Simmons, Experiments in the Exclusion of *Ephestia figulilella* Gregson From Drying Fruit. *Journal of Economic Entomology.* 27, pp. 1072-1975, 1934.

Dormaar, J. F. and B. D. Schaber, Burning of Alfalfa Stubble for Insect Control as it Affects Soil Chemical Properties. *Canadian Journal of Soil Science.* 72, pp. 169-175, 1992.

Dowd, P. F., R. J. Bartlet, and D. T. Wicklow, Novel Insect Trap Useful in Capturing Sap Beetles (Coleoptera: Nitidulidae) and Other Flying Insects. *Journal of Economic Entomology.* 85, pp. 772-778, 1992.

Dowdy, A. K. and M. A. Mullen, Multiple Stored-product Insect Pheromone Use in Pitfall Traps. *Journal of Stored Products Research.* 34, pp. 75-80, 1998.

Drake, C. J. and H. M. Harris, Asparagus Insects in Iowa. *Iowa State College of Agriculture and Mechanical Arts. Agricultural Experiment Station Circular No. 134.*, 1932.

Dunkel, F., R. Sterling, and G. Meixel, Underground Bulk Storage of Shelled Corn in Minnesota. *Tunneling and Underground Space Technology.* 2, pp. 367-371, 1987.

Ebeling, W. and R. J. Pence, Relation of Particle Size of the Penetration of Subterranean Termites Through Barriers of Sand or Cinders. *Journal of Economic Entomology.* 50, pp. 690-692, 1957.

Ebeling, W., Sorptive Dusts for Pest Control. *Annual Review of Entomology.* Annual Reviews, Inc., Palo Alto, CA. 16, pp. 123-158, 1971.

Ebeling, W., *Urban Entomology*. Division of Agricultural Sciences, University of California-Berkley, Berkley, CA, 1978.

El Kahdy, E. A. and A. M. Hekal, Irradiation Disinfestation of Pulses and Resistance of Packaging Films to Insect Penetration, in *Insect Disinfestation of Food and Agricultural Products by Irradiation. Proceedings of the Final Research Co-ordination Meeting, Beijing, China*. International Atomic Energy Agency, Vienna, pp. 59-68, 1991.

Elder, W. B., T. F. Ghaly, and G. R. Thorpe, Grain Refrigeration Trials in Australia. *Controlled Atmosphere and Fumigation in Grain Storages: Proceedings of an International Symposium Practical Aspects of Controlled Atmosphere and Fumigation in Grain Storages Held from 11 to 22 April 1983 in Perth, Western Australia*. Developments in Agricultural Engineering 5. Elsevier, Amsterdam, pp. 623-646, 1983.

Erdman, M., Tool of the Future. *Pest Control*. May 1994, pp. 78-79, 1994.

Evans, D. E., Some Biological and Physical Constraints to the Use of Heat and Cold for Disinfesting and Preserving Stored Products, in *Proceedings of the 4th International Working Conference on Stored-product Protection (formerly Stored-product Entomology), Tel Aviv, Israel, September 21-26 1986*, Donahaye, E. and S. Novarro, Eds. Agricultural Research Organization, Volcani Center, Bet Dagan, Israel, pp. 149-164, 1987.

Evans, D. E., The Influence of Relative Humidity and Thermal Acclimation on the Survival of Adult Grain Beetles in Cooled Grain. *Journal of Stored Products Research*. 19, pp. 173-180, 1983.

Evans, D. E., The Survival of Immature Grain Beetles at Low Temperatures. *Journal of Stored Products Research*. 23, pp. 79-83, 1987.

Evans, W. G., Infrared Receptor in *Melanophila acuminata. Nature*. 4928, p. 211, 1964.

Eyer, J. R., A Four Year Study of Codling Moth Baits in New Mexico. *Journal of Economic Entomology*. 24, pp. 998-1001, 1931.

Farias-Larios, J. and M. Orozco-Santos, Color Polyethylene Mulches Increase Fruit Quality and Yield in Watermelon and Reduce Insect Pest Populations in Dry Tropics. *Gartenbauwissenschaft*. 62, pp. 255-260, 1997b.

Farias-Larios, J. and M. Orozco-Santos, Effect of Polyethylene Mulch Colour on Aphid Populations, Soil Temperature, Fruit Quality, and Yield of Watermelon Under Tropical Conditions. *New Zealand Journal of Crop and Horticultural Science*. 25, pp. 369-374, 1997a.

Faustini, D. L., A. V. Barak, W. E. Burkholder, and J. Leosmartinez, Combination Type Trapping for Monitoring Stored-Product Insects – A Review. *Journal of the Kansas Entomological Society*. 63, pp. 539-547, 1990.

Ficht, G. A. and T. E. Hienton, Some of the More Important Factors Governing the Flight of European Corn Borer Moths to Electric Traps. *Journal of Economic Entomology*. 34, pp. 599-604, 1941.

Ficht, G. A. and T. E. Hienton, Studies of the Flight of European Corn Borer Moths to Light Traps: A Progress Report. *Journal of Economic Entomology*. 32, pp. 520-526, 1939.

Ficht, G. A., T. E. Hienton, and J. M. Fore, The Use of Electric Light Traps in the Control of the European Corn Borer. *Agricultural Engineering*. 21, pp. 87-89, 1940.

Fields, P. G. and W. E. Muir, Physical Control, in *Integrated Management of Insects in Stored Products*. Subramanyam, B. and D. W. Hagstrum, Eds. Marcel Dekker, Inc., New York, 1996.

Fields, P. G., The Control of Stored-product Insects and Mites with Extreme Temperatures. *Journal of Stored Products Research*. 28, pp. 89-118, 1992.

Finch, S., Influence of Trap Surface on the Number of Insects Caught in Water Traps in Brassica Crops. *Entomologia Experentia et Applicata*. 59, pp. 169-173, 1991.

Finch, S., The Effectiveness of Traps Used Currently for Monitoring Populations of the Cabbage Root Fly (*Delia radicum*). *Annals Applied Biology*. 116, pp. 447-454, 1990.

Fleisher, S. J. and M. J. Gaylor, Seasonal Abundance of *Lygus Lineolaris* (Heteroptera: Miridae) and Selected Predators in Early Season Uncultivated Hosts: Implications for Managing Movement into Cotton. *Environmental Entomology*. 16, pp. 379-389, 1987.

Flint, H. M. and E. L. Kressin, Gamma Irradiation of the Tobacco Budworm: Sterilization, Competitiveness, and Observations on Reproductive Biology. *Journal of Economic Entomology*. 61, 477-483, 1968.

Flint, W. P., The Chinch Bug. *Journal of Economic Entomology*. 28, pp. 333-341, 1935.

Flint, W. P., M. D. Farrar, and W. E. McCauley, Chinch Bug Barriers and Repellents. *Journal of Economic Entomology*. 28, pp. 410-414, 1935.

Foil, L. D. and J. A. Hogsette, Biology and Control of Tabanids, Stable Flies and Horn Flies. *Revue Scientifique et Technique de l'Office International des Epizooties*. 13, pp. 1125-1158, 1994.

Foster, W. A. and R. G. Hancock, Nectar-related Olfactory and Visual Attractants for Mosquitoes. *Journal of the American Mosquito Control Association*. 10, pp. 288-289, 1994.

Fowler, H. G., Traps for Collecting Live *Euphasiopteryx depleta* (Diptera: Tachinidae) at a Sound Source. *Florida Entomologist*. 71, pp. 654-656, 1988.

Frick, T. B. and D. W. Tallamy, Density and Diversity of Nontarget Insects Killed by Suburban Electrical Insect Traps. *Entomological News*. 102, pp. 77-82, 1996.

Frings, H. and M. Frings, Uses of Sounds by Insects. *Annual Review of Entomology*, Annual Reviews, Inc., Palo Alto, California, pp. 87-106, 1958.

Frings, H., Factors Determining the Effects of Radio-frequency Electromagnetic Fields on Insects and Materials They Infest. *Journal of Economic Entomology*. 45, pp. 396-408, 1952.

Frost, S. W., Bait Pails as a Possible Control for the Oriental Fruit Moth. *Journal of Economic Entomology*. 19, pp. 441-450, 1926.

Frost, S. W., Continued Studies of Baits for Oriental Fruit Moth. *Journal of Economic Entomology*. 21, pp. 339-348, 1928.

Frost, S. W., Fourth Contribution to a Study of Baits, with Special Reference to the Oriental Fruit Moth. *Journal of Economic Entomology*. 22, pp. 101-108, 1929.

Frost, S. W., L. E. Dills, and J. E. Nicholas, The Effects of Infrared Radiation on Certain Insects. *Journal of Economic Entomology*. 37, pp. 287-290, 1944.

Fujii, Y., M. Nogushi, Y. Imamura, and M. Tokoro, Using Acoustic Emission Monitoring to Detect Termite Activity in Wood. *Forest Products Journal*. 40, pp. 34-36, 1990.

Gahukar, G. T., Control of Cotton Insect and Mite Pests in Subtropical Africa: Current Status and Future Needs. *Insect Science and Its Application*. 12, pp. 313-338. 1991.

Garman, P., An Electric Sterilizer for Killing Insects in Milled Cereals. *Connecticut State Entomologist Thirtieth Report 1930, Connecticut Agricultural Experiment Station Bulletin 327*. 1931.

Gauthier, N. L., P. A. Logan, L. A. Tewskbury, C. F. Hollingsworth, D. C. Weber, and R. G. Adams, Field Bioassay of Pheromone Lures and Trap Designs for Monitoring Adult Corn Earworm (Lepidoptera: Noctuidae) in Sweet Corn in Southern New England. *Journal of Economic Entomology*. 84, pp. 1833-1836, 1991.

Gerhardt, P. D. and D. L. Lindgren, Penetration of Various Packaging Films by Common Stored-product Insects. *Journal of Economic Entomology*. 47, pp. 282-287, 1954.

Gilbert, D., Insect Electrocutor Traps, in *Insect Management for Food Storage and Processing*. American Association of Cereal Chemists, St. Paul, MN, pp. 89-108, 1984.

Girish, G. K., Studies on the Preservation of Food Grains Under Natural Airtight Storage. *Controlled Atmosphere Storage of Grains: An International Symposium Held From 12–15 May, 1980 at Castelgandolfo (Rome) Italy.* Elsevier Scientific Publications, New York, 1980.

Gold, C. S., J. A. Wightman, and M. P. Pimbert, Effects of Mulches on Foraging Behavior of *Microtermes obsei* and *Odontotermes* Spp. in India. *Insect Science and Its Application.* 12, pp. 297-303, 1991.

Golob, P., Current Status and Future Perspectives for Inert Dusts for Control of Stored Product Insects. *Journal of Stored Products Research.* 33, pp. 69-79, 1997.

Goodenough, J. L. and J. W. Snow, Increased Collection of Tobacco Budworm by Electric Grid Traps as Compared with Blacklight and Sticky Traps. *Journal of Economic Entomology.* 66, pp. 450-453, 1973.

Goodenough, J. L., Adult Screworms: Comparison of Captures in Wind Oriented and Electrocutor Grid Traps. *Journal of Economic Entomology.* 72, pp. 419-422, 1979.

Grace, J. K., J. R. Yates, C. H. M. Tome, and R. J. Oshiro, Termite-resistant Construction: Use of a Stainless Steel Mesh to Exclude *Coptotermes formosanus* (Isoptera: Rhinotermitidae). *Sociobiology.* 28, 365-372, 1996.

Gras, P. W. and M. L. Bason, Biochemical Effects of Storage Atmospheres on Grain Quality. *Fumigation and Controlled Atmosphere Storage of Grain: Proceedings of an International Conference, Singapore, 14-18 February 1989, ACIAR Proceedings No. 25.* pp. 83-91, 1989.

Green, T. A. and R. J. Prokopy, Visual Monitoring Trap for the Apple Blotch Leafminer Moth, *Phyllonorycter crataegella* (Lepidoptera: Gracillariidae). *Environmental Entomology.* 15, pp. 562-566, 1986.

Hall, R. D. and K. E. Doisy, Walk-through Trap for Control of Horn Flies (Diptera: Muscidae) on Pastured Cattle. *Journal of Economic Entomology.* 82, pp. 530-534, 1989.

Hallman, G. J. and Q. H. Zhang, Inhibition of Fruit Fly (Diptera: Tephritidae) Development by Pulsed Electric Field. *Florida Entomologist.* 80, pp. 239-248, 1997.

Hamilton, D. W. and L. F. Steiner, Light Traps and Codling Moth Control. *Journal of Economic Entomology.* 32, pp. 867-872, 1939.

Hanan, J. J., W. D. Holley, and K. L. Goldsberry, *Greenhouse Management.* Springer-Verlag, New York, 1978.

Haniotakis, G., M. Kozyrakis, T. Fitsakis, and A. Antonidaki, An Effective Mass Trapping Method for the Control of *Dacus oleae* (Diptera: Tephritidae). *Journal of Economic Entomology.* 84, pp. 564-569, 1991.

Hardee, D. D., A. A. Weathersbee III, J. M. Gillespie, G. L. Snodgrass, and A. R. Quisumbing, Performance of Trap Designs, Lures, and Kill Strips for the Boll Weevil (Coleoptera: Curculionidae). *Journal of Economic Entomology.* 89, pp. 170-174, 1996.

Hardee, D. D., O. H. Lindig, and T. B. Davisch, Suppression of Populations of Boll Weevils Over a Large Area in West Texas with Pheromone Traps in 1969. *Journal of Economic Entomology.* 64, pp. 928-933, 1971.

Harding, W. G., J. G. Hartsock, and G. G. Rohwer, Blacklight Trap Standards for General Insect Surveys Recommended by the Entomological Society of America. *Entomological Society of America Bulletin.* 12, pp. 31-32, 1966.

Harper, A. M. and T. P. Story, Reliability of Trapping in Determining the Emergence Period and Sex Ratio of the Sugarbeet Root Maggot *Tetanops myopaeformis* (Roder) (Diptera: Otitidae). *Canadian Entomologist.* 94, pp. 268-271, 1962.

Harris, H. M. and G. C. Decker, Paper Barriers for Chinch Bug Control. *Journal of Economic Entomology.* 27, pp. 854-857, 1934.

Hartsack, A. W. Jr., J. P. Hollingsworth, R. L. Ridgeway, and H. H. Hunt, Determination of Trap Spacings Required to Control an Insect Population. *Journal of Economic Entomology.* 64, pp. 1090-1100, 1971.

Hartsack, A. W., J. A. Witz, and D. R. Buck, Moth Traps for the Tobacco Budworm. *Journal of Economic Entomology.* 72, pp. 519-522, 1979.

Hartsack, A. W., J. P. Hollingsworth, and D. A. Lindquist, A Technique for Measuring Trapping Efficiency of Electric Insect Traps. *Journal of Economic Entomology.* 61, pp. 546-552, 1968.

Harvey, F. L., The Apple Maggot. *Maine Agricultural Experiment Station Report 1889.* 1890.

Harwood, R. F. and M. T. James, *Entomology in Human and Animal Health.* Macmillian Publishing Company, New York. 548 pp, 1979.

Haseman, L., Controlling Horn and Stable Flies. *University of Missouri College of Agriculture, Agricultural Experiment Station Bulletin 254.* 1927.

Haseman, L., Peach "Stop Back" and Tarnished Plant-bug. *Journal of Economic Entomology.* 6, pp. 237-240, 1913.

Hata, T. Y., A. H. Hara, B. K. S. Hu, R. T. Kaneko, and V. L. Tenbrink, Excluding Pests From Red Ginger Flowers with Insecticides and Pollinating, Polyester, or Polyethylene Bags. *Journal of Economic Entomology.* 88, pp. 393-397, 1995.

Headlee, T. J. and R. C. Burdette, Some Facts Relative to the Effect of High Frequency Radio Waves on Insect Activity. *Journal of New York Entomological Society.* 37, pp. 59-64, 1929.

Heath, R. R., N. D. Epsky, A. Guzman, B. D. Dueben, A. Manukian, and W. L. Meyer, Development of a Dry Plastic Insect Trap with Food-Based Synthetic Attractant for the Mediterranean and Mexican Fruit Flies (Diptera: Tephritidae). *Journal of Economic Entomology.* 88, pp. 1307-1315, 1995.

Heathcote, G. D., The Comparison of Yellow Cylindrical, Flat and Water Traps, and of Johnson Suction Traps for Sampling Aphids. *Annals of Applied Biology.* 45, pp. 133-139, 1957.

Heather, N. W., Commodity Disinfestation Treatments With Heat, in *Stored Product Protection Proceedings of the 6th International Working Conference on Stored-product Protection, 17-23 April 1994, Canberra, Australia.* Highley, E., E.J. Wright, H.J. Banks, and B.R. Champ, Eds. CAB International, Wallingford, U.K., pp. 1199-1200, 1994.

Heather, N. W., R. J. Corcoran, and C. Banos, Disinfestation of Mangoes with Gamma Irradiation Against Two Australian Fruit Flies (Diptera: Tephritidae). *Journal of Economic Entomology.* 84, pp. 1304-1307, 1991.

Herms, W. B. and J. K. Ellsworth, Field Tests of the Efficacy of Colored Light in Trapping Insect Pests. *Journal of Economic Entomology.* 27, pp. 1055-1067, 1934.

Herms, W. B., A Field Test of the Effect of Artificial Light on the Behavior of the Codling Moth *Carpocapsa pomonella* L. *Journal of Economic Entomology.* 22, pp. 78-88, 1929.

Hess, A. D., The Biology and Control of the Round-headed Apple-tree Borer, *Saperda candida* Fabricius. *New York State Agricultural Experiment Station Bulletin No. 688.* 1940.

Hickling, R. and W. Wei, Sound Transmission in Stored Grain. *Applied Acoustics.* 45, pp. 1-8, 1995.

Hickling, R., W. Wei, and D. W. Hagstrum, Studies of Sound Transmission in Various Types of Stored Grain for Acoustic Detection of Insects. *Applied Acoustic.* 50, pp. 263-278, 1997.

Highland, H. A. and R. Wilson, Resistance of Polymer Films to Penetration by Lesser Grain Borer and Description of a Device for Measuring Resistance. *Journal of Economic Entomology.* 74, pp. 67-70, 1981.

Highland, H. A., Penetration of Packaging Films by the Cowpea Weevil (Coleoptera: Bruchidae). *Journal of Entomological Science.* 21, pp. 33-37, 1986.

Highland, H. A., Post-irradiation Protection From Infestation by Insect Resistant Packaging, in *Insect Disinfestation of Food and Agricultural Products by Irradiation: Proceedings of the Final Research Co-ordination Meeting, Beijing, China*. International Atomic Energy Agency, Vienna, pp. 51-57, 1991.

Highland, H. A., Vacuumized and Unvacuumized Polyethylene or Polyester Film Bags: Insect Survival and Resistance to Insect Penetration. *Journal of Economic Entomology*. 81, pp. 955-958, 1988.

Highley, E. E. J. Wright, H. J. Banks, and B. R. Champ, *Stored Product Protection Proceedings of the 6th International Working Conference on Stored-product Protection, 17-23 April 1994, Canberra, Australia*. Volumes 1 and 2. CAB International, Wallingford, U.K., 1994.

Himelick, E. B. and D. W. Ceplecha, Dutch Elm Disease Eradication by Pruning. *Journal of Arboriculture*. 2, pp. 81-84, 1976.

Hinds, W. E. and H. Spencer, Sugarcane Borer Control Aided Through Utilization of Infested and Trap Corn. *Louisiana Experiment Station Bulletin No. 198*, Louisiana State University, 1927.

Hinkle, N. C., M. K. Rust, and D. A. Reierson, Biorational Approaches to Flea (Siphonaptera: Pulicidae) Suppression: Present and Future. *Journal of Agricultural Entomology*. 14, pp. 309-321, 1997.

Hjalten, J. and P. W. Price, The Effect of Pruning on Willow Growth and Sawfly Population Densities. *Oikos*. 77, pp. 549-555, 1996.

Holloway, T. E., W. E. Haley, U. C. Loftin, and C. Heinrich, The Sugar-cane Moth Borer in the United States. *United States Department of Agriculture Technical Bulletin No. 41.*, 1928.

House, G. J. and M. D. R. Alzugaray, Influence of Cover Cropping and No-tillage Practices on Community Composition of Soil Arthropods in a North Carolina Agroecosystem. *Environmental Entomology*. 18, pp. 302-307, 1989.

Howard, L. O., The Shade-tree Insect Problem in the Eastern United States, in *Yearbook of the United States Department of Agriculture 1895*. Government Printing Office, Washington, D.C., 1896.

Howard. L. O., Progress in Economic Entomology in the United States, in *Yearbook of the United States Department of Agriculture 1899*. United States Printing Office, Washington, D.C., 1900.

Hoy, C. W., J. A. Wyman, T. T. Vaughn, D. A. East, and P. Kaufman, Food, Ground Cover, and Colorado Potato Beetle (Coleoptera: Curculionidae) Dispersal in Late Summer. *Journal of Economic Entomology*. 89, pp. 963-969, 1996.

Huber, L. L. and J. S. Houser, A Comparison of Certain Materials Used as Chinch Bug Barriers. *Journal of Economic Entomology*. 28, pp. 414-416, 1935.

Huifen, F., L. Jingren, and H. Xin, Study on Feasibility of Insect Control with Gamma Ray. *Radiation Physics and Chemistry*. 42, pp. 625-627, 1993.

Hunter, A. J. and P. A. Taylor, Refrigeration Aeration for the Preservation of Bulk Grain. *Journal of Stored Product Research*. 16, pp. 123-131, 1980.

Ingram, J. W. and E. K. Bynum, The Sugarcane Borer. *United States Department of Agriculture Farmer's Bulletin No. 1884*, 1941.

Isely, D. and H. H. Schwardt, The Rice Water Weevil. *University of Arkansas Agricultural Experiment Station Bulletin No. 299*, 1934.

James, D. G., R. J. Bartlet, and C. J. Moore, Trap Design Effect on Capture of *Carpophilus* Spp. (Coleoptera: Nitidulidae) Using Synthetic Aggregation Pheromones and a Coattractant. *Journal of Economic Entomology*. 89, pp. 648-653, 1996.

Jay, E. and R. D'Orazio, Progress in the Use of Controlled Atmospheres in Actual Field Situations in the United States. *Controlled Atmospheres and Fumigation in Grain Storages: Proceedings of an International Symposium Practical Aspects of Controlled Atmosphere and Fumigation in Grain Storages held from 11 to 22 April 1983 in Perth, Western Australia, Developments in Agricultural Engineering 5.* Elsevier, Oxford, U.K., pp 3-14, 1984.

Joffe, A., The Effect of Physical Disturbance of "Turning" of Stored Maize on the Development of Insect Infestations. I. Grain Elevator Studies. *South African Journal of Agricultural Science.* 6, pp. 55-64, 1963.

Judd, G. R., J. H. Borden, and A. D. Wynne, Visual Behaviour of the Onion fly, *Delia antiqua*: Antagonistic Interaction of Ultraviolet and Visible Wavelength Reflectance. *Entomologia Experimentalis et Applicata.* 49, pp. 221-234, 1988.

Kahn, M. C. and W. Offenhauser, The First Field Tests of Recorded Mosquito Sounds Used for Mosquito Destruction. *American Journal of Tropical Medicine.* 29, pp. 811-825, 1949.

Kaakeh, W. and G. W. Bennett, Evaluation of Trapping and Vacuuming Compared with Low-Impact Insecticide Tactics for Managing German Cockroaches in Residences. *Journal of Economic Entomology.* 90, pp. 976-982, 1997.

Kamel, A. H., Underground Storage in Some Arab Countries. *Controlled Atmosphere Storage of Grains: An International Symposium Held From 12-15 May, 1980 at Castelgandolfo (Rome) Italy.* Elsevier Scientific Publications, New York, 1980.

Kerbel, E. L., F. G. Mitchell, and G. Mayer, Effect of Postharvest Heat Treatments for Insect Control on the Quality and Market Life of Avocados. *HortScience.* 22, pp. 92-92, 1987.

Kiritani, K., Pest Management in Rice. *Annual Review of Entomology.* Annual Reviews, Inc., Palo Alto, California. 24, pp 279-312, 1979.

Kitch, L. W. and G. Ntoukam, Airtight Storage of Cowpea in Triple Plastic Bags (Triple-bagging). *Agronomic Research Institute of Cameroon (IRA), Maroua Research Center, CRSP Cowpea Storage Project, Technical Bulletin,* 1991b.

Kitch, L. W. and G. Ntoukam, Le Stockage Du Niebe Dans La Cendre. *Fiche Technique, Collaborative Research Support Program.* Purdue University, West Lafayette, Indiana, 1991a.

Kitch, L. W., G. Ntoukam, R. E. Shade, J. L. Wolfson, and L. L. Murdock, A Solar Heater for Disinfesting Stored Cowpeas on Subsistence Farms. *Journal of Stored Products Research.* 28, pp. 262-267, 1992.

Klassen, W., Eradication of Introduced Arthropod Pests: Theory and Historical Practice. *Miscellaneous Publications of the Entomological Society of America No. 73,* 1988.

Kline, D. L., Introduction to Symposium on Attractants for Mosquito Surveillance and Control. *Journal of the American Mosquito Control Association.* 10: 253-257, 1994.

Klotz, J. H. and B. L. Reid, The Use of Spatial Cues for Structural Guide-line Orientation in *Tapinoma sessile* and *Camponotus pennsylvanicus* (Hymenoptera: Formicidae). *Journal of Insect Behavior.* 5, pp. 71-82, 1992.

Koppenhofer, A. M., K. V. Seshureddy, and R. A. Sikora, Reduction of Banana Weevil Populations with Pseudostem Traps. *International Journal of Pest Management.* 40, pp. 300-304, 1994.

Koptur, S., Facultative Mutualism Between Weedy Vetches Bearing Extrafloral Nectaries and Weedy Ants in California. *American Journal of Botany.* 66, pp. 1016-1020, 1979.

Kovacs, E., I. Kiss, A. Boros, N. Horvath, J. Toth, P. Gyulai, and A. Szalma, Disinfestation of Different Cereal Products by Irradiation. *Radiation Physics and Chemistry.* 28, pp. 545-548, 1986.

Kovacs, E., Irradiation Disinfestation of Wheat, Dried Wheat Products and Mushrooms, in *Insect Disinfestation of Food and Agricultural Products by Irradiation. Proceedings of the Final Research Co-ordination Meeting, Beijing, China.* International Atomic Energy Agency, Vienna, pp. 69-88, 1991.

Kring, J. B. and D. J. Schuster, Management of Insects on Pepper and Tomato with UV Reflective Mulches. *Florida Entomologist.* 73, pp. 119-129, 1992.

Krishnamurthy, T. S., N. Muralidharan, and M. K. Krishnakumari, Disinfesting Food Commodities in Small Storages Using Carbon Dioxide Rich Atmospheres. *International Pest Control.* 35, pp. 153-156, 1993.

Lambert, L. and L. G. Heatherly, Influence of irrigation on Susceptibility of Selected Soybean Genotypes to Soybean Looper. *Crop Science.* 35, pp. 1657-1660, 1995.

Landolt, P. J., Attraction of *Mocis latipes* (Lepidoptera: Noctuidae) to Sweet Baits in Traps. *Florida Entomologist.* 78, pp. 523-530, 1995.

Larson, A. O. and C. K. Fisher, The Bean Weevil and the Southern Cowpea Weevil in California. *United States Department of Agriculture Technical Bulletin No. 593,* 1938.

Latta, R., Vapor-Heat Treatment for the Control of Narcissus Bulb Pests in the Pacific Northwest. *United States Department of Agriculture Technical Bulletin No. 672,* 1939.

Lawton, J. K., Vacant Niches and Unsaturated communities: A Comparison of Blacken Herbivores at Sites on Two Continents. *Journal of Ecology.* 51, pp. 573-595, 1982.

Lemaster, R. L., F. C. Beall, and V. R. Lewis, Detection of Termites with Acoustic Emission. *Forest Product Journal.* 47, pp. 75-79, 1997.

Lenz, M. and S. Runko, Protection of Buildings, Other Structures and Materials in Ground Contact From Attack by Subterranean Termites (Isoptera) with a Physical Barrier-A Fine Mesh of High Grade Stainless Steel. *Sociobiology.* 24, pp. 1-16, 1994.

Leong, E. C. W. and S. H. Ho, Response of *Liposcelis bostrychophia* and *L. entomophila* (Psocoptera) to Carbon Dioxide. *Stored Product Protection Proceedings of the 6th International Working Conference on Stored-product Protection, 17-23 April 1994, Canberra, Australia* Volume 1. CAB International, Wallingford, U.K., pp. 108-112, 1994.

Lessard, F. F. and J. M. Le Torc'h, Practical Approach to Purging Grain with Low Oxygen Atmosphere for Disinfestation of Large Wheat Bins Against the Granary Weevil, *Sitophilus granarius,* in *Proceedings of the 4th International Working Conference on Stored-product Protection (formerly Stored-product Entomology), Tel Aviv, Israel, September 21-26 1986,* Donahaye, E. and S. Novarro, Eds. Agricultural Research Organization, Volcani Center, Bet Dagan, Israel, pp. 208-217, 1987.

Lewis, T., A comparison of Water Traps, Cylindrical Sticky Traps and Suction Traps for Sampling Thysanopteran Populations at Different Levels. *Entomologia Experimentalis et Applicata.* 2, pp. 204-215, 1959.

Lewis, V. R., M. I. Haverty, D. S. Carver, and C. Fouche, Field Comparison of Sand or Insecticide Barriers for Control of *Reticulitermes* spp. (Isoptera: Rhinotermitidae) Infestations in Homes in Northern California. *Sociobiology,* 28, pp. 327-336, 1996.

Liang, W. and M. Huang, Influence of Citrus Orchard Ground Cover Plants on Arthropod Communities in China: A Review. *Agriculture, Ecosystems and Environment.* 50, pp. 29-37, 1994.

Liburd, O. E., R. A. Casagrande, and S. R. Alm, Evaluation of Various Color Hydromulches and Weed Fabric on Broccoli Insect Populations. *Journal of Economic Entomology.* 91, pp. 256-262, 1998.

Lillie, T. H. and J. Goddard, Operational Testing of Electrocutor Traps for Fly Control in Dining Facilities. *Journal of Economic Entomology.* 80, pp. 826-829, 1987.

Lindgren, B. S. and R. G. Fraser, Control of Ambrosia Beetle Damage by Mass Trapping at a Dryland Log Sorting Area in British Columbia. *The Forestry Chronicle.* 70, pp. 159-163, 1994.

Liquido, N. J., Reduction of Oriental Fruit Fly (Diptera, Tephritidae) Populations in Papaya Orchards by Field Sanitation. *Journal of Economic Entomology.* 10, pp. 153-170, 1994.

Locatelli, D. P. and E. Daolio, Effectiveness of Carbon Dioxide Under Reduced Pressure Against Some Insects Infesting Packaged Rice. *Journal of Stored Products Research.* 29, pp. 81-87, 1993.

Longstaff, B. C., Environmental Manipulation as a Physiological Control Measure, in *Proceedings of the 4th International Working Conference on Stored-product Protection (formerly Stored-product Entomology), Tel Aviv, Israel, September 21-26 1986,* Dona- haye, E. and S. Novarro, Eds. Agricultural Research Organization, Volcani Center, Bet Dagan, Israel, pp. 47-61, 1987.

Loschiavo, S., Effect of Disturbance of Wheat on Four Species of Stored Product Insects. *Journal of Economic Entomology.* 71, pp. 888-893, 1978.

Lowry, W. L., A. J. Chapman, F. T. Wratten, and J. P. Hollingsworth, Tests of the Dielectric Treatment of Cotton Seed for Destroying Pink Bollworms. *Journal of Economic Ento- mology.* 47, pp. 1022-1023, 1954.

Lurie, S., E. Fallik, J. D. Klein, F. Kozar, and K. Kovacs, Postharvest Heat Treatment of Apples to Control San Jose Scale (*Quadraspidiotus perniciosus* Comstock) and Blue Mold (*Penicillium expansum* Link) and Maintain Fruit Firmness. *Journal of the American Society for Horticultural Science.* 123, pp. 110-114, 1998.

Lyon, W. F. and J. A. Steele, Mosquito Pest Management. *The Ohio State University Extension Bulletin 641,* 1998.

Mafra-Neto, A. and M. Habib, Evidence that Mass Trapping Suppresses Pink Bollworm Populations in Cotton Fields. *Entomologia Experimentalis et Applicata.* 81, pp. 315-323, 1996.

Maier, D. E., W. H. Adams, J. E. Throne, and L. J. Mason, Temperature Management of the Maize Weevil, *Sitophilus zeamais* Motsch (Coleoptera: Curculionidae), in Three Loca- tions in the United States. *Journal of Stored Products Research.* 32, pp. 255-273, 1996.

Mallis, A., *Handbook of Pest Control.* 6th ed. Franzak and Foster Company. Cleveland, OH, pp. 1101, 1982.

Mangan, F., R. Degregorio, M. Schonbeck, S. Herbert, K. Guillard, R. Hazzard, E. Sideman, and G. Litchfield, Cover Cropping Systems for Brassicas in the Northeastern United States. 2. Weed, Insect and Slug Incidence. *Journal of Sustainable Agriculture.* 5, pp. 15-26, 1995.

Mangan, R. L. and S. J. Ingle, Forced Hot-Air Quarantine Treatment for Grapefruit Infested with Mexican Fruit Fly (Diptera: Tephritidae). *Journal of Economic Entomology.* 87, pp. 1574-1579, 1994.

Mangan, R. L. and S. J. Ingle, Forced Hot-Air Treatment of Mangoes Infested with West Indian Fruit Fly (Diptera; Tephritidae). *Journal of Economic Entomology.* 85, pp. 1859-1864, 1992.

Mankin, R. W., J. S. Sun, D. Shuman, and D. K. Weaver, Shielding Against Noise Interfering With Quantitation of Insect Infestations by Acoustic Detection Systems in Grain Eleva- tors. *Applied Acoustics.* 50, pp. 309-323, 1997.

Manoto, E. C., L. R. Blanco, A. B. Mendoza, and S. S. Resilva, Disinfestation of Copra Desiccated Coconut and Coffee Beans Using Gamma Irradiation, in *Insect Disinfestation of Food and Agricultural Products by Irradiation. Proceedings of the Final Research Co-ordination Meeting, Beijing, China.* International Atomic Energy Agency, Vienna, pp. 105-126, 1991.

Mansour, M. and G. Franz, Gamma Radiation as a Quarantine Treatment for the Mediterranean Fruit Fly (Diptera: Tephritidae). *Journal of Economic Entomology*. 89, pp. 1175-1180, 1996.

Marlatt, C. L., Insect Control in California. *Yearbook of the United States Department of Agriculture 1896*, 1897.

Marlatt, C. L., The True Clothes Moths. *United States Department of Agriculture Circular No. 36*. 1908.

Masiunas, J. B., D. M. Eastburn, V. N. Mwaja, and C. E. Eastman, The Impact of Living and Cover Crop Mulch Systems on Pests and Yields of Snap Beans and Cabbage. *Journal of Sustainable Agriculture*. 9, pp. 61-89, 1997.

Mason, L. J. and C. A. Strait, Stored Product Integrated Pest Management with Extreme Temperatures, in *Lethal Temperatures in Integrated Pest Management*. Hallman, G. and D. Denlinger, Eds. Westview Press, Denver, Colorado. pp. 139-175, 1998.

Mason, L. J., Alternative Methods for Suppression of Pantry Pests. *Journal of Agricultural Entomology*. 14, pp. 323-331, 1997.

Mason, L. J., R. A. Rulon, and D. E. Maier, Chilled Versus Ambient Aeration and Fumigation of Stored Popcorn. Part 2: Pest Management. *Journal of Stored Product Research*. 33, pp. 51-58, 1997.

Matthews-Gehringer, D. and J. Hough-Goldstein, Physical Barriers and Cultural Practices in Cabbage Maggot (Diptera: Anthomyiidae) Management on Broccoli and Chinese Cabbage. *Journal of Economic Entomology*. 81, pp. 354-360, 1988.

Matthiessen, J. N. and S. E. Learmonth, Enhanced Survival and Reproduction of Whitefringed Beetle (Coleoptera: Curculionidae) with Irrigation of Pasture in a Dry Summer Environment. *Journal of Economic Entomology*. 85, pp. 2228-2233, 1992.

Mbata, G., C. Reichmuth, and T. Ofuya, Comparative Toxicity of Carbon Dioxide to Two *Callosobruchus* Species, in *Stored Product Protection Proceedings of the 6th International Working Conference on Stored-product Protection, 17-23 April 1994, Canberra, Australia*. Highley, E., E.J. Wright, H.J. Banks, and B.R. Champ, Eds. CAB International, Wallingford, U.K., pp. 120-122, 1994.

McHugh, J. J. and R. E. Foster, Reduction of Diamondback Moth (Lepidoptera: Plutellidae) Infestation in Head Cabbage by Overhead Irrigation. *Journal of Economic Entomology*. 88, pp. 162-168, 1995.

Mensah, R. K., Yellow Traps Can Be Used to Monitor Populations of *Coccinella tyransversalis* (F.) and *Adalia bipunctata* (L.) (Coleoptera: Cocinellidae) in Cotton Crops. *Australian Journal of Entomology*. 36, pp. 377-381, 1997.

Metcalf, R. L. and R. A. Metcalf,. *Destructive and Useful Insects: Their Habits and Control*. 5th ed., McGraw-Hill, Inc., New York. 1993.

Miller, H. C., S. B. Silverborg, and R. J. Capinera, Dutch Elm Disease: Relation of Spread and Intensification to Control by Sanitation in Syracuse, New York. *Plant Disease Reporter*. 53, pp. 551-555, 1969.

Misener, G. G., G. Boiteau, and L. P. McMillan, A Plastic-lining Trenching Device for the Control of Colorado Potato Beetle: Beetle Excluder. *American Potato Journal*. 70, pp. 903-908, 1993.

Mogi, M., Effect of Intermittent Irrigation on Mosquitos (Diptera, Culicidae) and Larvivorous Predators in Rice Fields. *Journal of Medical Entomology*. 30, pp. 309-319, 1993.

Moy, J. H., Efficacy of Irradiation vs Thermal Methods as Quarantine Treatments for Tropical Fruits. *Radiation Physics and Chemistry*. 42, pp. 269-272, 1993.

Mueller, D., L. Pierce, H. Benezet, and V. Krischik, Practical Application of Pheromone Traps in Food and Tobacco Industry. *Journal of the Kansas Entomological Society*. 63, pp. 548-553, 1990.

Muir, W. E., G. Yacuik, and R. N. Sinha, Effects of Temperature and Insect and Mite Population of Turning and Transferring Farm-stored Wheat. *Canadian Agricultural Engineering.* 19, 25-28, 1977.

Mullen, M. A., Low Frequency Sound Affecting the Development of the Indian Meal Moth. *Journal of Georgia Entomological Society.* 8, pp. 320-321, 1973.

Mullen, M. A., Rapid Determination of the Effectiveness of Insect Resistant Packaging. *Journal of Stored Products Research.* 30, pp. 95-97, 1994.

Muma, M. H., The Influence of Cover Crop Cultivation on Population of Injurious Insects and Mites in Florida Citrus Groves. *Florida Entomologist.* 44, pp. 51-68, 1961.

Murdock, L. L. and R. E. Shade, Eradication of Cowpea Weevil (Coleoptera: Bruchidae) in Cowpeas by Solar Radiation. *American Entomologist.* 37, pp. 228-231, 1991.

Nakagawa, S., R. Prokopy, T. Wong, J. Ziegler, S. M. Mitchell, T. Unago, and E. J. Harris, Visual Orientation of *Ceratitis capitata* Flies to Fruit Models. *Entomologia Experimentalis et Applicata.* 24, pp. 193-198, 1978.

Nakakita, H. and H. Ikenaga, Action of Low Temperature on Physiology of *Sitophilus zeamais* Motschulsky and *Sitophilus oryzae* (L.) (Coleoptera: Curculionidae) in Rice Storage. *Journal of Stored Products Research.* 33, pp. 31-38, 1997.

Nakakita, H. and K. Kawashima, A New Method to Control Stored-product Insects Using Carbon Dioxide with High Pressure Followed by Sudden Pressure Loss, in *Stored Product Protection Proceedings of the 6th International Working Conference on Stored-product Protection, 17-23 April 1994, Canberra, Australia.* Highley, E., E.J. Wright, H.J. Banks, and B.R. Champ, Eds. CAB International, Wallingford, U.K., pp. 126-129, 1994.

National Academy of Sciences, *Principles of Plant and Animal Pest Control, Volume 3. Insect-Pest Management and Control. Publication 1695.* National Academy of Sciences, Washington, D.C., p. 243, 1969.

Nelson, S. O. and B. H. Kantack, Stored-grain Insect Control Studies with Radio-frequency Energy. *Journal of Economic Entomology.* 59, pp. 588-594, 1966.

Nelson, S. O., Dielectric Properties of Grain and Seed in the 1 to 50 MC Range. *Transactions of the American Society of Agricultural Engineers.* 8, pp. 38-43, 1965.

Nelson, S. O., Electromagnetic and Sonic Energy for Insect Control. *Transactions of the American Society of Agricultural Engineers,* 9, pp. 398-405, 1966.

Nelson, S. O., Electromagnetic Energy, in *Pest Control Biological, Physical, and Selected Chemical Methods.* Kilgore, W. W. and R. L. Doutt, Eds. Academic Press, New York. pp. 89-145, 1967.

Nelson, S. O., Insect-control Studies with Microwaves and Other Radiofrequency Energy. *Bulletin of the Entomological Society of America.* 19, pp. 157-163, 1973.

Nelson, S. O., Potential Agricultural Applications for RF and Microwave Energy. *Transactions of the American Society of Agricultural Engineering.* 30, pp. 818-822, 1987.

Nelson, S. O., Radiation Processing in Agriculture. *Transactions of the American Society of Agricultural Engineers.* 5, pp. 20-30, 1962.

Nelson, S. O., Review and Assessment of Radio-frequency and Microwave Energy for Stored-grain Insect Control. *Transactions of the American Society of Agricultural Engineers.* 39, pp. 1475-1484, 1996.

Nettles, W. C., An Usual Outbreak of Stable Fly and Its Control. *Journal of Economic Entomology.* 27, pp. 1197-1198, 1934.

Neuenschwander, P., Beneficial Insects Caught by Yellow Traps Used in Mass Trapping of the Olive Fly, *Dacus oleae. Entomologia Experimentalis et Applicata.* 32, pp. 286-296, 1991.

Newton, J., Insect and Packaging — A Review. *International Biodeterioration.* 24, pp. 175-187, 1988.

Nichol, A. A., A Study of the Fig Beetle, *Cotinis texana* Casey. *University of Arizona College of Agriculture, Agricultural Experiment Station Technical Bulletin No. 55*, 1935.

Nickson, P. J., J. M. Desmarchelier, and P. Gibbs, Combination of Cooling with a Surface Application of Dryacide to Control Insects, in *Stored Product Protection Proceedings of the 6th International Working Conference on Stored-product Protection, 17-23 April 1994, Canberra, Australia.* Highley, E., E.J. Wright, H.J. Banks, and B.R. Champ, Eds. CAB International, Wallingford, U.K., pp. 185-188, 1994.

Norris, M. J., The Feeding Habits of the Adult Lepidoptera Heteroneura. *Transactions of the Royal Entomological Society of London.* 85, pp 61-90, 1933.

Ntoukam, G. and L. W. Kitch, Solar Heaters for Improved Cowpea Storage. *Agronomic Research Institute of Cameroon, Maruca Research Center, CRSP Cowpea Storage Project. Technical Bulletin.* 1991.

Nwanze, K. F., Components for the Management of Two Insect Pests of Pearl Millet in Sahelian West Africa. *Insect Science and Its Application.* 12, pp. 673-678, 1991.

Ofuya, T. I. And C. Reichmuth, Mortality of the Bean Bruchid, *Acanthoscelides obtectus* (Say), in Some High Carbon Dioxide Atmospheres. *Zeitschrift fur Pflanzenkrankheiten und Pflanzenschutz.* 100, pp. 165-169, 1993.

Orozco, S. M. and Z. O. Lopez, Effect of Transparent Mulch on Insect Populations, Virus Diseases, Soil-temperature, and Yield of Cantaloupe in a Tropical Region. *New Zealand Journal of Crop and Horticultural Science.* 23. pp. 199-204, 1995.

Orozco, S. M., O. Lopez, O. Perez, and F. Delgadillo, Effect of Transparent Mulch, Floating Row Covers and Oil Sprays on Insect Populations, Virus Diseases and Yield of Cantaloupe. *Biological Agriculture and Horticulture.* 10. pp. 298-234, 1994.

Osborn, H., The Hessian Fly in the United States. *United States Department of Agriculture, Bureau of Entomology Bulletin 16.* 1898.

Page, W. W., Destruction of Eggs as an Alternative to Chemical Control of the Grasshopper Pest *Zonocerus variegatus* (L.) (Orthoptera: Pyrgmorphidae) in Nigeria. *Bulletin of Entomological Research.* 68, pp. 575-581, 1978.

Parrott, P. J. and D. L. Collins, Some Further Observations on the Influence of Artificial Light Upon Codling Moth Infestations. *Journal of Economic Entomology.* 28, pp. 99-103, 1935.

Patkar, K. L., C. M. Usha, H. S. Shetty, N. Paster, and J. Lacey, Biogeneration of Carbon Dioxide for Use in Modified Atmosphere Storage of Sorghum Grains, in *Stored Product Protection Proceedings of the 6th International Working Conference on Stored-product Protection, 17-23 April 1994, Canberra, Australia.* Highley, E., E.J. Wright, H.J. Banks, and B.R. Champ, Eds. CAB International, Wallingford, U.K., pp. 144-147, 1994.

Pearce, M. J., *Termites Biology and Pest Management.* CAB International. Wallingford, U. K. 172 pp, 1997.

Pelletier, Y., G. C. Misener, and L. P. McMillan, Steam as an Alternative Control Method for the management of Colorado Potato Beetles. *Canadian Agricultural Engineering.* 40, pp. 17-21, 1998.

Permual, D., and G. Le Patourei, Laboratory Evaluation of Acid Activated Kaolin to Protect Stored Paddy Against Infestation by Stored Product Insects. *Journal of Stored Products Research.* 26, pp. 149-153, 1990.

Peterson, D. G. and W. A. Brown, Studies of the Responses of the Female *Aedes* Mosquito. Part III. The Response of *Aedes Aegypti* (L.) to a Warm Body and Its Radiation. *Bulletin of Entomological Research.* 42, pp. 535-541, 1951.

Pettit, R. H. and R. Hutson, Pests of Apple and Pear in Michigan. *Michigan State College Agricultural Experiment Station Circular Bulletin No. 137.* 1931.

Pettit, R. H., Insects New or Unusual in Michigan. *Michigan State Agricultural College Experiment Station Bulletin 244.* 1906.

Pettit, R. H., Insects of 1907. *Michigan State Agricultural College Experiment Station Bulletin 251*. 1908.

Phatak, S. C., An Integrated Sustainable Vegetable Production System. *HortScience*. 27, pp. 738-741, 1992.

Phillips, A. D. G. and T. D. Wyatt, Beyond Origami: Using Behavioural Observations as a Strategy to Improve Trap Design. *Entomologia Experimentalis and Applicata*. 62, pp. 67-74, 1992.

Phillips, T. W., Semiochemicals of Stored-product Insects: Research and Applications. *Journal of Stored Products Research*. 33, pp. 17-30, 1997.

Pickens, L. G. and G. D. Mills, Jr., Solar-powered Electrocuting Trap for Controlling House Flies and Stable Flies (Diptera: Muscidae). *Journal of Medical Entomology*. 30, pp. 872-877, 1993.

Pickens, L. G. and R. W. Thimijan, Design Parameters That Affect the Performance of UV-emitting Traps in Attracting House Flies (Diptera: Muscidae). *Journal of Economic Entomology*. 79, pp. 1003-1009, 1986.

Pickens, L. G., Battery Powered, Electrocuting Trap for Stable Flies (Diptera: Muscidae). *Journal of Medical Entomology*. 28, pp. 822-830, 1991.

Pickens, L. G., Factors Affecting the Distance of Scatter of House Flies (Diptera: Muscidae) From Electrocuting traps. *Journal of Economic Entomology*. 82, pp 149-151, 1989.

Pickens, L. G., G. D. Mills, Jr., and R. W. Miller, Inexpensive Tap for Capturing House Flies (Diptera: Muscidae) in Manure Pits of Caged-layer Poultry Houses. *Journal of Economic Entomology*. 87, pp. 116-119, 1994.

Pickett, J. A., L. J. Wadhams, and C. M. Woodcock, Developing Sustainable Pest Control From Chemical Ecology. *Agricultural Ecosystems and Environment*. 64, pp. 149-156, 1997.

Pittendrigh, B. R, J. E. Huesing, R. E. Shade, and L. L. Murdock, Monitoring of Rice Weevil, *Sitophilus oryzae*, Feeding Behavior in Maize Seeds and the Occurrence of Supernumerary Molts in Low Humidity Conditions. *Entomologia Experimentalis et Applicata*. 82, pp. 1-7, 1997.

Platt, R. R., G. W. Cuperus, M. E. Payton, E. L. Bonjour, and K. N. Pinkston, Integrated Pest Management Perceptions and Practices and Insect Populations in Grocery Stores in South Central United States. *Journal of Stored Products Research*. 34, pp. 1-10, 1998.

Price, P. W., C. E. Bouton, P. Gross, B. A. Mcperon, J. N. Thompson, and A. E. Weiss, Interactions Among Three Trophic Levels: Influences of Plants on Interactions Between Insect Herbivores and Natural Enemies. *Annual Review Ecology and Systematics*. Annual Reviews, Inc., Palo Alto, CA. 11, pp. 41-56, 1980.

Prokopy, R. J. and E. O. Owens, Visual Detection of Plants by Herbivorous Insects. *Annual Review of Entomology*. Annual Reviews, Inc., Palo Alto, CA. 28, pp. 337-364, 1983.

Prokopy, R. J., Apple-maggot Control by Sticky Red Spheres. *Journal of Economic Entomology*. 68, pp. 197-198, 1975.

Prokopy, R. J., Visual Responses of Apple Maggot Flies, *Rhagoletis pomonella* (Diptera: Tephritidae): Orchard Studies. *Entomologia Experimentalis et Applicata*. 11, pp. 403-423, 1968.

Rahim bin Muda, A., H. Osman, A. Sivaprogasm, N. O. Mohd, A. Radziah, S. Kaaruzzaman, and L. Karmariah, Irradiation of Stored Cocoa Beans, in *Insect Disinfestation of Food and Agricultural Products by Irradiation. Proceedings of the Final Research Co-ordination Meeting, Beijing, China*. International Atomic Energy Agency, Vienna, pp. 135-151, 1991.

Raupp, M. J., C. S. Koehler, and J. A. Davidson, Advances in Implementing Integrated Pest Management for Woody Landscape Plants. *Annual Review of Entomology*. Annual Reviews, Inc., Palo Alto, California. 37, pp. 561-585, 1992.

Reed, W. D. and J. P. Vinzant, Control of Insects Attacking Stored Tobacco and Tobacco Products. *United States Department of Agriculture Circular No. 635.* 1942.

Reed, W. D., A. W. Morrill, Jr., and E. M. Livingstone, Experiments with Suction Light Traps for Combating the Cigarette Beetle. *Journal of Economic Entomology.* 27, pp. 796-801, 1934.

Reichmuth, C. and R. Wohlgemuth, Carbon Dioxide Under High Pressure at 15 Bar and 20 Bar to Control the Eggs of the Indianmeal moth *Plodia interpunctella* (Hubner) (Lepidoptera: Pyralidae) as the Most Tolerant Stage at 25°C. *Stored Product Protection Proceedings of the 6th International Working Conference on Stored-product Protection, 17-23 April 1994, Canberra, Australia.* Highley, E., E.J. Wright, H.J. Banks, and B.R. Champ, Eds. CAB International, Wallingford, U.K., pp. 163-172, 1994.

Reichmuth, C., Low Oxygen Content to Control Stored Product Insects. *Proceedings of 4th International Working Conference on Stored-product Protection, Tel Aviv, Israel,* pp. 194-207, 1986.

Reissig, W. H., Performance of Apple Maggot Traps in Various Apple Free Canopy Positions. *Entomologia Experimentalis et Applicata.* 68, pp. 534-538, 1975.

Reynolds, A. H., R. J. Prokopy, T. A. Green, and S. E. Wright, Apple Maggot Fly (Diptera: Tephritidae) Response to Perforated Red Spheres. *Florida Entomologist.* 79, pp. 173-179, 1996.

Riedl, H., W. W. Barnett, W. W. Coates, R. Coviello, J. Joos, and W. H. Olson, Walnut Husk Fly (Dipterea: Tephritidae): Evaluation of Traps for Timing of Control Measures and for Damage Predictions. *Journal of Economic Entomology.* 82, pp. 1191-1196, 1989.

Ripp, B. E., Controlled Atmosphere and Fumigation in Grain Storages. *Proceedings of an International Symposium Practical Aspects of Controlled Atmosphere and Fumigation in Grain Storages held from 11 to 22 April 1983 in Perth, Western Australia, Developments in Agricultural Engineering 5,* Elsevier, Oxford, U.K., 1984.

Robacker, D. C., D. S. Moreno, and D. A. Wolfenbarger, Effects of Trap Color, Height, and Placement Around Trees on Capture of Mexican Fruit Flies (Diptera: Tephritidae). *Journal of Economic Entomology.* 83, pp. 412-419, 1990.

Roelofs, W. L., E. H. Glass, J. Tette, and A. Coineau, Sex Pheromone Trapping for Red-banded Leaf roller Control: Theoretical and Practical. *Journal of Economic Entomology.* 63, pp. 1162-1167, 1970.

Rogers, C. E., Extraflora Nectar: Entomological Implications. *Bulletin Entomological Society of America.* 31, pp. 15-20, 1985.

Rusell, E. P., Enemies Hypothesis: A Review of the Effect of Vegetational Diversity on Predatory Insects and Parasitoids. *Environmental Entomology.* 18, pp. 590-599, 1989.

Rust, M. K., P. O. Paine, and D. A. Reierson, Evaluation of Freezing to Control Wood-Destroying Insects (Isoptera: Coleoptera). *Journal of Economic Entomology.* 90, pp. 1215-1221, 1997.

Rutz, D. A., G. A. Scoles, and G. G. Howser, Evaluation of Fly-Electrocuting Black Light Devices in Caged-layer Poultry Facilities. *Poultry Science.* 67, pp. 871-877, 1987.

Ryan, J., M. F. Ryan, and F. McNaeidhe, The Effect of Interrow Plant Cover on Populations of the Cabbage Root Fly, *Delia brassicae* (Wiedemann). *Journal of Applied Ecology.* 17, pp. 31-40, 1980.

Samways, M. J., Spatial Distribution of *Scirtothrips avranti* Faure (Thysanoptera: Thripidae) and Threshold Level for One Per Cent Damage on Citrus Fruit Based on Trapping with Fluorescent Yellow Sticky Traps. *Bulletin of Entomological Research.* 76, pp. 649-659, 1986.

Sanderson, J. P. and R. T. Roush, Monitoring Insecticide Resistance in Greenhouse Whitefly (Homoptera: Aleyrodidae) with Yellow Sticky Cards. *Journal of Economic Entomology.* 85, pp. 634-641, 1992.

Schaber, B. D. and T. Entz, Effect of Annual and Biennial Burning of Seed Alfalfa (Lucerne) Stubble on Populations of Lygus (*Lygus* spp.) and Alfalfa Plant Bug (*Adelphocoris Lineolatus* (Goeze)) and Their Predators. *Annals of Applied Biology.* 124, pp. 1-9, 1994.

Schaber, B. D. and T. Entz, Effect of Annual and or Biennial Burning of Seed Alfalfa Stubble on Populations of Alfalfa Weevil and Pea Aphid. *Annals of Applied Biology.* 119, pp. 425-431, 1991.

Schaber, B. D. and T. Entz, Effect of Spring Burning on Insects in Seed Alfalfa Fields. *Journal of Economic Entomology.* 81, pp. 688-672, 1988.

Schal, C., Relation Among Efficacy of Insecticides, Resistance Levels, and Sanitation in the Control of the German Cockroach (Dictyoptera: Blattellidae). *Journal of Economic Entomology.* 81, pp. 536-544, 1988.

Seifert, S. and C. E. Snipes, Influence of Flame Cultivation on Mortality of Cotton (*Gossypium hirsutum*) Pests and Benefical Insects. *Weed Technology.* 10, pp. 544-549, 1996.

Service, M. W., *Medical Entomology for Students.* Chapman and Hall, London. 278 pp, 1996.

Sexton, J. M. and T. D. Schowalter, Physical Barriers to Reduce Damage by *Lepesoma lecontei* (Coleoptera: Curculionidae) to Conelets in a Douglas-fir Seed Orchard in Western Oregon. *Journal of Economic Entomology.* 84, pp. 212-214, 1991.

Sharp, J. L. and G. J. Hallman, Hot-air Quarantine Treatment for Carambolas Infested with Caribbean Fruit Fly (Diptera: Tephritidae). *Journal of Economic Entomology.* 85, pp. 168-171, 1992.

Sharp, J. L. and R. C. McGuire, Control of Caribbean Fruit Fly (Diptera: Tephritidae) in Navel Orange by Forced Hot Air. *Journal of Economic Entomology.* 89, pp. 1181-1185, 1996.

Sharp, J. L. and W. P. Gould, Control of Carribbean Fruit Fly (Diptera: Tephritidae) in Grapefruit by Forced Hot Air and Hydrocooling. *Journal of Economic Entomology.* 87, pp. 131-133, 1994.

Sharp, J. L. and H. Picho-Martinez, Hot-water Quarantine Treatment to Control Fruit Flies in Mangoes Imported into the United States from Peru. *Journal of Economic Entomology,* 83, pp. 1940-1943, 1990.

Sharp, J. L., J. J. Gaffney, J. I. Moss, and W. P. Gould, Hot-Air Treatment Device for Quarantine Research. *Journal of Economic Entomology.* 84, pp. 520-527, 1991.

Shayesteh, N. and N. N. Barthakur, Mortality and Behaviour of Two Stored-product Insect Species During Microwave Irradiation. *Journal of Stored Products Research.* 32, pp. 239-246, 1996.

Shejbal, J., *Controlled Atmosphere Storage of Grains: An International Symposium Held From 12 to 15 May 1980 at Castelgandolfo (Rome) Italy,* Elsevier/North-Holland, Amesterdam, Holland, 1980.

Shuman, D., D. K. Weaver, and R. W. Mankin, Quantifying Larval Infestation With an Acoustical Sensor Array and Cluster Analysis of Cross-correlation Outputs. *Applied Acoustics,* 50, pp. 279-296, 1997.

Sigaut, F., Significance of Underground Storage in Traditional Systems of Grain Production. *Controlled Atmosphere Storage in Grains: An International Symposium Held From 12 to 15 May 1980 at Castelgandolfo (Rome) Italy, Developments in Agricultural Engineering, 1.* Elsevier Scientific Publishing, Amsterdam, Holland, pp. 3-13, 1980.

Sivinski, J., Colored Spherical Traps for Capture of Caribbean Fruit Fly, *Anastrepha suspensa. Florida Entomologist.* 73, pp. 123-128, 1990.

Smart, M. R. and A. W. A. Brown, Studies on the Responses of the Female *Aedes* Mosquito. Part VII — The Effect of Skin Temperature, Hue and Moisture on the Attractiveness of the Human Hand. *Bulletin of Entomological Research*, 47. pp. 89-100, 1956.

Smith, A. W., R. B. Hammond, and B. R. Stinner, Influence of Rye-cover Crop Management on Soybean Foliage Arthropods. *Environmental Entomology*. 17, pp. 109-114, 1988.

Soemartaputra, M. H., Z. I. Purawanto, R. Chosdu, R. S. Haryadi, and A. Rahayu, Irradiation Disinfestation of Dry Leaf Tobacco and Coffee Beans, in *Insect Disinfestation of Food and Agricultural Products by Irradiation. Proceedings of the Final Research Co-ordination Meeting, Beijing, China.* International Atomic Energy Agency, Vienna, pp. 153-168, 1991.

Sorenson. W. C., *Brethern of the Net American Entomology, 1840-1880.* University of Alabama Press, Tuscaloosa, AL. 357 pp., 1995.

Southwood, T. R. E., W. F. Jepson, and H. F. Van Emden, Studies on the Behavior of *Oscinella frit* L. (Diptera) Adults of the Panicle Generation. *Entomologia Experimentalis et Applicata.* 4, pp. 196-210, 1961.

Spangler, H. G., Sound and the Moths That Infest Beehives. *Florida Entomologist.* 71, pp. 467-477, 1988.

Stanley, J. M., F. R. Lawson, and R. Gentry, Area Control of Tobacco Insects with Blacklight Radiation. *Transactions of the American Society of Agricultural Engineers.* 7, pp. 125, 1964.

Stern, V. M., A. Mueller, V. Sevacherian, and M. Way, Lygus Bug Control Through Alfalfa Interplanting. *California Agriculture.* 23, pp. 8-10, 1969.

Stern, V., Environmental Control of Insects Using Trap Crops, Sanitation, Prevention, and Harvesting, in *CRC Handbook of Pest Management in Agriculture. Volume 1.* Pimentel, D., Ed. CRC Press, Inc., Boca Raton, FL. pp, 199-207, 1981.

Storey, C. L., Exothermic Inert-atmosphere Generators for Control of Insects in Stored Wheat. *Journal of Economic Entomology.* 66, pp. 511-514, 1973.

Strong, R. G. and D. E. Sbur, Protection of Wheat Seed with Diatomaceous Earth. *Journal of Economic Entomology.* 56, pp. 372-374, 1963.

Su, N and R. H. Scheffrahn, Economically Important Termites in the United States and Their Control. *Sociobiology*, 17, pp. 77-94, 1990.

Su, N. and R. H. Scheffrahn, Penetration of Size-Particle Barriers by Field Populations of Subterranean Termites (Isoptera: Rhinotermitidae). *Journal of Economic Entomology.* 85, pp. 2275-2278, 1992.

Su, N., R. H. Scheffrahn, and P. M. Ban, Uniform Size Particle Barrier: A Physical Exclusion Device Against Subterranean Termites (Isoptera: Rhinotermitidae). *Journal of Economic Entomology.* 84, pp. 912-916, 1991.

Subramanyam, B., C. L. Swanson, N. Madamanchi, and S. Norwood, Effectiveness of Insecto. a New Generation Diatomaceous Earth Formulation in Suppressing Several Stored-Grain Insect Species, in *Proceedings of the 6th International Working Conference on Stored-Product Protection.* Highley, E., E. J. Wright, H. J. Banks, and B. R. Champ, Eds. Canberra, Australia, 1994.

Suss, L. and D. P. Locatelli, Effectiveness of Carbon Dioxide Under Reduced Pressure Against Some Insects Infesting Dried Fruit, in *Stored Product Protection Proceedings of the 6th International Working Conference on Stored-product Protection, 17-23 April 1994, Canberra, Australia.* Highley, E., E.J. Wright, H.J. Banks, and B.R. Champ, Eds. CAB International, Wallingford, U.K., pp. 194-200, 1994.

Sutherst, R. W. and R. S. Tozer, Control of the Buffalo Fly (*Haematobia irritans exigua*, de Meijere) on Dairy and Beef Cattle Using Traps. *Australian Journal of Agricultural Research.* 46, pp. 269-284, 1995.

Syms, P. R. and L. J. Goodman, The Effect of Flickering U-V Light Output on the Attractiveness of an Insect Electrocutor Trap to the House-Fly, *Musca Domestica*. *Entomologia Experimentalis et Applicata*. 43, pp. 81-85, 1987.

Tabashnik, B. E. and R. F. L. Mau, Suppression of Diamondback Moth (Lepidoptera: Plutellidae) Oviposition by Overhead Irrigation. *Journal of Economic Entomology*. 79, pp. 189-191, 1986.

Tamashiro, M, J. R. Yates, R. T. Yamamoto, and R. H. Ebesu, Tunneling Behavior of the Formosan Subterranean Termite and Basalt Barriers. *Sociobiology*. 19, pp. 163-170, 1991.

Thorpe, R. H., Hygienic Design Considerations for Chilled Food Plants, in *Chilled Foods A Comprehensive Guide*. Dennis, C. and M. Strinfer, Eds. Ellis Horwood, New York, 1992.

Timmon, G. M. and D. A. Potter, Influence of Pheromone Trap Color on Capture of Lilac Borer Males. *Environmental Entomology*. 10, pp. 756-759, 1981.

Titus, E. G., On the Life History of the Alfalfa Weevil. *Journal of Economic Entomology*. 3, pp. 459-470, 1910.

Toba, H. H. and H. R. Moffitt, Controlled Atmosphere Cold Storage as a Quarantine Treatment for Nondiapausing Codling Moth (Lepidoptera, Tortricidae) Larvae in Apples. *Journal of Economic Entomology*. 84, pp. 1316-1319, 1991.

Tozer, R. S. and R. W. Sutherst, Control of Horn Fly (Diptera: Muscidae) in Florida with an Australian Trap. *Journal of Economic Entomology*. 89, pp. 415-420, 1996.

Trumble, R. B. and R. F. Denno, Light Intensity, Host Plant Irrigation, and Habitat Related Mortality as Determinants of the Abundance of Azalea Lace Bug (Heteroptera: Tingidae). *Environmental Entomology*. 24, pp. 898-908, 1995.

Tuncbilek, A. S., Susceptibility of the Saw-toothed Grain Beetle, *Oryzaephilus surinamensis* (L.) to Gamma Radiation. *Journal of Stored Products Research*. 33, pp. 331-334, 1997.

Uchida, G. K., W. A. Walsh, C. Encarnacion, R. I. Vargas, J. D. Stark, J. W. Beardsley, and D. O. McInnis, Design and Relative Efficiency of Mediterranean Fruit Fly (Diptera: Tephritidae) Bucket Traps. *Journal of Economic Entomology*. 89, pp. 1137-1142, 1996.

Udayagiri, S., C. E. Mason, and J. D. Pisek, *Coleomegilla maculata, Coccinella septempunctata* (Coleoptera: Coccinellidae), *Chrysoperla carnea* (Neuroptera: Chrysopidae), and *Macrocentrus grandii* (Hymenoptera: Braconidae) Trapped on Colored Sticky Traps in Corn Habitats. *Environmental Entomology*. 26, pp. 983-988, 1997.

Ulrichs, C., Effects of Different Speed of Build Up and Decrease of Pressure with Carbon Dioxide on the Adults of the Tobacco Beetle *Lasioderma serricorne* (Fabricius) (Coleoptera: Anobiidae), in *Stored Product Protection Proceedings of the 6th International Working Conference on Stored-product Protection, 17-23 April 1994, Canberra, Australia*. Highley, E., E.J. Wright, H.J. Banks, and B.R. Champ, Eds. CAB International, Wallingford, U.K., pp. 214-216, 1994.

Valentin, N. and F. Preusser, Insect Control by Inert Gases in Museums, Archives and Libraries. *Restaurator*, 11, pp. 22-33, 1990.

Valentin, N., Comparative Analysis of Insect Control by Nitrogen, Argon, and Carbon Dioxide in Museum, Archive and Herbarium Collections. *International Biodeterioration and Biodegradation*. 32, pp. 263-278, 1993.

Van Steekelenburg, N., Novel Approaches to Integrated Pest and Disease Control in Glasshouse Vegetables in the Netherlands. *Pesticide Science*. 36, pp. 359-362, 1992.

Vernon, R. S. and J. S. Broatch, Responsiveness of *Delia* spp. (Diptera: Anthomyiidae) to Colored Sticky Traps in Flowering and Rosette Stage Canola. *Canadian Entomologist*. 128, pp. 1077-1085, 1996.

Wagner, D. L., W. Peacock, J. L. Carter, and S. E. Talley, Spring Caterpillar Fauna of Oak and Blueberry in a Virginia Deciduous Forest. *Annals of the Entomological Society of America*. 88, pp. 416-426, 1995.

Waite, M. B., W. W. Gilbert, N. A. Cobb, W. R. Beattie, F. E. Brooks, J. E. Graf, and W. L. McAfee, Diseases and Pests of Fruits and Vegetables. *United States Department of Agriculture Yearbook 1925.* pp. 453-600, 1926.

Washburn, F. L., Grasshoppers and Other Injurious Insects of 1911 and 1912. *Fourteenth Report of the State Entomologist of Minnesota to the Governor for the Years 1911 and 1912.* 1912.

Washburn, F. L., Injurious Insects of 1909 and 1910. *Thirteenth Report of the State Entomologist of Minnesota to the Governor for the Years 1909 and 1910.* 1910.

Watson, D. B. and I. A. Neale, Some Effects of Electric Fields on Living Creatures. *International Journal of Electrical Engineering Education,* 24, pp. 273-279, 1987.

Watson, D. B., R. D. Jamieson, and D. F. Solloway, Towards Insect Control by Electric Fields. *New Zealand Journal of Technology.* 2, pp. 167-170, 1986.

Webber, H. H., R. P. Wagner, and A. G. Pearson, High-frequency Electric Fields as Lethal Agents for Insects. *Journal of Economic Entomology.* 39, pp. 487-498, 1946.

Webster, F. M. Bringing Applied Entomology to the Farmer. *Yearbook of the United States Department of Agriculture 1913.* Government Printing Office, Washington, D.C. 1914.

Weintraub, P. G., Y. Arazi, and A. R. Horowitz, Management of Insect Pests in Celery and Potato Crops by Pneumatic Removal. *Crop Protection,* 15, pp. 763-769, 1996.

Whitcomb, W. H. and R. M. Marengo, Use of Pheromones in the Boll Weevil Detection and Control Program in Paraguay. *Florida Entomologist.* 69, pp. 153-156, 1986.

White, G. D. and S. R. Loschiavo, Factors Affecting the Survival of the Merchant Grain Beetle (Coleoptera: Cucujidae) and the Confused Flour Beetle (Coleoptera: Tenebrionidae) Exposed to Silica Aerogel. *Journal of Economic Entomology.* 82, pp. 960-969, 1989.

White, G. G., Temperature Changes in Bulk Stored Wheat in Subtropical Australia. *Journal of Stored Products Research.* 24, pp. 5-11, 1988.

White, N. D. G. and D. S. Jayas, Control of Insects and Mites with Carbon Dioxide in Wheat Stored at Cool Temperatures in Nonairtight Bins. *Journal of Economic Entomology.* 84, pp. 1933-1942, 1991.

White, N. D. G. and D. S. Jayas, Effectiveness of Carbon Dioxide in Compressed Gas or Solid Formulation for the Control of Insects and Mites in Stored Wheat and barley. *Phytoprotection.* 74, pp. 101-111, 1993.

White, N. D. G., D. S. Jayas, and C. J. Demianyk, Movement of Grain to Control Stored-product Insects and Mites. *Phytoprotection.* 78, pp. 75-84, 1997.

Wilcox, J., K. W. Gray, and D. C. Mote. The Strawberry Crown Moth. *Oregon State Agricultural College, Agricultural Experiment Station Bulletin 296,* 1932.

Wilson, J. Report of the Secretary, *Yearbook of the Department of Entomology 1910.* Government Printing Office, Washington, D.C., 1911.

Wingfield, M. J. and W. J. Swart, Integrated Management of Forest Tree Disease in South Africa. *Forest Ecology and Management.* 65, pp. 11-16, 1994.

Wudtke, A. and R. Reichmuth, Control of the Common Clothes Moth *Tineola bisselliella* (Hummel) (Lepidoptera:Tineidae) and Other Museum Pests with Nitrogen, in *Stored Product Protection Proceedings of the 6th International Working Conference on Stored-product Protection, 17-23 April 1994, Canberra, Australia.* Highley, E., E.J. Wright, H.J. Banks, and B.R. Champ, Eds. CAB International, Wallingford, U.K., pp. 251-254, 1994.

Wyss, E., The Effects of Artifical Weed Strips on Diversity and Abundance of the Arthopod Fauna in a Swiss Experimental Apple Orchard. *Agricultural Ecosystems and Environment.* 60, pp. 47-59, 1996.

Yothers, M. A., Biology and Control of Tree Hoppers Injurious to Fruit Trees in the Pacific Northwest. *United States Department of Agriculture Technical Bulletin No. 402,* 1934.

Yothers, M. A., Tree Hoppers and Their Control in the Orchard of the Pacific Northwest. *United States Department of Agriculture Circular No. 106*, 1931.

Young, W. R. and G. L. Teetes, Sorghum Entomology. *Annual Review of Entomology.* Annual Reviews, Inc., Palo Alto, CA. 22, pp. 193-218, 1977.

Zehnder, G. W. and J. Hough-Goldstein, Colorado Potato Beetle (Coleoptera: Chrysomelidae) Population Development and Effects on Yield of Potatoes With and Without Straw Mulch. *Journal of Economic Entomology.* 83, pp. 1982-1987, 1990.

Zoebisch, T. G. and D. J. Schuster, Influence of Height of Yellow Sticky Cards on Captures of Adult Leafminer (*Liriomyza Trifolii*) (Diptera: Agromyzidae) in Staked Tomatoes. *Florida Entomologist.* 73, pp. 505-507, 1990.

Chemical Control

CHAPTER **3**

Ecologically Based Use of Insecticides

Clive A. Edwards

CONTENTS

1-56670-478-2/00/$0.00+$.50
© 2000 by CRC Press LLC

3.1   INTRODUCTION

3.1.1   History of Insecticide Usage

The use of chemicals to control pests which harm crops and animals, annoy humans, and transmit diseases of both animals and humans is not a new practice. Hemlock and aconite were suggested for pest control in ancient Egyptian records as far back as 1200 B.C. Homer described how Odysseus fumigated the hall, house, and the court with burning sulfur to control pests. As long ago as A.D. 70, Pliny the Elder recommended the use of arsenic to kill insects, and the Chinese used arsenic sulfide for the same purpose as early as the 16th century. By the early 20th century, inorganic chemicals, such as lead arsenate and copper acetoarsenite, were in common use to control insect pests.

However, until 50 years ago, most arthropod pests, diseases, and weeds were still controlled mainly by cultural methods. The era of synthetic chemical pesticides truly began about 1940 when the organochlorine and organophosphorus insecticides were discovered. These chemicals and others that were developed subsequently, seemed to be so successful in controlling pests that there was extremely rapid adoption of their use and the buildup of a large multibillion-dollar agrochemical industry. There are currently more than 1600 pesticides available in the U.S. (Hayes and Lawes, 1991) and their worldwide use is still increasing (Edwards, 1994); about 4.4 million tons of pesticides are used annually with a total value of more than $20 billion (Environmental Protection Agency, 1989). The United States accounts for more than 30% of this market, exporting about 450 million pounds and importing about 150 million pounds.

In the early years of the rapid expansion of the use of insecticides, the effectiveness of these chemicals on a wide range of insect pests was so spectacular that they were applied widely and often indiscriminately in most developed countries. Indeed, aerial spray of forests and urban areas was quite common. There was little anxiety concerning possible human, ecological or environmental hazards until the late 1950s and early 1960s, when attention was attracted to the issue by the publication of Rachel Carson's book *Silent Spring* (1962), followed shortly after by *Pesticides and the Living Landscape* (Rudd, 1964). Although these publications tended to overdramatize the potential hazards of insecticides to humans and the environment, they effectively focused public attention on relevant issues. These concerns included the acute and chronic toxicity of many insecticides to humans, domestic animals, and wildlife; their phytotoxicity to plants; the development of new pest species after extensive pesticide use; the development of resistance to these chemicals by pests; the persistence of many insecticides in soils and water; and their capacity for global transport and environmental contamination.

In response to the recognition of such potential and actual environmental and human hazards from insecticides, most developed countries and relevant international agencies such as FAO and WHO set up complex registration systems, developed monitoring organizations, and outlined suites of regulatory requirements that had to be met before a new pesticide could be released for general use. Data were requested on toxicity to mammals and other organisms, pesticide degradation pathways, and fate. Monitoring programs were instituted to determine residues of insecticides in soil, water, and food, as well as in flora and fauna in the U.S. and Europe (Carey, 1979). Indeed, the registration demands have currently become so expensive to fulfill, that the development of new selective insecticides that are more environmentally acceptable has been discouraged since the registration period may take as much as six years. However, in many developing countries, many pesticides are still used without adequate registration requirements or suitable regulatory precautions, so potential environmental problems are often much greater in these countries. For instance, in a recent survey, Wiktelius and Edwards (1997) reported residues of organochlorine insecticides in the African fauna at greater levels than in the U.S. and European fauna in the 1970s. In the U.S., there are still many environmental problems that result from the extensive use of pesticides in spite of regulatory supervision.

3.1.2   Different Groups of Insecticides

The many insecticides from different chemical groups in current use vary greatly in structure, toxicity, persistence and environmental impact. They include the following:

Organochlorine Insecticides — These insecticides, which are very persistent in soil and are toxic to a range of arthropods, were used extensively in the 25 years after the Second World War. They include compounds as dichlorodiphenyltrichloroethane (DDT), benzene hexachloride (BHC, lindane), chlordane, heptachlor, toxaphene, methoxychlor, aldrin, dieldrin, endrin, and endosulfan, all of which are relatively non-soluble, have a low volatility, and are lipophilic. Many of them do not have very high acute mammalian toxicities but their persistence, and their tendency to become bioconcentrated into living tissues and move through food chains, has meant that, with the exception of lindane, their use has been largely phased out other than in certain developing countries. However, many soils and rivers are still contaminated with DDT, endrin, and dieldrin (White et al., 1983a; White and Krynitsky, 1986), the most persistent of these compounds, and there are still reports of organochlorine residues in wildlife (Riseborough, 1986), so the residues of these chemicals still present an environmental hazard. Unfortunately, most monitoring for these chemicals in developed countries was phased out after their use was banned or restricted, so we are not certain of the amounts still present in the environment in the U.S. or Europe. The environmental impact of these insecticides was considerable; hence their phasing out in most developed countries.

Organophosphate Insecticides — Some of the organophosphate insecticides were first developed as nerve gases during the Second World War. These, and others discovered later, include: carbophenothion, chlorfenvinphos, chlorpyrifos, diazinon, dimethoate, disulfoton, dyfonate, ethion, fenthion, fonofos, malathion, menazon, methamidophos, mevinphos, parathion, phosphamidon, phorate, thionazin, toxaphene, and trichlorfon. Although most of these chemicals are much less persistent than the organochlorines, many of them have much higher mammalian toxicities and greater potential to kill birds, fish, and other wildlife. Some of them are systemic, including dimethoate, disulfoton, mevinphos, phorate, methamidophos, and phosphamidon, and although this makes them much more selective they can be taken up into plants from where they may be consumed by vertebrates. They have sometimes caused severe local environmental problems, particularly in contamination of water and local kills of wildlife, but most of their environmental effects have not been drastic, although they can contaminate human food if suitable regulatory precautions are not observed.

Carbamate Insecticides — Typical carbamate insecticides include: aldicarb, bendicarb, carbaryl, carbofuran, methomyl, and propoxur. They tend to be rather more persistent than the organophosphates in soil and they vary considerably in their mammalian toxicity, which ranges from relatively low to comparatively high LD_{50}s. However, most carbamates are broad-spectrum toxicants affecting a range of quite different groups and phyla of organisms, so some of them have the potential for considerable environmental impact, particularly in soils, where they may influence populations of nematodes, earthworms and arthropods quite drastically.

Pyrethroid Insecticides — These are synthetic insecticides of very low mammalian toxicity and persistence, closely related to the botanical pyrethrins. They include allethrin, cyalothrin, cypermethrin, deltamethrin, fenvalerate, permethrin, and resmethrin, as well as many other related compounds. Since they are very toxic to insects they can be used at low dosages. However, since they affect a broad range of insects they may kill beneficial species as well as pests, lessen natural biological control, and increase the need for chemical control measures. Their main environmental impacts occur because they are broad-spectrum toxicants that are very toxic to fish and other aquatic organisms.

Avermectins — These are 16-membered macrocyclic lactones produced by the soil actinomycete *Streptomyces avermitilis*. They include: abamectin, spinosyn and ivermectin. To date, no significant environmental impacts have been recorded for them other than that they are relatively persistent in animal manures when used to control animal pests and soils, and can slow down organic matter degradation significantly.

Other Synthetic Chemical Insecticides — Recently, three new groups of insecticides have been developed. The first are the *formamidines* which include chlordimeforin and amitraz, which have a broad spectrum of activity. They also include *phenopyrazoles*, such as fipronil, introduced in 1987, which is effective against a wide range of insect taxa, but is relatively persistent with a half-life of 3 to 7 months in soil.

The third group is the *nitroguanidines*, such as imidacloprid, which has a wide spectrum of activity against various groups of insects, is systemic in plants, and is also quite persistent with a half-life in soil of about 5 months, but probably is relatively immobile and does not move into groundwater. Their overall environmental impacts are still unknown.

Insect Growth-Regulating Chemicals — A number of insecticides that kill insects by interfering with their molting process have been developed. One of the earliest was diflubenzuron, which kills mosquitoes and lepidopterour larvae, gypsy moths, and cotton boll weevils. These include: methoprene, used for mosquito control; kinoprene, which inhibit metamorphosis in Homoptera; and benzoylphenyl urea, which interferes with chitin formation and is relatively selective. A recent introduction is halofenozide. They have low mammalian toxicities, are relatively selective, and seem likely to have little environmental impact on the vertebrate fauna, soil, or water.

Biopesticides (Entomopathogens) — *Bacillus thuringiensis*, a bacteria widely accepted as an insect biocontrol agent most effective against Lepidoptera, has been in use for more than 40 years to control pests without harming humans, animals, and many beneficial insects. However, it has been marketed commercially in developing countries with mixed results, and it is not clear that the preparations developed for use in the U.S. or in Europe are suitable for use elsewhere. Genetically engineered strains of *B. thuringiensis* that are specific for different groups of insects have been produced to solve the problem of overspecificity. However, there are many concerns about insect resistance developing as a result of the widespread and careless application of *B. thuringiensis*, and it is important that this organism be seen as only one of the alternative control options. There are a number of other bacterial pesticides, including *Bacillus popullae* (and the closely-related *Bacillus lentimorbus*) and *Beavaria basicana* which control soil-inhabiting pests. Viruses such as nuclear polyhedrosis virus to control moths, cotton boll worm, and tobacco budworm, as well as pine sawfly bacilovirus have been used to control aerial pests. A protozoan *Nosema locustae* has been used to control grasshoppers. None of the biopesticides have been reported to have any serious environmental impacts.

Botanical Insecticides — Three broad categories of natural plant products are used to control insect pests: *botanical insect control agents*, such as pyrethrin and rotenone (produced from leguminous plants); *repellents and antifeedants*, such as *asarones* from *Acorus calamus*, *azadirachtin* from *neem*, and other isolates; and whole neem plants that are effective in the protection of stored grain in developing countries.

Although botanical insecticides have been in use for a long time, we do not understand the mode of action of many of them. The effects of azadirachtin, *Azadirachta indica*, have been known in India for millennia, but the pesticide has been characterized chemically only recently, although it has been synthesized in the laboratory. Although many botanical pesticides are known only to local farmers and to a handful of medicinal plant specialists, they often are available locally and could be produced and used by farmers themselves as a cottage industry.

If botanical pest control agents are to be more widely used, many ecological and environmental problems will have to be overcome. For example, the best known products — pyrethrin and rotenone — are not persistent and they affect pests and beneficial species alike. Neem is more of a systemic repellent and antifeedant (rather than a lethal toxin) that affects plant-feeding insects, but it has no apparent effect on wasps or bees. None of the botanical insecticides appear to have any environmental impact.

Entomophilic Nematode Products — Eelworms or nematodes belonging to the families Steinernematidae, Heterorhabditae and Mermithidae parasitize insects. They have symbiotic bacteria such as *Xenorhabdus* in their intestines; this can cause septicemia in insects and kill them in 24 to 48 hours. They are sensitive to environmental conditions and although they can be used as biological insecticides in much the same way as chemicals, they seem to be most effective in controlling soil-inhabiting insects. They have not been reported to cause any environmental problems and hold considerable promise for future development commercially.

3.1.3 General Concepts on Insecticide Use and Environmental Impacts

During the last 20 years, two new concepts have been developed progressively. The first of these is *ecotoxicology*, or *environmental toxicology*; a field in which holistic studies are made of the environmental impacts of toxic substances (including insecticides) in both natural and man-made environments, the environmental risks are assessed, and measures to prevent or minimize environmental damage are made (Truhart, 1975; Duffus, 1980; Butler, 1978). One of the best reference sources is the three-volume set *Handbook of Pesticide Toxicology*, edited by J. Wayland Hayes, Jr. and Edward R. Laws, Jr. (1991), and *Fundamentals of Aquatic Ecotoxicology: Effects, Environmental Fate and Risk Assessment* (Rand, 1995). There has also been great progress in the area of *agroecology*, which aims to understand the ecological processes that drive agricultural ecosystems. Such ecological knowledge is an important key to being able to minimize the amounts of synthetic insecticides used to manage pests (Carroll et al., 1990).

3.2 IMPACT OF INSECTICIDES ON THE ENVIRONMENT

Insects are living organisms, so the insecticides that are designed to control them are of necessity broad spectrum biocides. Indeed, some of the organophosphate insecticides that are effective insect control agents were developed originally during the Second World War as human nerve-gas agents. However, insecticides have a wide range in mammalian toxicity; toxic doses (L.D.$_{50}$) range from amounts as low as 1 mg/kg in the diet of a vertebrate animal to very large amounts needed to kill a mammal. They also differ greatly in persistence; some insecticides, particularly the organochlorine insecticides, are extremely stable compounds and persist in the environment for many years; others break down within a few hours or days.

There is increasing pressure, from national and international pesticide registration authorities, on insecticide manufacturers to provide comprehensive data about the environmental behavior of insecticides and on the acute toxicity of their chemicals to humans, rats or mice, fish, aquatic crustacea, plants, and other selected organisms. However, such data can only indicate the possible field toxicity of a particular insecticide to related organisms which may actually differ greatly from the test organism in their susceptibility to particular chemicals. No data at all are available on the toxicity of most insecticides to the countless species of untested organisms in the environment. Some of these species at potential hazard may include endangered species or species that may play important roles in dynamic biological processes or food chains.

There has been some progress in recent years in developing predictive models of the likely toxicity of a particular insecticide to different organisms, based on data on the behavior and toxicity of related compounds; the structure of the chemical; its water solubility and volatility; its lipid/water partition coefficient; and other properties (Moriarty, 1983). Edwards et al. (1996) have been using a microcosm technology to forecast environmental effects on soil ecosystems and Metcalf (1977) used an aquatic model ecosystem to forecast effects on aquatic ecosystems.

Different groups of living organisms vary greatly in their susceptibility to insecticides, but we are gradually accumulating a data bank identifying which chemicals present the greatest potential acute toxic hazard to the various groups of organisms. The characteristics of some of these organisms and their relative susceptibility to insecticides will be summarized briefly.

3.2.1 Effects of Insecticides on Microorganisms

The numbers of microorganisms in all of the physical compartments of the environment are extremely large and they have immense diversity in form, structure, physiology, food sources, and life cycles. This diversity makes it almost impossible to assess or predict the effects of insecticides upon them. Moreover, the situation is even more complex because microorganisms can utilize many insecticides as food sources upon which to grow; indeed, microorganisms are the main agents of degradation of many insecticides.

We still know relatively little of the complex ecology of microorganisms in soil and water, which makes it difficult to assess the impact of insecticides upon them.

Clearly, microorganisms can utilize many substances as food sources and are involved in complex food chains. Most of the evidence available indicates that if an ecological niche is made unsuitable for particular microorganisms by environmental or chemical factors, some other microorganism that can withstand these factors will fill the niche (functional redundancy). Moreover, unless an insecticide is very persistent, any effect it may have on particular microorganisms is relatively transient, so populations usually recover in 2 to 8 weeks after exposure, particularly if the chemical is transient.

Since there are such enormous numbers of microorganisms, it is impossible to test the acute toxicity of insecticides to them individually, and it is possible to generalize only in the broadest terms, as to the acute toxicity of insecticides to particular soil- and water-inhabiting microorganisms, based on tests on groups of organisms.

Most of those workers who have reviewed the effects of insecticides on **soil microorganisms** (Parr, 1974; Brown, 1978; Edwards, 1989; Domsch, 1963, 1983) have reported that insecticides have relatively small environmental impacts on microorganisms.

There are relatively few data on the toxicity of insecticides to **microorganisms in aquatic environments** (Parr, 1974). Much of the microbial activity is limited to the bottom sediments, and this is where insecticide residues in aquatic systems become concentrated through runoff and erosion from agricultural land (Rand, 1995).

There is a considerable literature on the effects of insecticides on **aquatic algae** that are a major part of the phytoplankton in aquatic systems. Herbicides such as simazine and terbutryn can have drastic effects on these organisms (Gurney and Robinson, 1989).

3.2.2   Effects of Insecticides on Aerial and Soil-Inhabiting Invertebrates

The kinds of invertebrates that inhabit soil or live above ground are extremely diverse, belonging to a wide range of taxa. There are extremely large numbers of species, with many species still to be described, and their overall populations are enormous. We know most about the effects of insecticides on insect pests, beneficial insects, and invertebrate predators that live on or are associated with plants and how they affect populations and communities. The main generality is that the broader the spectrum of activity of the insecticide, the greater its impact on beneficial invertebrates is likely to be.

We still know relatively little of the biology and ecology of many of the invertebrate species that inhabit soil. Thompson and Edwards (1974) reviewed the effects of insecticides on soil and aquatic invertebrates, but there have been few comprehensive reviews of the effects of insecticides on particular groups of invertebrates, an exception being a review of the effects of pesticides on earthworms (Edwards and Bohlen, 1991). Because of the diversity of the invertebrate fauna it is extremely difficult to make any generalizations on the acute toxicity of pesticides in individual species.

Soil-Inhabiting Invertebrates — A review of the effects of insecticides on soil-inhabiting invertebrates (Edwards and Thompson, 1973) reported that there are relatively few data on the acute toxicity of insecticides to individual species of soil-inhabiting invertebrates; most studies have involved studying the effects of insecticides on mixed populations on invertebrates in soil in the laboratory or field. More recently, Edwards and Bohlen (1992) made a comprehensive review of the effects of more than 200 pesticides on earthworms. Hence, it is possible to make some empirical assessments of the susceptibility of different groups of earthworms and other soil-inhabiting invertebrates to different groups of insecticides.

Nematodes — Nematodes, which are extremely numerous in most soils, and include parasites of plants and animals as well as free-living saprophagous species, are not susceptible to most insecticides. Insecticides have little direct effect on nematodes, although there is evidence that insecticides can have indirect effects on nematode populations, e.g., they can decrease communities of nematodes from fungivorous, bactivorous, and predator species and increase those for plant parasitic species (Yardim and Edwards, 1998).

Mites (Acarina) — Populations of mites are extremely large both above and below ground, and occur in most soils. The different taxa differ greatly in susceptibility to insecticides. The more active predatory species of mites tend to be more susceptible to pesticides than the sluggish saprophagous species. This has led to upsurges in mite populations and creation of new mite pests such as red spider mites with the confirmed extensive use of chemical insecticides. Similar effects have been reported for mite communities in soils.

Springtails (Collembola) — These arthropods, which are closely related to insects, are extremely numerous in most soils. They are susceptible to many insecticides, but their susceptibility to different insecticides has not been well documented and is extremely difficult to predict. There seems to be a strong positive correlation between the degree of activity in springtails and their susceptibility to insecticides. The main predators of springtails are mesostigmatid mites, and there have been many reports of upsurges in springtail populations in response to the use of organochlorine and organophosophate insecticides.

Pauropods (Pauropoda) — These very small animals, which are common in many soils but occur in smaller number than mites or springtails, seem to be extremely sensitive to many insecticides. Little is known about their feeding habits or ecological importance, but they are common in soils and are excellent indicators of the overall effects of insecticides in soils.

Symphylids (Symphyla) — Related to millipedes and centipedes, sometimes pests and other times saprophagous or even predators, these arthropods are common in many soils worldwide, and tend not to be very susceptible to insecticides; moreover, they are repelled by insecticides and can penetrate deep into the soil, where their exposure to these chemicals is minimized until the insecticide residues break down or disappear and they can return to the surface soil strata.

Millipedes (Diplopoda) — These common soil-inhabiting arthropods, which live mainly on decaying organic matter and sometimes tender young seedlings, can be serious pests of seeding sugar beet and cucumbers. They are intermediate in their susceptibility to insecticides between that of pauropods and that of symphylids. Since they live on or near the soil surface, they are exposed to many insecticides that occur as surface residues as they move over the soil surface, so are quickly eliminated.

Centipedes (Chilopoda) — These predatory invertebrates, which are often important predators of soil-inhabiting pests, are common in most soils and tend to be very susceptible to many insecticides. Since they are very active, their exposure to insecticide residues is considerable as they move through contaminated soil and they are relatively sensitive to many insecticides.

Earthworms (Lumbricidae) — These are probably the most important of invertebrates in breaking down organic matter and in the maintenance of soil structure and fertility. Because of their importance in soils, and because of their selection as key-indicator organisms for soil contamination (Edwards, 1983), there is a great deal more information available on the acute toxicity of insecticides to them than to any other group of soil-inhabiting invertebrates (Edwards and Bohlen, 1992). The insecticides that are acutely toxic to earthworms include endrin, heptachlor, chlordane, parathion, phorate, aldicarb, carbaryl, bendiocarb, and benomyl. This is a relatively small number of insecticides out of the more than 200 that have been tested for acute toxicity. All carbamates tested seem to be particularly toxic to earthworms.

Molluscs (Mollusca) — No insecticides, with the exception of methiocarb, are toxic to slugs or snails, probably because of their protective coating of mucus. However, lipophilic insecticides such as the organochlorines can bioconcentrate into the tissues of molluscs, as can some organophosphate insecticides such as diazinon and carbamates such as aldicarb. This can affect birds that feed on the molluscs.

Insects (Insecta) — Different species of soil-inhabiting insects and larvae can be pests, predators of pest, or are important in breaking down organic matter in soil; indeed, some species may act in two or more of these capacities. They are susceptible to many insecticides, but the variability in susceptibility between species to different chemicals is much too great for any general ecological trends to emerge. However, there is some tendency for the more active predatory species to be more susceptible to insecticides than the more nonmotile species, leading to upsurges in populations of some pests after sustained insecticide use.

Many insecticides also affect aerial insects, including bees. Bees are extremely important, not only in providing honey but also in pollination of crops. Data on acute toxicity of insecticides to bees have been required by most pesticide registration authorities, but this does not avoid considerable bee mortality in the field which can create a significant environmental impact.

3.2.3   Effects on Aquatic Invertebrates

In general, aquatic invertebrates are much more susceptible to insecticides than soil-inhabiting invertebrates, particularly if the insecticide is water-soluble. Lethal doses of an insecticide can be picked up readily as water passes over the respiratory surfaces of the aquatic invertebrates, and it is difficult for aquatic invertebrates to escape such exposure. A single incident of spill of an insecticide can cause drastic and widespread kills of invertebrates.

Although it is much easier to assess the acute toxicity of insecticides to individual species of aquatic invertebrates than to species of soil organisms in laboratory tests, not many aquatic species have been tested extensively in this way. The aquatic invertebrate species that have been tested most commonly have been *Daphnia pulex*, *D. magna*, *Simocephalus*, mosquito larvae (*Aedes*), *Chironomus* larvae, stonefly nymphs (*Pteroarcys californica* and *Acroneuria pacifica*), (Jensen and Gaufin, 1964), mayfly nymphs (*Hexagenia*), caddis fly larvae (*Hydropsyche*), copepods (*Cyclops*), ostracods, and the amphipod *Gammarus*.

It is extremely difficult to differentiate between the different taxa of aquatic invertebrates in terms of their susceptibility to different groups and kinds of insecticides. However, there is little doubt that pyrethroid insecticides have the most effect on most aquatic invertebrates. There have been several reviews of the effects of insecticides on aquatic organisms (Muirhead-Thompson, 1971; Edwards, 1977; Thompson and Edwards, 1974; Parr, 1974; Brown, 1978) and a more recent comprehensive review of the impact of insecticides on aquatic organisms (Rand, 1995). The aquatic invertebrate fauna in terms of susceptibility to insecticides can be reviewed as follows:

Crustacea — Aquatic crustacea differ greatly in both size and numbers, and populations vary greatly seasonally. They include: small, swimming crustaceans such as *Cyclops* and *Daphnia*; intermediate-sized organisms such as shrimps and prawns; and larger invertebrates, including crabs and lobsters. They all seem to be relatively susceptible to most insecticides, and there have been many incidences of large-scale kills of crustacea by insecticides that reach aquatic systems from spraying or spills.

Molluscs and Annelids — These are bottom-living aquatic organisms such as oysters, clams, and other shellfish or small worms that live in the bottom mud or sediment in salt- and freshwater systems. Most of them tend to have much lower sensitivities to insecticides than the different groups of arthropods in aquatic systems, although they take up and bioconcentrate some persistent insecticides into their tissues as contaminated water passes over their gills, and these may eventually accumulate to a toxic level; however, reports of serious incidences of this kind are relatively rare, but this may be because of the difficulty of tracing the sources of contamination.

Insects — A wide range of insect larvae inhabit water, particularly fresh water. Some, such as mosquito larvae, are free-living in water, but the majority live on or in the bottom sediment. These include chironomid, mayfly, dragonfly, stonefly, and caddis fly larvae. These insects are very susceptible to many insecticides, particularly the more persistent ones which tend to concentrate in the bottom sediment and remain there for considerable periods.

The organochlorine insecticides are moderately toxic to not only insect larvae but also to many other aquatic invertebrates. Organophosphate and carbamate insecticides tend to be less toxic than organochlorines to insect larvae, but very toxic to some species. Carbamates are probably the least toxic. The most toxic group of insecticides to aquatic invertebrates in general is the pyrethroids, which have a broad spectrum of activity and affect most species of aquatic invertebrates. For instance, Anderson (1989) reported that pyrethroids were very toxic to mosquitoes, blackfly, and chironomid larvae, and Day (1989) reported the same for zooplankton.

3.2.4   Effects on Fish

All aquatic organisms tend to be much more susceptible to insecticides than terrestrial ones. There are many reasons for this, but the most important is that the contamination can spread rapidly through an aquatic system, and there is no escape for fish or other organisms. In most developed countries, reports of fish kills by insecticides are very common, particularly in summer (Muirhead-Thompson, 1971; Rand, 1995). These can be caused both through direct toxicity of an insecticide, or because insecticides kill many of their food organisms. There have been no good estimates of overall losses of fish due to insecticides, but there is little doubt that such losses must be enormous worldwide.

There is a considerable data bank on the acute toxicity of insecticides to fish since, in the developed countries, a major requirement before a pesticide can be registered is to provide data on its acute toxicity to fish. However, these data tend to be confined to assays on a relatively few species of fish that are easy to breed and culture, and may not always be relevant to field populations.

Pesticides applied to agricultural land can fall out from aerial sprays on to water, or eventually reach aquatic systems such as rivers or lakes through drainage, or by runoff and soil erosion. Another source of contamination of water is the disposal of insecticides and their containers in aquatic systems and industrial effluents from pesticide factories. Fish are particularly susceptible to poisonous chemicals since they are exposed to such chemicals in solution, in the water in which they live, or in suspension absorbed on to sediments, as the water passes over the fishes' gills.

3.2.5   Effects on Amphibians and Reptiles

There are relatively few documented data on the effects of insecticides on amphibians and reptiles, although there has been considerable speculation recently that deformities reported in frogs in parts of the U.S. and elsewhere are due to insecticides. Wiktelius and Edwards (1997) reported that crocodiles and their eggs bioconcentrate organochlorine insecticides.

3.2.6 Effects on Birds

Birds are susceptible to many insecticides and we have a great deal of information on the acute toxicity of different insecticides to different species since not only are many insecticides tested for their effects on indicator bird species during the registration process, but also there are monitoring schemes for recording numbers of birds killed by insecticides in many countries (Riseborough, 1986; Hardy, 1990). Most incidents of toxicity of insecticides to birds are encountered through feeding on contaminated food, such as seed dressed with pesticides, plants treated with pesticides, or animals fed upon that have died from pesticides. It can be difficult to confirm, where large numbers of birds are found dead on agricultural land, whether they were killed by insecticides. Avian toxicity is sometimes a reason for registration of a pesticide to be refused, particularly if the use pattern, e.g., as a seed dressing, would expose large numbers of birds to the insecticide.

3.2.7 Effects on Mammals

All insecticides are tested for their acute toxicity to representative mammals during their development and registration phases. The species that are normally tested for acute toxic responses to insecticides are mice or rats, and there are comprehensive lists of the toxicity of virtually all pesticides to these animals. Any insecticide with a high mammalian toxicity is more difficult to register for general use in developed countries unless it is very effective in killing pest insects and provides good benefits. However, it is difficult to use such specific data on laboratory animals to predict harm to other mammals with quite different habits and susceptibilities in the field. Data on the relative toxicity of insecticides to mammals in the field is relatively scarce.

3.2.8 Effects on Humans

It is virtually impossible to obtain direct acute toxicity data for the effects of insecticides on human beings, although there have been a few studies on prison inhabitants and volunteers who received insecticides in their daily diet. Hence, data from animal toxicity is used to predict the potential acute toxicity of insecticides to human beings for insecticide regulatory and registration purposes. This is far from satisfactory as a toxicity index, since different groups of mammals have considerably different susceptibilities to insecticides. However, since insecticides can have important toxic effects on humans, such toxicity data are usually used with added safety factors to minimize adverse effects. Even with such precautions, it has been estimated that there are between 850,000 and 1.5 million insecticide poisonings of humans annually worldwide, from which between 3,000 and 20,000 people die. There have been many serious accidents in which many people have died from insecticides

(Hayes and Lawes, 1991) and there is considerable circumstantial evidence that farmers have a much greater incidence of illness that could be due to insecticides than the rest of the population. Certain groups of people with limited diets such as babies may have a greater susceptibility to insecticides.

There have been a number of books and major reviews of the environmental impacts of insecticides. These include: *Organic Pesticides in the Environment* (Rosen and Kraybill, 1966); *Pesticides in the Environment and their Effects on Wildlife* (Moore, 1966); *Pesticides and Pollution* (Mellanby, 1967); *Organochlorine Pesticides in the Environment* (Stickel, 1968); *Since Silent Spring* (Graham, 1970); *Persistent Pesticides in the Environment* (Edwards, 1973a); *Environmental Pollution by Pesticides* (Edwards, 1973b); *Pesticide Residues in the Environment in India* (Edwards et al., 1980); *Pollution and the Use of Chemicals in Agriculture* (Irvine and Knights, 1974); *Ecology of Pesticides* (Brown, 1978); *Use and Significance of Pesticides in the Environment* (McEwan and Stevenson, 1979); *Chemicals in the Environment: Distribution, Transport, Fate and Analysis* (Neely, 1980); *Ecotoxicology: The Study of Pollutants in Ecosystems* (Moriarty, 1983); *Wildlife Toxicology and Population Modelling: Integrated Studies of Agroecosystems* (Kendall and Lacher, 1994); *Introduction to Environmental Toxicology: Impacts of Chemicals Upon Ecological Systems* (Landis and Yu, 1995); *Fundamentals of Aquatic Toxicology: Effects, Environmental Fate and Risk Assessment* (Rand, 1995); and *The Pesticide Question: Environment, Economics and Ethics* (Pimentel and Lehman, 1993).

3.3 ECOLOGICAL PRINCIPLES INVOLVED IN JUDICIOUS INSECTICIDE USE

3.3.1 General Concepts

The judicious use of insecticides to avoid major environmental or human impacts involves a broad range of ecological concepts and principles associated with: predicting insect pest attacks; assessing how rapidly they will build up or decline; identifying cultural or biological inputs to insect management that will minimize pest damage and losses; and selecting the kind, dosage, and placement of an insecticide that will have minimal adverse ecological and human side-effects. Implicit in such judicious insecticide use are several strategies:

- *Develop suitable pest monitoring or forecasting techniques.* This involves the assessment of insect pest populations (numbers of eggs, larvae, insects, crop damage, etc.) or amount of damage, through suitable absolute or relative sampling or scouting programs.
- *Establish pest economic thresholds*, to identify those pest populations or incidence levels that can cause losses in crop value that significantly exceed the cost of management. It may be difficult to establish such levels for some insects, particularly since the buildup of most insect pest and predator populations is mainly climatically controlled, and affected by many other factors.
- *Develop a pest management strategy.* It is necessary to identify the least environmentally hazardous insecticide that can be used, identify the minimal dose and

placement that will be effective, and complement this with appropriate cultural and biological techniques to provide an overall pest management strategy. The general aim is to maintain insect pest numbers and damage at economically acceptable levels, with minimal use of hazardous insecticides. This overall strategy usually targets pest containment rather than eradication.

3.3.2   Forecasting Insect Pest and Predator Populations

The prediction of insect pest outbreaks is extremely difficult, since the buildup of populations of most pests and predators is seasonal and linked strongly with variable climatic factors. Additionally, the accurate estimation of insect egg, larva or adult insect populations on plants or in soil is both labor-intensive and difficult; since it is inevitably based on some form of selective sampling, so many inaccuracies are inevitably involved.

Estimates of insect pest densities are usually determined by some kind of field sampling. The different estimates of insect population densities in agricultural crops can be categorized into three main groups: **absolute sampling methods, relative sampling methods,** and **population indices**. Absolute sampling methods, which can only rarely be used because of the high labor requirements, provide actual counts of population density per unit area (e.g., number of larvae per m^2), of crop or field, whereas relative sampling methods usually use some other unit other than unit area (e.g., number of weevils per 20 net sweeps, numbers per leaf, or number of moths caught per pheromone, light or sticky trap). Population indices do not count insects at all; instead, they count insect products (e.g., exuviae or frass) or insect attack effects (e.g., plant damage). There is an essential trade-off between the level of precision necessary for making management decisions and the amount of time, in terms of labor costs, that has to be devoted to sampling. This usually dictates the type of sampling program that is adopted.

Obviously, absolute estimates of insect pest densities would be preferable for making wise management decisions, but it is generally much too costly to obtain and use such estimates. However, should absolute pest density estimates be established and correlated with amounts of pest damage, it is often possible to relate future relative sampling estimates to these absolute sampling estimates using regression analyses. This enables the relative sampling estimates obtained to be used to make management decisions more effectively.

3.3.2.1   Sampling Methods

The field or crop area to be sampled influences the number and kind of samples that need to be taken. If there are considerable variations in infestation within the same field, it may be necessary to divide the field into separate sections for sampling, so that each section can be assessed separately. The most practical size of the sample unit must be established by preliminary sampling before main sampling begins. The sample should be a biologically meaningful unit (e.g., a leaf), yet not so large as to be difficult to handle. For example, when sampling a cabbage plant for insects, it may not be necessary to examine the entire head, especially when the variability in the data is greater among plants than between plants. If the relevant pest complex is restricted to the leaves, then it is only necessary to sample these leaves.

It is rarely possible to examine every plant within a field, even in absolute sampling, so usually a subsample of the overall population is taken, consisting of a number of individuals taken from the entire insect population. The number of observations made in obtaining the sample is referred to as the *sample size*.

Because there can be considerable variability between sampling units, it is necessary to obtain a common measure of spread or variation among the samples, based on the standard deviation; if this is large compared to the mean, then more samples are needed, thus the level of sampling precision helps in making the decision to treat or not to treat with an insecticide.

Most sampling programs are based on taking random samples throughout the field; statistically, this is preferable, although it may not be the most efficient way to sample insect populations in a field. If the pest is not distributed randomly across the field, it may be more efficient to stratify the area to be sampled. Typically, this consists of a number of rows or a transect across the field. This type of sampling is called *stratified random sampling*. It is efficient, increases precision, and reduces the variance between sample units.

3.3.2.2 Sequential Sampling

One of the most efficient insect sampling methods which minimizes the amount of work required is sequential sampling, which has to be based on some pre-knowledge of the variance and overall distribution of an insect population in a field. The number of samples taken varies with the size of the infestation, and sampling stops once it is known that a pest population steadies at a certain density. Based on an extensive sampling and pest loss program, number tables can be constructed which can indicate whether any further sampling is needed to make a decision. If the accumulated number falls into a designated area of uncertainty, then more samples should be taken to confirm the action threshold. To determine the relationship between the total number of insects per plant and the number of samples to be taken, it is necessary to determine the level of error that is acceptable in assessing the damage threshold.

3.3.2.3 Relative Sampling

Many threshold predictions are based on relative sampling methods, because such assessments are more practical than obtaining absolute estimates of population densities. Relative sampling often involves using various different types of traps, such as pheromone traps, foliage-mimic traps, water or yellow pan traps, emergence traps, suction traps, light traps or sticky traps. Many insects attract mates by chemicals, and these chemicals or pheromones have been synthesized for many pest species and used inside traps. The baited traps are placed in areas where the pest insects commonly mate, and the insects attracted to these pheromone traps become trapped once inside. Aphids are attracted to yellow objects that act as "foliage mimics," e.g., cake pans painted yellow on the inside and filled with soapy water, used for monitoring flight times of winged aphids to allow early detection. Many insect pests can be assessed by trapping them on sheets, or on objects such as bottles coated with a sticky material. It is possible to relate pest damage to the number of insects trapped; in this way the presence of pests can be assessed quickly so management decisions can be based on such relative sampling methods.

Area assessments can be based on local or regional trapping programs. For instance, in England a network of light traps over the whole country are used by the Rothamsted Experimental Station for timing the population levels of many pest moth species. Similarly, a network of high-level automatic suction traps covering the whole of northern Europe allows forecasts of flight periods and population density of economic species of aphids to be assessed quickly, as warnings for the need to spray crops. The main problem is to relate the number of insects caught by traps to the economic threshold in individual fields; hence trapping still involves a degree of insurance insecticide treatment.

3.3.2.4 Population Indices

Threshold estimates based on amount of damage, cast insect skins, excreted frass, and other criteria are termed population indices. Such estimates often bear only poor relationships to absolute populations and their use is justified only if they have been fully validated and they provide considerable savings in time and labor.

3.3.3 Determining Insect Pest Thresholds for Economic Damage

There have been many discussions and definitions on the best methods of assessing insect pest thresholds, which are usually expressed as the *economic injury level* (EIL) or the *economic threshold level* (ETL). In economic terms, an insect is not a pest until its population reaches a density at which it has to be controlled to avoid serious crop losses. The EIL is the population density at which the financial benefits of control measures are equal to or greater than the cost of control. The ETL is the population density at which control tactics must be used to prevent the pest population from reaching the EIL. When absolute pest incidence levels have not been established, such as when a relative population estimate based on trap counts is used to measure pest densities, an *action threshold* (AT) is used. The action threshold is generally based on empirical observations obtained from working with growers and farmers, and often improves as the knowledge base increases.

The mere presence of an insect pest in a crop does not mean that damage will be caused. Many species of plants are able to compensate for the injury of small numbers of pests. In studying the effect on insect injury on crop yield, one must investigate the effect of a whole population of pests upon a population of plants. Because plants compete with each other for light, moisture, and nutrition, the loss of some plants to insects, such as in a wheat field, will result in neighboring plants producing more. In such cases, the presence of small numbers of insects will not cause any reduction in overall yields of a crop. Insects may also have a pruning effect upon plants by suppressing the growth of one organ and increasing the weight of others. Small infestations of the bean aphid decrease apical growth of the bean plant and increase yields. No damage occurs when small numbers of a pest attack leaves or roots, which provide more nutrients than the harvestable organs can assimilate. The twospotted spider mite can destroy up to 30% of the leaf area without affecting the yield of cucumbers. All this makes forecasting difficult.

Four main factors influence the economic injury threshold: control costs, crop market value, pest damage, and crop response to injury. The first two are relatively simple to estimate, but the latter two are difficult because they depend on complex biological processes about which little is known. All four factors vary, causing the economic injury level of pest density to be a variable or dynamic number.

The establishment of a reliable EIL or ETL estimate is difficult and must be based not only on population estimates but also on intensive studies of the pest and plant ecology, and how they are affected by weather, natural enemies, host plant resistance, plant growth, and environmental consequences of the control measure. Pest population densities tend to oscillate wildly over a long time-span, although they may eventually reach a general equilibrium. It is the magnitude of these oscillations that affects plant and insect ecology and the ability to calculate meaningful insect pest thresholds. If the general seasonal pest equilibrium is usually below the EIL, then control measures are rarely needed. However, some insects reach pest densities virtually every year and usually have to be controlled with insecticides or other measures. Finally, the general equilibrium of a pest may always be above the EIL. This is typical for most insects that are specific pests of a crop, and the only way to keep such populations below the EIL is to make regular applications of insecticides unless other control measures are available.

3.3.4 Cultural Inputs into Minimizing Pest Attack

Before the advent of modern insecticides, manipulation of farming practices was the main insect pest control tool available to farmers. Some of the cultural practices available to control insects were well established and most of them are still used extensively in many developing countries. Others are relatively new practices and need further testing to justify adoption. Although many cultural techniques used in the past were not scientifically based, most of those used currently have a firm ecological basis. The main cultural techniques available to minimize insect pest attacks include the following:

Tillage — It was widely thought that deep moldboard plowing had beneficial effects on pest insect populations; there are some situations in which it is true, particularly for long-lived soil-inhabiting pests such as wireworms and chafers. However, with the advent of conservation tillage and no-till, there is good evidence that although many insect pest problems are decreased by cultivations, others can be made more serious. In general, the overall trend from plowing to conservation tillage tends to decrease insect pest problems, although the effects of tillage differ considerably from pest to pest and crop to crop (Stinner and House, 1990; Edwards and Stinner, 1989).

Resistant Crop Varieties — The use of crop varieties that resist attack by arthropod pests has been a major tool in minimizing the use of insecticides and in developing pest management strategies. International Agricultural Research Centers have implemented large seed bank programs, of all related plant species, as the basis for developing crop varieties resistant to insect attack. This process may be accelerated through genetic engineering of new strains and varieties, particularly by techniques such as by the insertion of the *Bacillus thuringiensis* gene into crops. As much as 75% of U.S. cropland is planted with strains resistant to some form of pests. Although resistance can control some insect pests completely, insecticides may be needed to complement this component. There are many mechanisms of resistance to insects, including lack of synchronization with the insect life cycle, slowing down of insect development, and making the plant less attractive to insects. *Antibiosis* is an adverse effect of the plant on the pest insect through a deleterious direct effect or lack of a specific nutrient requirement. *Tolerance* is the ability of the plant to grow and yield in spite of the insect attack. A major problem is the possibility for the resistance mechanism to break down over time.

Rotations and Fallowing — The regular use of crop rotations has long been a major strategy in minimizing arthropod pest attacks and the current trend to monoculture or biculture has been the cause of much more serious pest incidence. Such rotations are essential in controlling most of the long-lived, soil-inhabiting insect pests effectively. They are also important in controlling annual insect pests which overwinter on residual foliage or in soils. Much more research is needed, however, to identify those rotations that minimize attack by different insect pests effectively and to define the important role that fallow periods play in this approach to pest management.

Cropping Patterns — It has long been understood that host plant diversity decreases arthropod pest attack on plants. Crop diversity provides alternatives to the populations of susceptible plants and maximizes the potential of populations of natural enemies to develop and provide alternative hosts and habitats for both pests and predators. Much more research is needed on the effects of undersowing, and amalgamation of tree growing with annual crops (agroforestry). For instance, *intercropping* in alternate rows or strip cropping in alternate rows of the tractor have been shown to lessen insect pest attacks significantly. There is accumulating evidence that such cropping patterns are very effective in controlling insect pest attacks (Edwards et al., 1992). Small areas of a preferred crop may be planted near a major crop to act as a trap. When pest insects are attracted to the trap crop, they may be sprayed with insecticides.

Timing of Farm Operations — The attacks by many pest insects on crops can often be minimized by careful studies of their life cycles and changes in the timing of farm operations, such as sowing and harvesting, to put the incidence of pest and crop out of synchronization and thereby reduce pest attacks and avoiding seasonal carryover of pests from crop to crop.

3.3.5   Biological Inputs into Minimizing Insect Pest Attack

Some of the biological tools available to control insect pests, including the use of insect pathogens and the use of entomophilic nematodes have been discussed in earlier sections. Additional to these are the following possible biological inputs:

3.3.5.1   Insect Attractants

These chemical stimuli which are usually referred to as *semiochemicals* cause a behavioral response by the pest insect to attract it to the source of the chemical. For instance, food or ovipositional chemicals, or *kairomones*, which are produced by plants or microorganisms, can attract the insect to its host. An example is that carbon dioxide which is produced by roots in the plant rhizosphere can attract soil-inhabiting pests or larvae towards the roots on which they feed.

Many insects are attracted to the other sex of the same species by chemicals commonly called *pheromones*. Such sexual attractants have been identified and synthesized for a wide range of insect pests and are in common use to assess pest incidence or trap pests. Many insects are also attracted to general odors such as that from alcohol or to baits such as bran on which they can feed. The use of these attractants can supplement chemical control in several ways. They can be used to assess when the insect pests are most active, to optimize the timing of insecticide sprays, or the attractants can be combined with an insecticide so that all the insects attracted to the attractant are killed.

3.3.5.2   Parasites and Predators

All insect pests have a range of natural enemies. It is possible, although expensive, to complement insect control by chemicals releasing specific parasites and predators of the pest insect into the crop. However, it is preferable from an economic aspect to select insecticides that are either very specific to the pest or are systemic in the plant, so as to minimize their effects on natural enemies of the pest and maximize natural ecological control measures. Broad spectrum insecticides used on a regular long-term basis tend to eliminate populations of most parasites and predators and increase the dependence on regular insecticide treatments. Mites provide an excellent example of such interactions. Many plant-feeding mites are kept in check by predatory mites: when broad-spectrum acaracides or insecticides that are toxic to predatory mites are used, such as DDT and many organophosphate insecticides, the pest mite populations tend to increase on a long-term basis; indeed, it was the widespread use of DDT in orchards that brought fruit tree red spider mites to full pest status. As a practical example, populations of predators and parasites have been maintained in cereal fields in Europe by leaving the margins of the crop untreated with insecticides (Brown et al., 1992).

3.3.5.3   Alternative Hosts

Many insect pests feed on more than one species of plant. To attract pests away from the crop, the provision of alternative hosts through activities such as the controlled growth of weeds, or by planting trap plants or crops to attract insect pests away, can supplement insecticide control. For instance, populations of pest predators were much greater in maize crops populated with broad-leaved weeds, or grasses, than when these weeds were controlled by herbicides (Edwards et al., 1992).

3.3.5.4 Plant Nutrition

Many insect pests, particularly sucking insects such as aphids and leafhoppers, are attracted to succulent foliage when crops are very well fertilized or even over-fertilized with nitrogen (van Einden, 1966) or potassium (Kitchen et al., 1990) as can occur and populations build up more rapidly. Keeping fertilizer amendments slightly below the optimal can make a useful complement to the use of insecticides in the management of insect pests. The balance between obtaining optimal crop growth and minimizing insect attack is a difficult one to maintain.

3.3.6 Insecticide Inputs into Minimizing Insect Pest Attack

The kinds of insecticides used and the ways in which insecticides are used are critical in promoting ecologically based insecticide use.

3.3.6.1 Choice of Insecticide

Choosing the best insecticide is one of the most critical decisions in insect pest management. The more specific the insecticide is to the target pest, the less damage is caused to populations of parasites and predators. There have been extensive attempts in Europe to make information on the relative toxicity of the most commonly used insecticide to parasites and predators generally available, if this information is not included on the insecticide label. The International Organization for Biological Control has been working towards this end for the last 20 years and has accumulated an extensive data bank on the toxicity of a wide range of insecticides to natural enemies (Hassan et al., 1991).

Whenever possible, pesticides of low mammalian toxicity should be used. It seems certain that most of the high mammalian-toxic organochlorines and organophosphates still in current use will be phased out. Many of the more recent pesticides such as the pyrethroid insecticides not only have low mammalian toxicities but are also effective at very low doses that minimize most of their environmental effects. Whenever possible, pesticides specific to particular pests or groups of pests that have the minimum side effects should be used. There is considerable legislative pressure in most countries to restrict pesticides that have been demonstrated to be sources of actual or potential hazards to wildlife or to humans that persist in the environment. There are economic problems because the development of new selective insecticides is expensive, particularly in obtaining registration for use, which unfortunately mediates against the development of highly selective insecticides.

3.3.6.2   Frequency of Insecticide Use

If the same or closely related insecticides are used very frequently against a particular insect pest, there is a great potential for the insect pest to develop resistance to the insecticide that can be cross-linked with resistance to other insecticides. Such resistance develops most rapidly with pest insects that go through multiple generations in a season, such as aphids, since at least 50 generations are usually needed before resistance develops. Once an insect strain has developed resistance to one group of insecticides it can extend this resistance to other insecticides quite quickly. Insecticide resistance is indicated in the field when two or three times the recommended rate of insecticide is needed to control the insect pest in successive seasons, although levels of resistance up to 100 times have been reported. Currently, more than 500 species of insects have developed resistance to insecticides. Resistance to insecticides is transferred genetically and is based on a variety of mechanisms including detoxification, effects or penetration of the chemical through the cuticle. It is unfortunate that natural enemies of insect pests tend to develop resistance to insecticides more slowly than the pests.

3.3.6.3   Mammalian Toxicity

Insecticides with a high mammalian toxicity should be used only when no less toxic effective alternative insecticides are available. Toxic insecticides are not only hazardous to the applicator, but require stringent protective clothing and can be subject to drift to other areas on windy days. They may also cause hazard to the consumer if they leave residues on food crops when applied close to harvest or are taken up systemically into such crops. Such human hazards can be increased because toxic doses of some insecticides can be accumulated and stored in the tissues progressively.

The kinds of strategies to minimize human hazard depends upon the nature and cause of the exposure, the pesticides involved, and the resources available. Lowering dietary exposure, for instance, may require lengthening the time between spraying and harvesting the crop, thus allowing the pesticide to break down. Protecting field workers, on the other hand, may entail changing safety procedures or using better equipment. Pesticide residues in food and water remain an important concern — especially where persistent, fat-soluble organochlorine chemicals are still in use — and one that can be answered mainly by eliminating or severely curtailing their use. But because most evidence suggests that occupational exposure is a more severe problem than dietary intake, the emphasis should be on ways to reduce risks to exposed workers as well as to consumers.

3.3.6.4   Insecticide Persistence

Although persistence of an insecticide on plants or in soil means that fewer treatments need to be applied to control the pest, it can also increase environmental impacts by extending the period of exposure and affecting wildlife and natural enemies of pests over a much larger period. Ideally, the persistence of the insecticide used should match that of the insect pest to be controlled, as closely as possible. It is easier to increase persistence through appropriate formulations such as granules than to accelerate the loss of a relatively persistent insecticide once it has been applied.

3.3.6.5 Environmental Impact

As newer insecticides are developed, our knowledge of their potential environmental impact is increasing because of the extensive data bank on environmental issues required by national and international pesticide registration and regulatory authorities. The kind of data that allow more environmentally desirable insecticides to be chosen offers extensive knowledge of the toxicity of the insecticide to soil and aquatic invertebrates, fish, birds, and mammals. A further factor is to be aware of any potential of the insecticide to move into groundwater, persist in the environment, or bioconcentrate into the tissues of invertebrates and vertebrates.

3.3.6.6 Minimal Effective Dose

Most insecticides have recommendations for the dosage rates that should be applied for control of specific pests. Such dosage rates have usually been developed from the results of an extensive series of field trials. However, these recommendations almost invariably have an insurance factor built into them and in many instances, particularly with low pest populations, lower insecticide doses are quite effective even if applied at doses as low as half the recommended application rate.

3.3.6.7 Insecticide Placement and Formulation

It has been calculated that on average less than 50% of insecticide applications reach their desired target and even less actually reaches the pest insect. Clearly, placement of the insecticide as precisely as possible is a major factor in insect control and increases the potential to use lower doses vary considerably. Insecticide placement can be improved in a number of ways. For instance, there is a current trend towards the application of ultra-low volume concentrated sprays applied in small quantities of carrier in order to decrease dosages per unit area. However, this can result in greater persistence of residues on the plant. Another technique being adopted increasingly are charged sprays which are attracted to the leaf surface. For soil-inhabiting pests, placement of the insecticide around the base of the plants or in the row can improve insect pest control or using slow-release granules. Seed dressings are the best way of controlling many soil pests. The development of such techniques is progressing rapidly and holds considerable promise for lessened ecological and environmental impact.

3.4 CONCLUSIONS ON THE ECOLOGICAL AND ECONOMIC ASPECTS OF INSECTICIDE USE

Despite the widespread use of insecticides in the U.S. (about 100 million kg/annum) with a ten-fold increase in use between 1945 and 1989, the total crop losses due to insect damage have doubled from 7 to 13% over the same period (Pimentel et al., 1991). This has been due mainly to dramatic changes in agricultural practices, particularly the widespread adoption of continuous production of crops such as corn which has resulted in a four-fold increase in corn losses to insects, despite about 1000-time increases in insecticide use in corn production. Clearly, with more attention paid to rotations and ecologically-based and judicious use of insecticides, such increases in insecticide use would not have been necessary.

The continued use of extensive insecticides for controlling insects is under question, since progressive increases in costs of these chemicals are not correlated with increasing financial returns to farmers. There is little doubt that if the costs of the environmental impacts of insecticides were subtracted from the economic benefits, insecticide use would be much less attractive, particularly if the users or the producers of the insecticides were required to pay these costs. Examples of such costs include those of extracting insecticides from contaminated drinking water; provision of land for disposal of highly contaminated material dredged from rivers and waterways; loss of fish productivity in contaminated freshwater such as the Great Lakes; losses of crustacea that provide human food in contaminated estuaries; and effects on crop yields through decreased pollination by bees and other insects.

In developed countries, there is a demand for fruits and vegetables that are "cosmetically" attractive and have no blemishes. However, there is increasing question as to whether the cost of achieving this in both financial terms as well as in the accompanying potential contamination of the food by insecticides is worthwhile. There is an increasing pressure by consumers for "clean" and uncontaminated foods. This in turn is putting increased demand on the insecticide industry to produce chemicals with low mammalian toxicity that can be used at low doses with little environmental impact. There are increasing costs in the production of insecticides based on finite supplies of oil and increasing requirements for data to prove their environmental safety. Paradoxically, as more environmental data are required for each insecticide, the cost of producing a new insecticide goes up. This economic pressure on the agrochemical industry makes it uneconomical to develop insecticides that are highly specific for certain organisms, because of the limited market for such chemicals.

Most of the assessments of the benefits of insecticide use are based on direct crop yields and economic returns and are termed risk/benefit analyses. However, these do not take account of the indirect costs associated with insecticide use, in terms of environmental impacts and human health — costs which are not paid for by the insecticide user. There has been much discussion on the subject of risk-benefit evaluations. In particular, such assessments are extremely difficult to make accurately because of all the variables involved and the intangibles in valuing environmental and human health. How can crop yield increases be compared to hazards to wildlife or even endangered species? There is no valid way to compare economic benefits resulting from abundant and inexpensive food with the esthetics of loss of populations of birds or other wildlife. It is easier to pass judgment on crop yield benefits gained at a cost of off-site pollution of groundwater and drinking water by insecticides.

Moreover, the cost/benefit analyses assume that the farmer knows, with certainty, factors such as the EIL, the effects of alternate management tactics, and the effects of weather, soil fertility, and the soil pH on yield. It also assumes that each pest control decision is independent of every other decision on other inputs, such as cultural practices. Many grower practices, in particular insecticide applications, are a form of insurance aimed at minimizing risk. Generally, growers select plant cultivars and cultural practices independently of their susceptibility to pests. Most farmers are most interested in expenditures that reduce uncertainty or risk as far as possible. However, any insurance use of insecticides should be linked to some probability function rather than some formulated schedule.

Since most insecticides are used by farmers on a routine insurance basis to minimize risk, scant attention is usually paid to the potential for increasing profits through lowering variable costs, by the judicious use of much less insecticides on a field-by-field basis. It is important to educate farmers in understanding that insecticide use can be minimized by practices such as those described earlier in this chapter, using ecological principles for decisions on insecticide use, complemented with cultural and biological inputs, and by using insecticides only when it can be forecast that significant crop losses are likely to occur.

On an international scale, there have been major attempts in recent years to decrease insecticide use in various countries by up to 50%. For instance, in Ontario, Canada, a 10-year program to reduce insecticide use by 50% in 2002 is currently in progress (Surgeoner and Roberts, 1993). In Denmark, a program to reduce overall insecticide use by 50% began in 1985. Similarly, in Sweden a project to reduce insecticide use aimed at an overall decrease of 50% between 1988 and 1993, which was to be followed by a second target for a further 50% reduction (Petterson, 1993). The Netherlands began a program in 1989 to reduce pesticide use by 50% in 10 years. Various studies have suggested that it is feasible to reduce insecticide use in the U.S. by 35 to 50% without decreasing crop yields significantly (Pimentel et al., 1993). Pimentel et al. (1993) concluded that decreasing pesticide use by approximately 50% in the U.S. would add some additional costs in use of alternative pest control tactics, but the overall cost of food production would increase only by 0.5%. They estimated that extra pest control measures would add $818 million annually, compared with the $4.1 billion already being spent, but that this would be more than offset by a significant decrease in the direct environmental and public health costs which he estimated at a total of $8.1 million.

Achieving such decreases in insecticide use would involve considerable education of farmers into the diverse ecological and cultural techniques available for decreasing the use of insecticides that were discussed earlier. It would also require considerable changes in current insecticide regulations and in current government policies, e.g., commodity and price support programs. Finally, it would need considerably more research into the ecological aspects of insecticide use and alternative insect control practices.

REFERENCES

Anderson, R.L., 1989. The toxicity of synthetic pyrethroids to freshwater invertebrates. *Environmental Toxicology and Chemistry*, 8 (5): 403-410.

Brown, A.W.A., 1978. *Ecology of Pesticides*. Wiley and Sons, New York, Chichester, Brisbane, Toronto, 525 pp.

Brown, R.A., Jepson, P.C., and Sotherton, N.W. (Eds.) 1992. Interpretation of Pesticide Effects on Beneficial Arthropods. Association of Applied Biologists. *Aspects of Applied Biology*, 31, 208 pp.

Butler, G.C. (Ed.) 1978. *Principles of Ecotoxicology. Scope 12*. Wiley and Sons, Chichester, 350 pp.

Carey, A.E., 1979. Monitoring pesticides in agricultural and urban soils in the United States. *Pesticides Monitoring Journal*, 13 (1): 23-27.

Carroll, C.R., Vandermeer, J.H., and Rosset, P.M. (Eds.) 1990. *Agroecology*. McGraw-Hill Publishing Co., New York, 641 pp.

Carson, R., 1962. *Silent Spring*. Hamish Hamilton, London, 304 pp.

Day, K.E., 1989. Acute chronic and sublethal effects of synthetic pyrethroids in freshwater. *Environmental Toxicology and Chemistry*, 8 (5): 411-416.

Domsch, K.H., 1963. Influence of plant protection chemicals on the soil microflora: Review of the literature. *Mitt. Biologische Bundesanstalt Land-Forstirwch. Berlin Dahlem*, 105: 5.

Domsch, K.H., 1983. An ecological concept for the assessment of the side-effects of agrochemicals on soil microorganisms. *Residue Reviews*, 86: 65.

Duffus, J.H., 1980. *Environmental Toxicology*. Resource and Environmental Science Series, Halsted Press, John Wiley and Sons, New York, 164 pp.

Edwards, C.A., 1973a. *Persistent Pesticides in the Environment*. 2nd Edition. CRC Press, Cleveland, Ohio, 170 pp.

Edwards, C.A., 1973b. *Environmental Pollution by Pesticides*. Plenum Press, London and New York, 542 pp.

Edwards, C.A., 1977. Nature and origins of pollution of aqautic systems by pesticides. In: *Pesticides in Aquatic Environments*. M.A.Q. Khan (Ed.). Plenum Press, New York, 11-37.

Edwards, C.A. 1983. Agrochemicals as environmental pollutants affecting human health with particular reference to developing countries. *World Health Organization Report EFP/EC/WP83.3*. 22 pp.

Edwards, C.A., 1985. Agrochemicals as environmental pollutants. In: *Control of Pesticide Residues in Food. A Directory of National Authorities and International Organizations*. Bengt. V. Hofsten., 1-19.

Edwards, C.A., 1989. The Impact of Herbicides on Soil Ecosystems. *Critical Reviews in Plant Sciences* 8 (3): 221-257.

Edwards, C.A., 1994. Pesticides as Environmental Pollutants. In: *World Directory of Pesticide Control Organizations*, G. Ekstrom (Ed.), British Crop Protection, Cornwall, 1-24.

Edwards, C.A. and Bohlen, P.J., 1992. The effect of toxic chemicals on earthworms. *Reviews of Environmental Contamination and Toxicology*, 125: 23-99.

Edwards, C.A., Brust, G.E., Stinner, B.R., and McCartney, D.A., 1992. Work in the United States on the use of cropping patterns to promote natural enemies of pests. In: *Interpretation of Pesticide Effects on Beneficial Arthropods*. R.A. Brown, P.C. Jepson and N.W. Sotherton (Eds.), Association of Applied Biologists, *Aspects of Applied Biology*, 31: 139-148.

Edwards, C.A., Knacker, T.T., Pokarzhevskii, A.A., Subler, S., and Parmelee, R., 1996. The Use of Soil Microcosms in Assessing the Effects of Pesticides on Soil Ecosystems. In: *Proceedings of International Symposium on Environmental Behavior of Crop Protection Chemicals*. IAEA, Vienna, 435-452.

Edwards, C.A., Krueger H.R., and Veeresh, G.K. (Eds.) 1980. *Pesticide Residues in the Environment in India.* University of Agricultural Science Press, Bangalore, India, 650 pp.

Edwards, C.A. and Stinner, B.R., 1989. Pest Management and Conservation Tillage. *Proceedings of First International Conference on Conservation Tillage.* Tuxtla Guttierez, Mexico, 181-192.

Edwards, C.A. and Thompson, A.R., 1973. Pesticides and the Soil Fauna. *Residue Reviews,* 45: 1-79.

Environmental Protection Agency, 1989. Pesticide Industry Sales and Usage: 1988 Market Estimates. EPA Economic Analysis Branch, Washington., D.C.

Graham, F. Jr. 1970. *Since Silent Spring.* Hamish Hamilton, London, 297 pp.

Gurney, S.E. and Robinson, G.G.C., 1989. The influence of two herbicides on the productivity, biomass and community composition of freshwater marsh periphyton. *Aquatic Botany,* 36 (1): 1-22.

Hardy, A.R., 1990. Estimating exposure: the identification of species at risk and routes of exposure. In: *Effects of Pesticides on Terrestrial Wildlife.* L. Somerville and C.H. Walker (Eds.). Taylor and Francis, London, 81-97.

Hassan, S.A., Bigler, F., Bogenshutz, H., Boller, E., Brun, J. Callis, J.N.M., Chwerton, P., Coremans-Pelseneer, J., Duso, C., Lewis, G.B., Monsour, F., Moreth, L., Ooman, P.A., Overmeer, W.P.J., Polgar, L., Reickman, W., Samsoe-Peterson, L., Stanbli, A., Sterk, G., Tavares, K., Tuset, J.J. and Viggiani, G., 1991. Results of the Fifth Joint Pesticide Testing Programme by IOBC/WPRS Working Group. Pesticides and Beneficial Organisms. *Entomophaga,* 36: 55-67.

Hayes, J. Wayland, Jr. and Lawes, Edward R., Jr., 1991. *Handbook of Pesticide Toxicology.* Vols. I, II, and III. Academic Press, San Diego, London, Sydney, Toronto. 1527 pp.

Irvine, D.E.G. and Knights, B., 1974. *Pollution and the Use of Chemicals in Agriculture.* Butterworths, London, 136 pp.

Kendall, R.J. and Lacher, T.E. (Eds) 1994. *Wildlife Toxicology and Population Modelling: Integrated Studies of Agroecosystems.* Lewis Publishers, Boca Raton, New York, London, Tokyo. 576 pp.

Kitchen, N.R., Bucholz, D.D., and Nelson, C.J., 1990. Potassium fertilizer and potato leaf hopper effects on alfalfa growth. *Agronomy Journal,* 82: 1069-1074.

Landis, W.G. and Yu, Ming-Ho. 1995. *Introduction to Environmental Toxicology: Impacts of Chemicals Upon Ecological Systems.* Lewis Publishers, Boca Raton, New York, London, Tokyo. 328 pp.

McEwen, S.L. and Stevenson, G.R., 1979. *Use and Significance of Pesticides in the Environment.* Wiley and Sons, New York, Chichester, Brisbane, Toronto.

Mellanby, K., 1967. *Pesticides and Pollution.* New Naturalist Series No. 50. Collins, London, 221 pp.

Metcalf, R.L., 1977. Model Ecosystem Studies of Bioconcentration and Biodegradation of Pesticides. In: *Pesticides in Aquatic Environments.* M.A.Q. Khan, Ed., Plenum Press, 127-144.

Moore, N.W., 1966. *Pesticides in the Environment and Their Effects on Wildlife.* Blackwell, Oxford, 311 pp.

Moriarty, F., 1983. *Ecotoxicology: The Studies of Pollutants in Ecosystems.* Academic Press, London, 233 pp.

Muirhead-Thompson, R.C., 1971. *Pesticides and Freshwater Fauna.* Academic Press, London and New York, 248 pp.

Neely, W.B., 1980. *Chemicals in the Environment: Distribution, Transport, Fate and Analysis.* Dekker, New York.

Parr, J.F., 1974. Effects of pesticides on microorganisms in soil and water. In: *Pesticides in Soil and Water,* W.D. Guenzi (Ed.). Soil Science Society of America, Madison, WI, 315 pp.

Pettersson, O., 1993. Swedish Pesticide Policy in a Changing Environment. In: *The Pesticide Question: Environment, Economics and Ethics*. D. Pimentel and H. Lehman (Eds.) Chapman and Hall, New York and London, 182-205.

Pimentel, D., McLaughlin, L., Zepp, A., Lakitan, B., Kraus, T., Kleinman, P., Vancini, F., Roach, W.J., Graap, E., Keeton, W.S., and Selig, G., 1991. Environmental and economic impacts of reducing U.S. agricultural pesticide use. D. Pimentel (Ed.). *Handbook on Pest Management in Agriculture*. Boca Raton: CRC Press, 679-718.

Pimentel, D., Acquay, H., Biltonen, M., Rice, P., Silva, M., Nelson, J., Lipner, V., Giordano, S., Horowitz, A., and D'Amore, M., 1993. Assessment of Environmental and Economic Impacts of Pesticide Use. In: *The Pesticide Question: Environment, Economics and Ethics*. D. Pimentel and H. Lehman (Eds.) Chapman and Hall, New York and London, 47-84.

Pimentel, D. and Lehman, H. (Eds.), 1993. *The Pesticide Question: Environment, Economics and Ethics*. Chapman & Hall, London and New York. 441 pp.

Rand, G.M. (Ed.) 1995. *Fundamentals of Aquatic Tocixology: Effects, Environmental Fate and Risk Assessment*. 1125 pp.

Riseborough, R.W. et al., 1968. Organochlorine pesticides in the atmosphere. *Nature*, London, 211, 259.

Riseborough, R.W., 1986. Pesticides and Bird Populations. In: *Current Ornithology*. R.F. Johnston (Ed.), Plenum Press, London and New York, 397-427.

Rosen, A.A. and Kraybill, H.F., 1966. *Organic Pesticides in the Environment*. Advances in Chemistry. Series 60. American Chemical Society, Washington, D.C., 309 pp.

Rudd, R.L., 1964. *Pesticides and the Living Landscape*. Faber and Faber, London, 320 pp.

Stickel, L.R., 1968. Organochlorine Pesticides in the Environment. *Report of the U.S.D.I. Bureau of Sports Fisheries and Wildlife*, 1, 119 pp.

Stinner, B.R. and House, G., 1990. Arthropods and other invertebrates in conservation tillage. *Annual Review of Entomology*, 39: 299-318.

Surgeoner, G.A. and Roberts, W., 1993. Reducing Pesticide Use by 50% in the Province of Ontario: Challenges and Progress. In: *The Pesticide Question: Environment, Economics and Ethics*. D. Pimentel and H. Lehman (Eds.), Chapman and Hall, New York and London, 206-222.

Thompson, A.R. and Edwards, C.A., 1974. Effects of Pesticides on Non-Target Invertebrates in Freshwater and Soil. *Soil Science Society of America Special Publication*, No. 8, Chapter 13, 341-386.

Truhart, R., 1975. Ecotoxicology — A new branch of toxicology. In: *Ecological Toxicology Research*. A.D. McIntyre and C.F. Mills (Eds.). Proceedings of NATO Science Communications Conference, Quebec. Plenum Press, New York, 323 pp.

van Einden, H.F., 1966. A comparison of reproduction of *Brevicoryne brassicae* and *Myzus persicae* on brussel sprout plants supplied with different rates of nitrogen and potassium. *Entomologia Experimentalis et Applicata*, 9: 444-460.

White, D.H. and Krynitsky, A.J., 1986. Wildlife in Some Areas of New Mexico and Texas Accumulate Elevated DDE Residues, 1983. *Archives of Environmental Contamination and Toxicology*, 15, 149-157.

White, D.H., Mitchell, C.A., and Prouty, R.M., 1983a. Nesting biology of laughing gulls and relation to agricultural chemicals. *Wilson Bulletin*, 95 (4): 540-551.

Wiktelius, S. and Edwards, C.A., 1997. Organochlorine Residues in the African Terrestrial and Aquatic Fauna 1970-1995. *Review of Environmental Contamination and Toxicology*, 151, 1-38.

Yardim, E.N. and Edwards, C.A., 1998. Effects of chemical pest disease and weed management on the trophic structure of nematode populations in tomato agroecosystems. *Applied Soil Ecology*, 7: 137-147.

CHAPTER 4

Selective Insecticides

Douglas G. Pfeiffer

CONTENTS

4.1   INTRODUCTION

For decades, the backbone of most pest management programs in agriculture have been broad spectrum insecticides, primarily organophosphates, carbamates, pyrethroids, and, to a lesser extent lately, organochlorines. These are mainly neurotoxins, differing in the specific mode of action. Organophosphates and carbamates are cholinesterase inhibitors, affecting the neurotransmitter acetyl cholinesterase at the nerve synapse. Organochlorines and pyrethroids affect the transport of ions across axonic membranes, affecting the transmission of electrical charges along the neuron.

The broad-spectrum activity of modern insecticides has had a two-edged effect. A single application could control a wide range of pests for a minimum monetary cost. This advantage is especially acute in crops with a wide range of pests. This advantage has not been without cost, however. Broad-spectrum sprays also can wipe out populations of beneficial predators, parasites, and pollinators. In some cropping systems, growers must now contend with a wide array of secondary pests which follow treatment for a relatively small group of primary pests. Furthermore, non-target toxicity of such pesticides has caused concerns for farm worker safety and food safety. Nevertheless, broad-spectrum pesticides have been more commercially desirable because of market size and cost or product registration.

These negative attributes of conventional insecticides have been prime motives for development of alternative tactics for IPM programs for years. Recently, the impetus has been bolstered by the passage of the Food Quality Protection Act (FQPA), and uncertainties regarding the Act's impact on a variety of registered materials.

Although this chapter deals primarily with insecticides, this category will for purposes of discussion encompass acaricides. Furthermore, points may be illustrated with other classes of pesticides (e.g., herbicides).

4.2   MODES OF SELECTIVITY

There are several ways in which selectivity can be attained. Perhaps the most obvious is by **differential toxicity** (physiological selectivity of Ripper et al., 1951), i.e., inherently greater toxicity to the target pest than to non-target organisms. This may be a marked absolute difference, such as the use of ovicidal acaricides hexythiazox or clofentezine, which are nontoxic to predators; or simply a relative difference, such as the use of oxamyl rather than methomyl for aphids. The former pesticide is less toxic to the coccinellid predator of spider mites, *Stethorus punctum* (LeConte) (Pfeiffer et al., 1998).

Selectivity can also be achieved by adjusting **use pattern** (ecological selectivity of Ripper et al., 1951), modifying either spatial or temporal aspects of application. Spatial selectivity is used where crop borders are sprayed to kill immigrants, thus sparing predators, or by spraying trap crops. Temporal selectivity is maintained when an otherwise widely toxic pesticide is applied at a time when nontarget species are not present in the crop. For example, the herbicide paraquat negatively affects populations of the predatory mite, *Neoseiulus fallacis* (Garman), when that predator is in the orchard ground cover in the spring (Pfeiffer, 1986). The herbicide glyphosate is likewise toxic, but may be applied in the fall when predators are in the tree canopy; this also is a desirable use timing for this herbicide, since this is when translocation to roots of nutrient stores (and herbicide) occurs. Until physiologically selective insecticides are sufficiently developed, older and less selective materials must be used in as selective a fashion as possible (Watson, 1975).

4.3   SELECTIVITY OF CONVENTIONAL INSECTICIDES

When organochlorine insecticides first appeared in agricultural use in the 1950s, there were dramatic disruptions of biological control systems. The mode of action of organochlorine insecticides is at the membrane of axons, affecting the movement of ions across the membranes during neurotransmission, and also on the enzyme ATPase in mitochondria (Cutkomp et al., 1982). This class of pesticides has been largely replaced for environmental reasons. In many crops, the most widely used insecticide class is now the organophosphates. After widespread use for the past

40 to 50 years, some organophosphates are now somewhat selective since some predators have acquired their own resistance, or predatory species with some degree of inherent tolerance have been selected. For example, the phytoseiid *Neoseiulus fallacis* (Garman) seems inherently better able to tolerate organophosphate sprays than other related species, and is often the predominant phytoseiid in sprayed orchards (Berkett and Forsythe, 1980). But development of resistance normally takes longer in predators than for their prey because when a pest survives because of resistance, it has an almost unlimited food supply, whereas if a resistant predator survives, it has the additional problem of a sharply reduced food supply.

Some newer classes of pesticides, such as pyrethroids, are extremely non-selective for insect populations. While significantly less toxic to vertebrates than to insects (Miller and Adams, 1982), pyrethroids are extremely detrimental to predatory populations. In apple orchards, pyrethroids are commonly recommended to be applied no later than bloom in order to avoid inducing spider mite outbreaks. Some progress has been made on breeding resistant strains of predatory mites to pyrethroids (Hoy et al., 1982). But it should be noted that even if resistant phytoseiids could be bred and released, many natural enemies are responsible for regulating other pests, and these biological control systems can also be disrupted by pyrethroids (Penman and Chapman, 1980).

4.4 SELECTIVITY OF NOVEL INSECTICIDES

Avermectins — The avermectins are a class of macrocyclic lactone pesticides originally derived from the soil microorganism, *Streptomyces avermitilis*. These compounds are experiencing widespread use in a variety of applications (Lasota and Dybas, 1991). Abamectin (avermectin B_1) is widely used as an insecticide; synthetic derivatives include the ivermectins. In fruit crops, avermectin is effective against spider mites, spotted tentiform leafminer, *Phyllonorycter blancardella* (Fabr.), pear psylla, *Cacopsylla pyricola* (Foerster), and a range of other pests (Dybas, 1989; Lasota and Dybas, 1991). In livestock, abamectins are used internally as anthelmentics as well as insecticide/acaricides (Benz et al., 1989). The targets of action for avermectin are receptors for the neurotransmitter γ-aminobutyric acid (GABA), at neuromuscular synapses (Clark et al., 1995). This is a common neurotransmitter in a wide range of organisms, including insects, arachnids, crustaceans, and nematodes, as well as the smooth muscles of the mammalian gut. While highly toxic to many arthropods and nematodes, the avermectins are only moderately toxic to mammals, and have low dermal toxicity (Lankas and Gordon, 1989). Impact on non-target organisms is low, partly because of the short half-life of avermectin in the environment (Wislocki et al., 1989). Further selectivity is conferred by the fact that this material is subject to translaminar absorption in young leaves. Beneficial species are thus partly protected, while phytophagous species are exposed to the pesticide. After the first few weeks following bloom in orchards, the foliage is less able to absorb the material. When avermectin was applied to apple trees in two sprays, a temporal pattern useful against San Jose scale crawlers, *Quadraspidiotus perniciosus* (Comstock), there was no detrimental effect on populations of the phytoseiid, *N. fallacis*.

But when applied in a season-long program, densities of predators were reduced (Pfeiffer, 1985). Coupled with this is the fact that the LC_{90} for predatory mites is often substantially higher than for phytophagous mites (Grafton-Cardwell and Hoy, 1983; El-Banhawy and El-Bagoury, 1985). Newer examples of the avermectin class offer a greater degree of selectivity to lepidopteran larvae (Lasota and Dybas, 1991)

Naturalytes — The insecticide spinosad is the first representative of this new class of pesticide chemistry. This product has two main components, spinosyns A and D. Spinosad is effective against a variety of lepidopterans, such as beet armyworm, *Spodoptera exigua* (Hubner) (Mascarenhas et al., 1996). Boyd and Boethel (1998) reported that spinosad is neither directly nor indirectly (through ingestion of contaminated prey) toxic to *Geocoris punctipes* (Say), *Nabis capsiformis* Germar, *Nabis roseipennis* Reuter, and *Podisus maculiventris* (Say), important hemipteran predators in soybean systems in the U.S. Spinosad is also effective against tephritid fruit flies, e.g., *Ceratitis capitata* Wied. (Adan et al., 1996). A further step in selectivity has been proposed for spinosad through incorporation into baits for Caribbean fruit fly, *Anastrepha suspensa*, by King and Hennessey (1996). Spinosad is subject to microbial degradation in the soil; its half-life is 9 to 17 days (Hale and Portwood, 1996).

Selective Acaricides — In recent years, two highly selective acaricides have been developed for spider mite control. While most acaricides work on adults or immature motile stages, the compounds hexythiazox and clofentezine work only on eggs of Tetranychidae. These compounds are not toxic even to other mites outside the family Tetranychidae. This includes the predatory family Phytoseiidae, and rust mites (Eriophyidae). A common species, the apple rust mite, *Aculus schlechtendali* (Nalepa), is thus allowed to survive to serve as an alternative food source for spider mites.

Care should be taken to avoid resistance to these acaricides. Herron et al. (1997) predicted that resistance could develop to clofentezine in twospotted spider mite in as few as four sprays. Dew may resuspend clofentezine and increase mortality of mite eggs (Rudd, 1997). Yamamoto et al. (1996a) tested various ratios of susceptible to resistant citrus red mite for reversion of hexythiazox resistance (50:50, 30:70, 10:90, and 2:98 S:R). Resistance reverted rapidly in the 50:50 and 30:70 ratios, but not sufficiently in the other ratios by the twelfth generation. In the field, resistance developed after 19 applications in 7 years and subsided over 33 months. Reversion was ascribed to incompletely recessive resistance and reproductive disadvantages associated with the resistant genotype. Yamamoto et al. (1996b) found that hexythiazox resistance would be highly heritable in a strain where the initial resistance level was moderate.

Insect growth regulators (IGR's) — The metamorphosis of insects can be the focus of action of pesticides in several ways. One class of insect growth regulators includes the chitin synthesis inhibitors (Marks et al., 1982). The pesticide diflubenzuron (Dimilin) is one such chitin synthesis inhibitor, affecting the manner in which cuticle is formed. When ecdysis is initiated after exposure to diflubenzuron, the inner cuticle pulls away from the exocuticle. Unfortunately, diflubenzuron has a fairly broad spectrum of activity. Where this material is used for gypsy moth control programs in forests in eastern North America, it should not be applied near streams or rivers so that it will not reach larger bodies of water with their susceptible crustacean populations.

More recently, more selective IGRs have appeared. Tebufenozide is an ecdysone agonist, mimicking 20-hydroxyecdysone, the insect hormone that induces molting. The material binds with the ecdysone receptor protein (Tomlin, 1994), causing a lethal premature molt. It has received Section 18 registrations in the U.S. for the past four years for leafrollers on apple. It is essentially nontoxic to predatory mites and Coleoptera in this system. Pfeiffer et al. (1996) found that parasitism of spotted tentiform leafminer was greater in tebufenozide-treated plots than in the conventionally treated control. This material is toxic only to larval Lepidopterans, with some selectivity even within the order. Tebufenozide has been effective against beet armyworm (Mascarenhas et al., 1996). Tebufenozide has been classified as a low-risk pesticide by the U. S. Environmental Protection Agency; this category is intended to speed the registration process of new pesticides intended to replace older, more non-selective pesticides.

Almost all activity of tebufenozide is limited to larvae of Lepidoptera; there is virtually no effect on other orders of insects. Smagghe et al. (1996b) found that tebufenozide was toxic to lepidopteran stored product pests, but not coleopteran species. Smagghe et al. (1996a) reported that although oriental cockroach reacted with hyperactivity and uncoordinated movements when injected with tebufenozide, behavior soon returned to normal and normal molting ensued. Butler et al. (1997) found that arthropod diversity (excluding Macrolepidoptera) was not reduced by tebufenozide treatment. However, Macrolepidoptera richness and abundance was reduced relative to the control by the IGR.

Halofenozide and methoxyfenozide are other examples of ecdysone agonists. The former has been reported to be effective against Japanese beetle, *Popillia japonica* Newman, at 3 ppm (Cowles and Villani, 1996); the latter is even more effective against European corn borer, *Ostrinia nubilalis* (Hubner), than was tebufenozide (Trisyono and Chippendale, 1997).

Pheromones as insecticides — Mating disruption is a pest management tactic that uses pheromones as insecticides, in the broad sense of FIFRA (a material that kills pests or mitigates their injury). Pheromone dispensers are placed in a field or orchard at a relatively high density (100–400/A, 250–1000/ha). This high density of pheromone dispensers, each releasing pheromone at high rates, disrupts the ability of male moths to orient to calling females. The exact mechanism is debated; several alternative mechanisms include false trail following, trail camouflage, and habituation/adaptation (Bartell, 1982; Cardé and Minks, 1995). There is evidence for various mechanisms in different systems; it is likely that most have some role under some conditions.

Advantages of mating disruption largely stem from the selectivity of the approach. Pheromones are non-toxic and so pose no hazard to farm workers, natural enemies, pollinators, or other non-target species. Consequently, there will be less induction of secondary pest outbreaks or pest resurgence. Selection pressure for pesticide resistance will be lessened. Disadvantages include lack of control of a wide variety of pests (the negative side of selectivity); the likelihood of control failure at high pest density; and the high cost of pheromone dispensers (this is more true for some pests, e.g., codling moth, than others, e.g., grape berry moth). The high cost of dispensers may be partially offset by economic benefits of increased biological control for both target and non-target pests. There may be increased survival of carabid beetles (Gronning, 1994), spiders (unpublished data), Aphidophaga (Knowles, 1997), and hymenopteran parasites (Biddinger et al., 1994) in mating disruption programs.

Another obstacle has been, in some cases, inadequate characterization of the pheromone composition of target species, e.g., a recent attempt to control dogwood borer, *Synanthedon scitula* (Harris), attacking apple in Virginia (Pfeiffer and Killian, 1999).

Most research on mating disruption has involved pests of orchards and vineyards in Europe and North America. Much attention has focused on codling moth, *Cydia pomonella* (L.), a key pest of apple worldwide (Moffitt and Westigard, 1984; Barnes et al., 1992). The cost of this disruption system is high. Adoption of the technique is proceeding in the Pacific Northwest of the U.S., but has lagged in the eastern states, where the pest complex is more diverse. Pest pressure from codling moth is often lower in the east, where the approach can be very successful (Pfeiffer et al., 1993a)

Oriental fruit moth, *Grapholita molesta* (Busck), is a key pest of stone fruits and an occasional pest of apple. Pioneering work was done on this pest in Australia (Rothschild, 1975; Vickers et al., 1985), and has been successfully used in the U.S. (Pfeiffer and Killian, 1988; Rice and Kirsch, 1990).

Most apple-growing areas of the world possess a leafroller complex, the specific composition of which varies geographically. One factor impeding adoption of codling moth mating disruption is the need to spray for mid- and late-season leafroller populations. Mating disruption programs for leafrollers would therefore have a more far-reaching effect than apparent from their control directly. Successful use of the technique has been made in the eastern U.S. (Pfeiffer et al., 1993b). But even here, there are several species involved: tufted apple bud moth, *Platynota ideausalis* (Walker), variegated leafroller, *Platynota flavedana* Clemens, and red-banded leafroller, *Argyrotaenia velutinana* Walker. Development of a general lea-froller blend that would control all the species in a region would improve the economic considerations. Promising results have been achieved in this area (Gronning et al., in press). Some species are more difficult targets for mating disruption than others, e.g., obliquebanded leafroller, *Choristoneura rosaceana* (Harris), in Michigan and New York (Agnello et al., 1996).

Peachtree borer, *Synanthedon exitiosa* (Say), and lesser peachtree borer, *Synanthedon pictipes* (Grote & Robinson), are important stone fruit pests in the southeastern U.S. Control provided by mating disruption against lesser peachtree borer was more effective than the most effective chemical treatment (Pfeiffer et al., 1991). Not only is this approach very effective, but is much safer than the relatively dangerous handgun application of sprays recommended for this complex, and is much more convenient than sprays, which must be applied immediately after harvest, a timing that conflicts with the labor-intensive apple harvest which occurs on most orchards with peaches. While peachtree borer disruption is registered by EPA, no registration has ever been granted for the mating disruption package for lesser peachtree borer.

Mating disruption has been effective for grape berry moth, *Endopiza viteana* Clemens, in North America (Dennehy et al., 1990), and for several grape tortricids in Europe (Descoins, 1990).

Mating disruption has also been developed for pests in other cropping systems. Commercial use is made against a complex of bollworms in U.S. and Egyptian cotton (Cardé and Minks, 1995), for tomato pinworm (Cardé and Minks, 1995), and also for diamondback moth (Cardé and Minks, 1995), a key pest of cabbage in many areas, one that is difficult to control because of resistance.

Pathogens — Entomopathogens offer great potential for selective tools for IPM, though relatively few have yet been registered. Tanada and Kaya (1993) listed seven bacteria, four viruses, two fungi, and one protozoan for a variety of pests. A brief description of the main groups of pathogens follows:

Bacteria — Perhaps the most widely used entomopathogen is the bacterium, *Bacillus thuringiensis* Berliner. This bacterium was discovered in 1902 in *Bombyx mori* and later characterized from *Ephestia kuehniella* Zell. The first use as an insecticide was in 1938 (Tomlin, 1994). Although infected insects may take several days to die, gut paralysis occurs quickly and no further damage occurs. Bacteria do not reproduce outside the host, and residual life on the plant surface is short. Commercial formulations do not contain viable bacteria. The most common toxin, the δ-endotoxin, is produced from crystal proteins in the bacterium. However, there is actually a variety of toxins produced by various strains of *B. thuringiensis*; the specific combination of toxins confer some differences in host spectrum (Tanada and Kaya, 1993). Most commercial use has been made of the subspecies *B. t. kurstaki*, which is toxic primarily to Lepidoptera. Even within that order, there is not uniform susceptibility. But despite this level of specificity, there may be negative environmental impacts from the use of *B. t. kurstaki*. Miller (1990) reported reduced diversity of forest Lepidoptera following the use of this pathogen applied against gypsy moth, *Lymantria dispar* (L.), in Oregon. In all three years of that study, the number of species was reduced; the total number of nontarget Lepidoptera was reduced in two of the three years.

A novel use of *B. thuringiensis* has involved the development of transgenic plants. The *B. thuringiensis* δ-endotoxin gene has been inserted into corn, tomato, potato, and tobacco (Leemans et al., 1990; Ebora and Sticklen, 1994). While this has often been initially quite effective, there has been controversy because of concerns over resistance. This results from the continual exposure of subeconomic populations of pests to the toxin. Some tactics could help prevent the development of this resistance, such as engineering plants that would express the toxin only at key life stages of the plant or in selected plant parts, using multiple toxins, mixtures of resistant and susceptible plants to provide harborage for susceptible populations, or combining with other control tactics such as biological control (Ebora and Sticklen, 1994; Gould et al., 1994). One company involved in the production of such resistant lines has published a set of guidelines that are intended to slow or prevent the development of *B. thuringiensis* resistance in target insects (Monsanto, 1997).

Some other subspecies of *B. thuringiensis* are used, some of which exhibit a spectrum of activity beyond the Lepidoptera. Examples of other subspecies are *B. t. aizawa* (used for caterpillars, especially diamondback moth), *B. t. isrealensis* (mosquito and black fly larvae), *B. t. tenebrionis* (Coleoptera, especially Colorado potato beetle), and *B. t. sandiego* (Coleoptera, especially Colorado potato beetle). Furthermore, recombinant DNA technology has allowed the blending of toxins from different subspecies, forming bacteria with wider host ranges (Tanada and Kaya, 1993).

Bacillus thuringiensis has developed a reputation for vertebrate safety, since most common formulations have been used for lepidopteran larvae. These products contain the δ-endotoxin, which is nontoxic to vertebrates. However the δ-endotoxin derived from some subspecies has a greater effect on vertebrates in toxicological studies, as does the β-endotoxin.

Bacillus popilliae has been used for control of larvae of Japanese beetle. Unlike *B. thuringiensis*, *B. popilliae* does replicate in the host after application. When used in an area-wide program, this bacterium can provide overall population suppression, rather than acting as a microbial insecticide at a particular site (Klein, 1995). New formulations of *B. popilliae* are under development to provide greater virulence.

Bacillus sphaericus, a spore-forming aerobic bacterium, is used for control of mosquito larvae The bacterium produces a crystal toxin, which is highly larvicidal. This toxin binds to brush-border membranes in the mid gut. The use of this agent was recently reviewed by Charles et al. (1996). Early strains were not sufficiently efficacious, but a strain from Indonesia imposed a high degree of mortality, mainly to the genera *Anopheles* and *Culex*.

Bacillus subtilis has been used as a seed treatment to protect against root-infecting phytopathogens (Tomlin, 1994).

Fungi — The interactions between fungal pathogens and insect hosts were recently reviewed by Hajek and St. Leger (1994). Most control efforts have used inundative augmentation (mycoinsecticides), although permanent establishment and conservation of natural populations have also been effective.

Beauveria spp. are Deutoeromycete fungi that have been used in practical application in IPM. Spores penetrate the insect cuticle by enzymatic action, requiring free water. Death of the insect takes 3 to 5 days, after which fresh spores are produced on the cadaver. The fungus *Beauveria bassiana* (Balsamo) was isolated from European corn borer, *Ostrinia nubilalis* in France (Tomlin, 1994). This pathogen has been used successfully in a granular formulation for European corn borer (Tomlin, 1994) and as a wettable powder formulation for diamondback moth, *Plutella xylostella* (L.) (Vandenberg et al., 1998). *B. bassiana* was more effective against diamondback moth than two other fungal pathogens, *Metarhizium anisopliae* Metschnikoff and *Paecilomyces fumosoroseus* (Wize) (Shelton et al., 1998). A related fungus, *Beauveria brongniartii*, has been used against cockchafers (*Melolontha melolontha*) in Europe, and for white grubs in sugar cane (Tomlin, 1994), This fungus was originally collected from the scarab, *Hoplochelus marginalis*, in Madagascar (Tomlin, 1994).

A case of successful use of fungi is represented by *Acremonium* endophytes in plants, reviewed by Breen (1994). These fungi are in the tribe Balansiae within the family Clavicipitaceae, and live almost entirely within the host plant. They differ widely in their host ranges. In pastures, these endophytes can pose a hazard to livestock, while in turf grass situations, resistance to insect feeding is conferred. Resistance is reported to various Homoptera, Hemiptera, Lepidoptera, and Coleoptera. Most susceptible pests feed on above-ground portions of the host plant. Some success has been reported against Japanese beetle (Potter et al., 1992).

The fungus *Ampelomyces quisqualis*, is a hyperparasite of powdery mildew fungi, *Erysiphaceae* spp. While this fungus may be used against the several mildew species in this genus, sprays are not compatible with fungicides (Tomlin, 1994).

The need for application of fungicides will hinder the adoption of entomopathogenic fungi. This is a good example of the need for higher levels of integration in IPM programs, i.e., the development of disease-resistant varieties could facilitate the development of arthropod IPM.

Viruses — Granulosis virus is used for codling moth (*Cydia pomonella* granulosis virus) commercially in Europe (Blommers, 1994). This Baculovirus was obtained from codling moth in Mexico. The virus is obtained from living codling moth larvae, and concentrated by centrifugation (Tomlin, 1994). There is also a granulosis virus for summerfruit tortrix (*Adoxophyes orana* granulosis virus) in Switzerland, where it has been registered since 1985 (Tomlin, 1994). These viruses have no toxicity or irritability to mammals (Tomlin, 1994).

Recombinant baculovirus insecticides were recently reviewed by Bonning and Hammock (1996). Baculoviruses constitute an important natural mortality factor in many insect populations, especially holometabolous insects. The length of time required to effect mortality (days to weeks) has been a limiting factor in their use in IPM. However, modern recombinant DNA technology has been used to modify these organisms (Bonning and Hammock, 1996). An early field test was described by Cory et al. (1994), where the alfalfa looper, *Autographa californica*, was controlled with a NPV (AcNPV) that expressed an insect-specific neurotoxin from a scorpion, *Androctonus australis* Hector. That study reported hastened mortality and reduced crop damage.

One concern over this new technology is that selectivity of the virus may be lessened. Heinz et al. (1995) reported that when a modified AcNPV was used in the field, two generalist predators, *Chrysoperla carnea* (Stephans) and *Orius insidiosus* (Say), and honey bee, *Apis mellifera* L., were not adversely affected.

The Baculoviridae contains two genera, *Nucleopolyhedrovirus* and *Granulovirus*. Viruses in this family are each specific to only a few species, usually from within the same family. Some baculoviruses induce insect hosts to climb to a high point before they die, facilitating dispersal of virus from the cadavers.

These viruses may act synergistically with conventional insecticides (McCutcheon et al., 1997). Recombinant AcNPV acted synergistically with cypermethrin and methomyl relative to the wild type virus (in terms of median lethal time). Other insecticides, while still showing a positive interaction (less time until mortality was achieved), the effect was antagonistic as opposed to synergistic, i.e., the effect was less than predicted.

One disadvantage of recombinant baculoviruses pointed out by Bonning and Hammock (1996) is that since the host insect dies relatively quickly, there is less chance for the virus to be replicated and recycled. This fact makes the derived strains of virus less competitive with wild-type viruses. At this time, the recombinant technology may be applied to only a small number of species. Another factor impeding development of this technology is a lack of a complete understanding of the host range of these viruses (Richards et al., 1998). Those authors discussed ways in which impacts of altered viruses may be assessed. Current *in vivo* techniques are too expensive for commercial production of viruses.

Nematodes — Various nematodes have been used as entomopathogens, especially in the families Steinernematidae and Heterorhabditidae; these organisms have been exempted from registration requirements by the U.S. Environmental Protection Agency (Klein, 1995). Some entomopathogenic nematodes have been used against a variety of pests, e.g., *Steinernema carpocapsae* (Weiser), but with variable results. Some are highly host specific, e.g., *S. scapterisci* Nguyen and Smart attacking mole crickets (Klein, 1995). Wider host ranges are sometimes reported than are naturally found, based on laboratory trials using high inoculation rates in Petri dishes (Lewis et al., 1996). Experiments in the field therefore sometimes yield discouraging results; Georgis and Gaugler (1991) recommended standardized testing to better predict the likelihood of successful biological control.

In some cases there may be interactions between treatments of *S. carpocapsae* and herbicides, though the mechanism is unclear (Gibb and Buhler, 1998). Synergism between an insecticide and entomopathogen has also been described for white grubs, using imidacloprid and the nematode, *Heterorhabditis bacteriophora* (Poinar) (Koppenhöfer and Kaya, 1998). In this case, the pesticide inhibits grooming, a primary method of defense against pathogens.

Protozoans — Little use has been made commercially of protozoan diseases of insects. An exception is *Nosema locustae* Canning, which has been used in grasshopper IPM programs in the U.S. and Africa (Klein, 1995).

REFERENCES

Adan, A, P. DelEstal, F. Budia, M. Gonzalez, and E. Vinuela. Laboratory evaluation of the novel naturally derived compound spinosad against *Ceratitis capitata. Pesticide Science.* 48, pp. 261-268, 1996.

Agnello, A. M., W. H. Reissig, S. M. Spangler, R. E. Charlton, and D. P. Kain. Trap response and fruit damage by obliquebanded leafroller (Lepidoptera: Tortricidae) in pheromone-treated apple orchards in New York. *Environmental Entomology.* 25, pp. 268-282, 1996.

Barnes, M. M., J. G. Millar, P. A. Kirsch, and D. C. Hawks. Codling moth (Lepidoptera: Tortricidae) control by dissemination of synthetic female sex pheromone. *Journal of Economic Entomology.* 85, pp. 1274, 1992.

Bartell, R. J. Mechanisms of communication disruption by pheromone in the control of Lepidoptera: A review. *Physiological Entomology.* 7, pp. 353-364, 1982.

Benz, G. W., R. A. Roncalli, and S. J. Gross. Use of ivermectin in cattle, sheep, goats, and swine. p. 215-229. In *Ivermectin and Abamectin.* Campbell, W. C. (ed.). Springer-Verlag, NY. 363 p., 1989.

Berkett, L. P. and H. Y. Forsythe. Predaceous mites (Acari) associated with apple foliage in Maine. *Canadian Entomologist.* 112, pp. 497-502, 1980.

Biddinger, D. L., C. M. Felland, and L. A. Hull. Parasitism of tufted apple bud moth (Lepidoptera: Tortricidae) in conventional and pheromone-treated Pennsylvania apple orchards. *Environmental Entomology.* 23, pp. 1568-1579, 1994.

Blommers, L. H. M. Integrated pest management in European apple orchards. *Annual Review Entomology.* 39, pp. 213-241, 1994.

Bonning, B. C. and B. D. Hammock. Development of recombinant baculoviruses for insect control. *Annual Review Entomology.* 41, pp. 191-210, 1996.

Boyd, M. L. and D. J. Boethel. Susceptibility of predaceous hemipteran species to selected insecticides on soybean in Louisiana. *Journal of Economic Entomology.* 91, pp. 401-409, 1998.

Breen, J. P. *Acremonium* endophyte interactions with enhanced plant resistance to insects. *Annual Review Entomology.* 39, pp. 401-423, 1994.

Butler, L., V. Kondo, and D. Blue. Effects of tebufenozide (RH-5992) for gypsy moth (Lepidoptera: Lymantriidae) suppression on nontarget canopy arthropods. *Environmental Entomology.* 26, pp. 1009-1015, 1997.

Cardé, R. T. and A. K. Minks. Control of moth pests by mating disruption: Successes and constraints. *Annual Review Entomology.* 40, pp. 559-585, 1995.

Casida, J. E. and G. B. Quistad. Golden age of insecticide research: Past present, or future? *Annual Review Entomology.* 43, pp. 1-16, 1998.

Charles, J. F., C. Nielsen-LeRoux, and A. Delcéluse. *Bacillus sphaericus* toxins: Molecular biology and mode of action. *Annual Review Entomology.* 41, pp. 451-472, 1996.

Clark, J. M., J. G. Scott, F. Campos, and J. R. Bloomquist. Resistance to avermectins: Extent, mechanisms, and management implications. *Annual Review Entomology.* 40, pp. 1-30, 1995.

Cowles, R. S. and M. G. Villani. Susceptibility of Japanese beetle, Oriental beetle, and European chafer (Coleoptera: Scarabaeidae) to halofenozide, an insect growth regulator. *Journal of Economic Entomology.* 89, pp. 1556-1565, 1996.

Cutkomp, L. K., R. B. Koch, and D. Desaiah. Inhibition of ATPases by chlorinated hydrocarbons. p. 45-69. In: *Insecticide Mode of Action.* Coats, J. R. (ed.) Academic, NY. 470 p., 1982

Cory, J. S., M. L. Hirst, T. Williams, R. S. Hails, and D. Goulson, B. M. Green, T. M. Carty, R. D. Possee, P. J. Cayley and D. H. L. Bishop. Field trial of a genetically improved baculovirus insecticide. *Nature.* 370, pp. 138-140, 1994.

Dennehy, T. J., W. L. Roelofs, E. F. Taschenberg, and T. N. Taft. Mating disruption for control of grape berry moth in New York vineyards. p. 223-240. In *Behavior-Modifying Chemicals for Pest Management: Applications for Pheromones and Other Attractants.* Ridgway, R. L., R. M. Silverstein, and M. N. Inscoe (eds.). Marcel Dekker, N.Y., 1990.

Descoins, C. Grape berry moth and grape vine moth in Europe. p. 213-222. In *Behavior-Modifying Chemicals for Pest Management: Applications for Pheromones and Other Attractants.* Ridgway, R. L., R. M. Silverstein, and M. N. Inscoe (eds.). Marcel Dekker, N.Y., 1990.

Dhadialla, T. S., G. R. Carlson, and D. P. Le. New insecticides with ecdysteroidal and juvenile hormone activity. *Annual Review Entomology.* 43, pp. 545-569, 1998.

Dybas, R. A. Abamectin use in crop protection. p. 287-310. In *Ivermectin and Abamectin.* Campbell, W. C. (ed.). Springer-Verlag, NY. 363 p., 1989.

Ebora, R. V. and M. B. Sticklen. Genetic transformation of potato for insect resistance. p. 509-521. In *Advances in Potato Pest Biology and Management.* Zehnder, G. W., M. L. Powelson, R. K. Jansson, and K. V. Raman (eds). American Phytopathological Press, St. Paul, MN. 655 p., 1994.

El-Banhawy, E. M. and M. E. El-Bagoury. Toxicity of avermectin and fenvalerate to the eriophyid gall mite *Eriophyes dioscoridis* and the predacious mite *Phytoseius finitimus* (Acari: Eriophyidae; Phytoseiidae). *International Journal of Acarology.* 11, pp. 237-240, 1985.

Georgis, R. and R. Gaugler. Predictability in biological control using entomopathogenic nematodes. *Journal of Economic Entomology* 84: 713-720. 1991.

Gibb, T. J. and W. G. Buhler. Infectivity of *Steinernema carpocapsae* (Rhabditida: Steinernematidae) in sterilized and herbicide-treated soil. *Journal of Entomological Science.* 33, pp. 152-157, 1998.

Gould, F. Sustainability of transgenic insecticidal cultivars: Integrating pest genetics and ecology. *Annual Review Entomology.* 43, pp. 701-726, 1998.

Gould, F., P. Follett, B. Nault, and G. G. Kennedy. p. 255-277. In *Advances in Potato Pest Biology and Management.* Zehnder, G. W., M. L. Powelson, R. K. Jansson, and K. V. Raman (eds). American Phytopathological Press, St. Paul MN 655 p., 1994.

Grafton-Cardwell, E. E. and M. A. Hoy. Comparative toxicity of avermectin B_1 to the predator *Metaseiulus occidentalis* (Nesbitt) (Acari: Phytoseiidae) and the spider mites *Tetranychus urticae* Koch and *Panonychus ulmi* (Koch) (Acari: Tetranychidae). *Journal of Economic Entomology*. 76, pp. 1216-1229, 1983.

Gronning, E. K. Mating disruption in apple orchards: Dispenser release rates, generic blends and community impact. M.S. thesis, Virginia Polytechnic Institute and State University, Blacksburg, 1994.

Gronning, E. K., D. M. Borchert, D. G. Pfeiffer, C. M. Felland, J. F. Walgenbach, L. A. Hull, and J. C. Killian. Effect of specific and generic pheromone blends on captures in pheromone traps of four leafroller species in mid-Atlantic apple orchards. *Journal of Economic Entomology*. In press.

Hajek, A. E. and R. S. St. Leger. Interactions between fungal pathogens and insect hosts. *Annual Review Entomology*. 39, pp. 293-322, 1994.

Hale, K. A. and D. E. Portwood. The aerobic soil degradation of spinosad — A novel natural insect control agent. *Journal of Environmental Science and Health. B. Pesticides, Contaminants and Agricultural Wastes*. 31, pp. 477-484, 1996.

Heinz, K. M., B. M. McCutcheon, R. Herrmann, M. P. Parrella, and B. D. Hammock. Direct effects of recombinant nuclear polyhedrosis viruses on selected nontarget organisms. *Journal of Economic Entomology*. 88, pp. 259-264, 1995.

Herron, G. A., S. A. Learmonth, J. Rophail, and I. Barchia. Clofentezine and fenbutatin-oxide resistance in the two-spotted spider mite, *Tetranychus urticae* Koch (Acari: Tetranychidae) from deciduous fruit tree orchards in Western Australia. *Experimental and Applied Acarology*. 21, pp. 163-169, 1997.

Hopkins, W. L. *Global Insecticide Directory*. Agricultural Chemical Information Service, First Ed. Indianapolis, IN. 183 pp., 1996.

Hoy, M. A., D. Castro, and D. Cahn. Two methods for large scale production of pesticide-resistant strains of the spider mite predator *Metaseiulus occidentalis* (Nesbitt) (Acari, Phytoseiidae). *Zietschrift für angewandte Entomologie*. 94, pp. 1-9, 1982.

King, J. R. and M. K. Hennessey. Spinosad bait for the Caribbean fruit fly (Diptera: Tephritidae). *Florida Entomologist*. 79, pp. 526-531, 1996.

Klein, M. G. Microbial control of turfgrass insects. p. 95-100. In: *Handbook of Turfgrass Insect Pests*. Brandenburg, R. L. and M. G. Villani (eds.). Entomological Society of America, Lanham, MD. 140 p., 1995.

Knowles, K. L. Impact of low-spray mating disruption programs on aphidophagous insect populations in Virginia apple orchards. M.S. thesis, Virginia Polytechnic Institute and State University, Blacksburg, 1997.

Koppenhöfer, A. M. and H. K. Kaya. Synergism of imidacloprid and an entomopathogenic nematode: A novel approach to white grub (Coleoptera: Scarabaeidae) control in turfgrass. *Journal of Economic Entomology*. 91, pp. 618-623, 1998.

Lankas, G. R. and L. R. Gordon. Toxicology. p. 89-112. In *Ivermectin and Abamectin*. Campbell, W. C. (ed.). Springer-Verlag, NY. 363 p., 1989.

Lasota, J. A. and R. A. Dybas. Avermectins, a novel class of compounds: Implications for use in arthropod pest control. *Annual Review Entomology*. 36, pp. 91-117, 1991.

Leemans, J., A. Reynaerts, H. Höfte, M. Peferoen, H. Van Mellaert, and H. Joos. Insecticidal crystal proteins from *Bacillus thuringiensis* and their use in transgenic crops. p 573-581. In *New Directions in Biological Control*. Baker, R. R. and P. E. Dunn (eds.). Alan R. Liss, N.Y., 1990.

Lewis, E. E., M. Ricci and R. Gaugler. Host recognition behaviour predicts host suitability in the entomopathogenic nematode *Steinernema carpocapsae* (Rhabditida: Steinernematidae). Parasitology 113: 573-579, 1996.

Marks, E. P., T. Leighton, and F. Leighton. Modes of action of chitin synthesis inhibitors. p. 281-313. In *Insecticide Mode of Action*. Coats, J. R. (ed.) Academic, N.Y. 470 p., 1982.

Mascarenhas V. J., B. R. Leonard, E. Burris, and J. B. Graves. Beet armyworm (Lepidoptera: Noctuidae) control on cotton in Louisiana. *Florida Entomologist*. 79, pp. 336-343, 1996.

McCutcheon, B. F., Hoover, K., Preisler, H. K., Betana, M. D., Herrman, R., Robertson, J. L., and Hammock, B. D., Interactions of recombinant and wild-type baculoviruses with classical insecticides and pyrethroid-resistant tobacco budworm (Lepidoptera: Noctuidae). *Journal of Economic Entomology*. 90, pp. 1170-1180, 1997.

Miller, J. C. Field assessment of the effects of a microbial pest control agent on nontarget Lepidoptera. *American Entomologist*. 36, pp. 135-139, 1990.

Miller, T. A. and M. E. Adams. Mode of action of pyrethroids. p. 3-27. In *Insecticide Mode of Action*. Coats, J. R. (ed.) Academic, NY 470 p., 1982.

Moffitt, H. R. and P. H. Westigard. Suppression of the codling moth (Lepidoptera: Tortricidae) population on pears in southern Oregon through mating disruption with sex pheromone. *Journal of Economic Entomology*. 77, p. 1513-1519, 1984.

Monsanto. Resistance management guide. NatureMark Potatoes. Monsanto. Boise, ID, 1997.

Penman, D. R. and R. B. Chapman. Woolly apple aphid outbreaks following use of fenvalerate in apple in Canterbury, New Zealand. *Journal of Economic Entomology*. 73, pp. 49-51, 1980.

Pfeiffer, D. G. Toxicity of avermectin B₁ to San Jose scale (Homoptera: Diaspididae) crawlers, and effects on orchard mites by crawler sprays versus full-season applications. *Journal of Economic Entomology*. 78, pp. 1421-1424, 1985.

Pfeiffer, D. G. Effects of field applications of paraquat on densities of *Panonychus ulmi* (Koch) and *Neoseiulus fallacis* (Garman). *Journal of Agricultural Entomology*. 3, pp. 322-325, 1986.

Pfeiffer, D. G., S. B. Ahmed, J. C. Killian and M. H. Rhoades. Use of mating disruption and tebufenozide against codling moth and leafrollers. 1996. Proc. 72nd Cumberland-Shenandoah Fruit Workers' Conf., Winchester VA. Nov. 21-22, 1996.

Pfeiffer, D. G. (Bulletin Coordinator), R. D. Fell, H. W. Hogmire, G. P. Dively, J. A. Barden, R. E. Byers, R. P. Marini, C. S. Walsh, R. F. Heflebower, R. J. Rouse, K. S. Yoder, A. R. Biggs, J. B. Kotcon, P. W. Steiner, J. F. Derr, C. L. Foy, M. J. Weaver, and J. F. Baniecki. 1998 Spray Bulletin for Commercial Tree Fruit Growers. Va. Coop. Ext. Serv. Publ. 456-419, 1998.

Pfeiffer, D. G., W. Kaakeh, J. C. Killian, M. W. Lachance and P. Kirsch. Mating disruption for control of damage by codling moth in Virginia apple orchards. *Entomologia Experimentalis et Applicata*. 67, pp. 57-64, 1993a.

Pfeiffer, D. G., W. Kaakeh, J. C. Killian M. W. Lachance, and P. Kirsch. Mating disruption to control damage by leafrollers in Virginia apple orchards. *Entomologia Experimentalis et Applicata*. 67, pp. 47-56, 1993b.

Pfeiffer, D. G. and J. C. Killian. Disruption of olfactory communication in oriental fruit moth and lesser appleworm in a Virginia peach orchard. *Journal of Agricultural Entomology*. 5, pp. 235-239, 1988.

Pfeiffer, D. G. and J. C. Killian. Dogwood borer (Lepidoptera: Sesiidae) flight activity and an attempt to control damage in 'Gala' apples using mating disruption. *Journal of Entomological Science*. 34, pp. 210-218, 1999.

Pfeiffer, D. G., J. C. Killian, E. G. Rajotte, L. A. Hull, and J. W. Snow. Mating disruption for reduction of damage by lesser peachtree borer (Lepidoptera: Sesiidae) in Virginia and Pennsylvania peach orchards. *Journal of Economic Entomology*. 84, pp. 218-223, 1991.

Pfeiffer, D. G. and S. W. Pfeiffer. Relative susceptibility to slide-dip application of cyhexatin in three populations of *Panonychus ulmi* (Koch) in Virginia apple orchards. *Journal of Agricultural Entomology*. 3, pp. 326-328, 1986.

Potter, D. A., C. G. Patterson, and C. T. Redmond. Influence of turfgrass species and tall fescue endophyte on feeding ecology of the Japanese beetle and southern masked chafer grubs (Coleoptera: Scarabaeidae). *Journal of Economic Entomology.* 85, pp. 900-909, 1992.

Rice, R. E. and P. Kirsch. Mating disruption of the oriental fruit moth in the United States. p. 193-211. In *Behavior-Modifying Chemicals for Pest Management: Applications for Pheromones and Other Attractants.* Ridgway, R. L., R. M. Silverstein, and M. N. Inscoe (eds.). Marcel Dekker, N.Y., 1990.

Richards, A., M. Mathews, and P. Christian, Ecological considerations for the environmental impact evaluation of recombinant baculovirus insecticides. *Annual Review Entomology.* 43, pp. 493-517, 1998.

Ripper, W. E., R. M. Greenslade, and G. S. Hartley. Selective insecticides and biological control. *Journal of Economic Entomology.* 44, pp. 448-459, 1951.

Rothschild, G. H. L. Control of oriental fruit moth (*Cydia molesta* (Busck) (Lepidoptera: Tortricidae)) with synthetic female pheromone. *Bulletin of Entomological Research.* 65, pp. 473-490, 1975.

Rudd, J. A. Effects of pesticides on spin down and webbing production by the two-spotted spider mite, *Tetranychus urticae* Koch (Acari: Tetranychidae). *Experimental and Applied Acarology.* 21, pp. 615-628, 1997.

Shelton, A. M., J. D. Vandenberg, M. Ramos, and W. T. Wilsey. Efficacy and persistence of *Beauveria bassiana* and other fungi for control of diamondback moth (Lepidoptera: Plutellidae) on cabbage seedlings. *Journal of Entomological Science.* 33, pp. 142-151, 1998.

Smagghe, G., J. A. Jacas, P. Del Estal, E. Vinuela, and D. Degheele. Tebufenozide: Effects of a nonsteroidal ecdysone agonist in the oriental cockroach *Blatta orientalis. Parasitica* 52, pp. 53-59, 1996a.

Smagghe, G., H. Salem, L. Tirry, and D. Degheele. Action of a novel insect growth regulator tebufenozide against different developmental stages of four stored product insects. *Parasitica* 52, pp. 61-69, 1996b.

Tanada, Y. and H. K. Kaya. *Insect Pathology.* Academic, N.Y. 666 p., 1993.

Tomlin, C. (ed.). *The Pesticide Manual.* Brit. Crop Protect. Council, Surrey, U.K. 1341 p., 1994.

Trisyono, A. and G. M. Chippendale. Effect of the nonsteroidal ecdysone agonists, methoxyfenozide and tebufenozide, on the European corn borer (Lepidoptera: Pyralidae). *Journal of Economic Entomology.* 90, pp. 1486-1492, 1997.

Vandenberg, J. D., A. M. Shelton, W. T. Wilsey, and M. Ramos. Assessment of *Beauveria bassiana* sprays for control of diamondback moth (Lepidoptera: Plutellidae) on crucifers. *Journal of Economic Entomology.* 91, pp. 624-630, 1998.

Vickers, R. H., G. H. L. Rothschild, and E. L. Jones. Control of the oriental fruit moth, *Cydia molesta* (Busck) (Lepidoptera: Tortricidae), at a district level by mating disruption with synthetic female pheromone. *Bulletin of Entomological Research.* 75, pp. 626-634, 1985.

Watson, T. F. Practical considerations in use of selective insecticides against major crop pests. p. 47-65. In *Pesticide Selectivity.* Street, J. C. (ed.). Marcel Dekker, N.Y. 197 p., 1975.

Wislocki, P. G., L. S. Grosso, and R. A. Dybas. Environmental aspects of abamectin use in crop protection. p. 182-200. In *Ivermectin and Abamectin.* Campbell, W. C. (ed.). Springer-Verlag, NY. 363 p., 1989.

Yamamoto, A., H. Yoneda, R. Hatano, and M. Asada. Stability of hexythiazox resistance in the citrus red mite, *Panonychus citri* (McGregor) under laboratory and field conditions. *Journal of Pesticide Science.* 21, pp. 37-42, 1996a.

Yamamoto, A., H. Yoneda, R. Hatano, and M. Asada. Realized heritability estimates of hexythiazox resistance in the citrus red mite, *Panonychus citri* (McGregor). *Journal of Pesticide Science.* 21, pp. 43-47, 1996b.

SECTION IV

Cultural Practices

CHAPTER 5

Using Cultural Practices to Enhance Insect Pest Control by Natural Enemies

N.A. Schellhorn, J.P. Harmon, and D.A. Andow

CONTENTS

1-56670-478-2/00/$0.00+$.50
© 2000 by CRC Press LLC

5.1   INTRODUCTION

Cultural control of insect pests includes any modification in the way a crop or livestock is produced that results in lower pest populations or damage. This includes both changes in production practices of the crop or livestock and changes in surrounding areas of production. Some pest management specialists define cultural controls as purposeful manipulation of production practices to reduce pest populations or damage, but the concept is used more broadly here to include any change in production practice that results in lower pest populations or damage, whether intentional or not.

Cultural controls are defined to exclude production practices that act directly on insect pests, such as insecticide application, biological control, genetic control, and behavioral modifiers. In some treatments of the topic, plant and animal resistance is included as a cultural control. Because both the genetics and the environment of the crop or livestock influence plant and animal resistance to pest attack, resistance is in part determined by cultural practices. Traditionally, resistance is treated as a separate pest control tactic, and it is excluded from the present discussion of cultural control.

Cultural controls include a diverse set of practices, including: sanitation; destruction of alternate habitats and hosts used by the pest; tillage; water management; plant or animal density; crop rotation and fallow; crop planting date; trap cropping; vegetational diversity; fertilizer use; and harvest time. Sanitation is the removal and destruction of crop or animal material to reduce pest density, including the destruction of crop residues and the disposal of animal wastes (Stern, 1991). Destruction of alternate habitats and hosts is usually aimed at overwintering habitats and hosts, and has met with limited success. Tillage is used to prepare soil for planting and to reduce weeds. The various forms of tillage have diverse effects on insect pests (Stinner and House, 1990). Water management, such as irrigation, can affect pest populations, but because of its importance for growth and development of crops and livestock, it has been little used as a pest control tactic (Pedigo, 1996). Plant and animal density has significant effects on pests (Teetes, 1991). Many pests become more abundant at higher plant or animal density, but some become rarer. Often, however, non-pest control considerations determine production densities, and the general effects of density are only partially understood. Crop rotation entails changing the crop in subsequent plantings, and crop fallow involves suppressing all plant growth on a field for a production season. Both practices can disrupt the normal life cycle of a pest, reducing its populations and damage (Brust and Stinner, 1991). Planting date has dramatic effects on pests, and prior to the advent of inexpensive, synthetic organic insecticides, was widely used to avoid pest attack (Teetes, 1991). The timing of other cultural practices, such as cattle dehorning and crop harvest, can also affect pests (Stern, 1991; Pedigo, 1996). Trap cropping involves planting a crop to attract pests, to divert them from the nearby main crop or to concentrate them for easy destruction (Hokkanen, 1991). Vegetational diversity involves using other plants in the crop field to reduce pest attack (Andow, 1991). This includes intercropping, strip cropping, and weedy culture. Nitrogen applications such as fertilizers can have large effects on insect populations and attack, because nitrogen is limiting to most insects that eat plants (Mattson, 1980). All of these direct cultural effects on insect pests have been evaluated for many decades and excellent reviews of most of these controls have been published recently. Here we focus on a less-evaluated factor: how cultural practices affect natural enemies of insect pests, concentrating on predators and parasitoids.

The effects of cultural practices on natural enemies and the potential consequent effects on insect pests are an indirect mechanism for cultural control. In some cases, these indirect effects could be discussed as a type of biological control, emphasizing the role of the natural enemies. In this chapter, however, the role of the cultural practices that can affect natural enemies will be emphasized to draw a more explicit link between the practices that humans can manipulate and the effects on the natural enemies. In the long run, it will be useful to identify these links so that reliable, sustainable insect pest control tactics can be developed.

Cultural practices can affect natural enemy population densities and species diversity. Either of these can influence the ability of the natural enemies to suppress pest populations. Increased density of a particular species or a greater number of natural enemy species can result in greater mortality of the target pest. There are numerous examples in the literature demonstrating that cultural practices can enhance natural enemy abundance, and possibly their efficiency; however, the majority are descriptive and usually only compare abundance in one production system to another. Understanding the population processes involved in the population changes is necessary to develop a general realization of how cultural practices can result in higher densities of parasitoids and predators. Colonization, reproduction, and longevity are three fundamental population processes that influence natural enemy density. By concentrating on these population processes it is possible to develop specific predictions for mechanisms by which cultural practices can affect natural enemy density.

The effects of cultural practices on natural enemy diversity are less commonly studied. Greater species diversity of natural enemies may result in reduced pest populations, because each species kills a part of the pest population that otherwise would have survived (Riechert et al., 1999; Schellhorn and Andow, 1999; but see Rosenheim, 1998). The interactions among natural enemies require further study to understand the role of natural enemy diversity on pests.

5.2 NATURAL ENEMY COLONIZATION

Natural enemy colonization may be higher in one location than another because: (1) there were more natural enemies near the location; (2) surrounding areas became less suitable and the natural enemies left these areas ending up in the location; or (3) the location became attractive to natural enemies and they accumulated there. The first hypothesis does not require that the natural enemies have a difference in preference among locations. If natural enemies are colonizing species (Southwood, 1962), or exhibit an oogenesis-flight syndrome (Johnson, 1960; Dingle, 1972), then they will disperse from habitats irrespective of the relative quality of the surrounding habitats. Under these circumstances, locations that are near large numbers of natural enemies will be colonized more readily than those farther away. The second two hypotheses require that there is a difference in preference. In the second, natural enemies are induced to leave a deteriorating area, and in the third, they are attracted to a particularly good area. The importance of preference in habitat selection is predicted by foraging theory (Kamil et al., 1987). Using population dynamics theory, the conditions under which natural enemies will become more abundant in the target habitat are developed in Andow (1996). In practice, these three hypotheses are often difficult to distinguish.

5.2.1   Hypothesis 1 — Natural Enemy Abundance is Increased Because the Spatial Proximity of Source Populations Results in Higher Colonization

Most agricultural crops do not by themselves have sufficient resources to keep and maintain high levels of natural enemies throughout the entire year. Parasites and predators use non-crops and non-crop habitats for overwintering sites, refuges, and more favorable microclimates, as well as additional prey, hosts, or food. Many natural enemies will move throughout the landscape to locate necessary habitats and resources. Cultural control tactics can be used to take advantage of this movement and increase the colonization of fields and crops that harbor pest species. If sufficient spatial and temporal synchrony is attained, natural enemy populations can increase in an area because of the proximity of nearby source populations, and the spatial structure of the habitats on the landscape.

Overwintering is a crucial part of the life cycle for most insects in temperate areas. Culture control tactics take advantage of this to directly reduce pests, for example, by sanitation and tillage. These same practices may also work to decrease natural enemy abundance. Overwintering might also be a key to abundant natural enemy populations. Adding overwintering sites such as hedge rows, grassy edges, non-crop habitats or other landscape modifications has been touted as a cultural control technique with great potential to increase enemy populations, strengthen the insect-enemy interaction, and increase the diversity of natural enemy species (Wratten and Thomas, 1990). By increasing natural enemy overwintering survival, colonization from these overwintering sites may be an important mechanism to increase densities of natural enemies associated with target crops, fields, and livestock.

Some artificial overwintering sites such as human-made boxes have found success in increasing the abundance of predators such as the green lacewing *Chrysoperla carnea* Stephens (Sengonca and Frings, 1989) and *Polistes* wasps (Gillaspy, 1971). Natural overwintering sites can be improved by management techniques. For example, adding leaves, grass, or other organic litter to the base of trees may lead to higher quality overwintering sites for the predaceous mite *Metaseiulus occidentalis* (Deng et al., 1988) and the coccinellid *Stethorus punctum punctum* (Fell and Hull, 1996). Where overwintering is associated with suppression of reproduction and the natural enemies continue to feed, planting specific vegetation and ensuring adequate food sources may be the key for reducing overwintering mortality (James, 1989).

Extensive research has been performed to determine how natural boundaries and edges surrounding agricultural fields influence aphid predators in cereal and grain crop systems in Europe. Studies have demonstrated how hedges and other boundary areas are crucial to the overwintering survival of species of carabid and staphylinid predators (Sotherton, 1984). By applying insecticides to these habitats, Sotherton (1984) was able to show a considerable reduction in the predator populations in adjoining crops the next spring. Other evidence for increased movement of natural enemies from overwintering sites include mark and recapture studies that have shown predators from edge habitats immigrate into nearby crop fields, and correlations between the number of predators in overwintering sites and the number of those predators in fields early in the growing season (Coombes and Sotherton, 1986). For some natural enemies such as species of ground beetles, progeny of overwintered adults have been shown to immigrate into adjacent fields and then have an affinity for returning to the same boundary areas as the previous generation (Coombes and Sotherton, 1986). In many systems, it may be important to look for changes in natural enemies' populations both within and between generations.

Maintaining field boundaries in an appropriate habitat can be an important way to increase colonization of a variety of natural enemy species into target fields, but it is important to consider numerous factors including the type of crop, field boundary, key predators, and disturbance schedule (e.g., pesticide applications, tillage, harvest). Each of these variables can have a significant effect on the timing and extent of predator colonization (Coombes and Sotherton, 1986; Thomas et al., 1991; Wallin, 1985). For example, Carillo (1985) showed that earwigs (Dermaptera) seem to have more limited movement through barley than they do through non-crop grasses. Wallin (1985) showed that different species of carabids used adjacent field boundaries at different times of the year for different purposes.

In some cases, overwintering sites that are separated from crop fields are necessary for the survival of the natural enemy. Minute solitary egg parasitoids, *Anagrus* spp., have been found to be an important mortality factor for the western grape leafhopper, *Erythroneura elegantula*, an economically significant pest of grapes in the western U.S. (Corbett and Rosenheim, 1996). *Anagrus* spp. require an egg host to overwinter; however, all of the major species of leafhoppers found in grapes overwinter in the adult stage. Therefore, other leafhoppers must be used as overwintering hosts of the parasitoids. *Anagrus* spp. can overwinter in the eggs of a native non-pest leafhopper found in wild blackberries and then move into vineyards the next year (Doutt and Nakata, 1973). Vineyards within 5.6 km of blackberries have been reported to benefit from parasitoids emigrating from the blackberry refuges (Doutt and Nakata, 1973). Kido et al. (1984) showed that *Anagrus* adults were also capable of parasitizing another leafhopper species, *Edwardsiana prunicola*, that overwinters as an egg in French prune tree orchards. They showed a correlation between grape leafhopper parasitism in vineyards and *Anagrus* dispersal during early spring from nearby French prune tree orchards that harbored *E. prunicola*. Laboratory studies revealed that parasitoids reared on one leafhopper species can readily parasitize the other species (Kido et al., 1984; Williams, 1984), so either alternative host can act as an overwintering refuge to increase the colonization of parasitoids to vineyards early in the growing season.

Significant correlation between the presence of French prune tree refuges and higher parasitoid abundance in grape vineyards has been found repeatedly (Kido et al., 1984; Murphy et al., 1996). To prove that these refuges were the source of parasitoids in grape vineyards, however, it was necessary to show that the overwintering parasitoids were indeed immigrating into adjacent vineyards. Corbett and Rosenheim (1996) used rare element labeling to mark overwintering parasitoids in the refuge and then track their movement by recapturing individuals in the vineyards the next year. This mark-recapture experiment demonstrated that parasitoids from the nearby refuges do colonize adjacent vineyards, yet the contribution colonists made to the total early season parasitoid population was relatively low and variable (1% and 34% of parasitoids in two experimental vineyards). By immigrating early in the season, even the smaller numbers of parasitoids from these refuges may be able to play a critical role in increasing parasitism and controlling populations of the western grape leafhopper (Murphy et al., 1998).

It is also possible that the prune tree refuge may increase parasitoid immigration in more subtle ways. Flying insects accumulate in sheltered regions downwind of natural or artificial windbreaks (Lewis and Stephenson, 1966). Because dispersing *A. epos* accumulate at a much greater rate downwind of prune tree refuges, it has been speculated that the French prune trees act both as a collection of overwintering hosts and as a natural windbreak which influences the colonization of dispersing parasitoids (Corbett and Rosenheim, 1996). Further research may be needed to determine optimal refuge size and placement in order to provide sufficient pest control.

Aphid parasitoids in grass and cereal crops provide another example of an association between higher colonization of natural enemies and the proximity of overwintering sites (Vorley and Wratten, 1987). Barley and early sown wheat (drilled before mid-October) provide a significant source of parasitoids that immigrate into later planted wheat fields. This was demonstrated both by trapping parasitoids in spatially oriented baffle traps, and by calculating the expected number of *Aphidius* spp. parasitoids and comparing it to the actual field surveys. The early sown fields may benefit the parasitoids in two ways. First, it creates an overwintering refuge with high densities of aphid hosts in the fall. The early sown fields also allow for the development of an aphid host early in the season, which in turn allows for parasitoid populations to build up when other hosts may be relatively scarce. Vorley and Wratten (1987) suggested that one early planted wheat field generated sufficient parasitoids in the spring to account for immigration into about 25 late planted fields. Early movement of parasitoids in the spring may coincide with the initial build up of aphids in the other fields, when parasitoids are capable of the greatest impact on aphid populations.

Natural enemy populations may benefit from managing landscapes to increase the temporal availability of habitats and food so that resources are available for natural enemies throughout the growing season. This has been studied for aphids and their parasitoids on a variety of weeds and other non-crop hosts (Perrin, 1975; Stary and Lyon, 1980; Müller and Godfrey, 1997). Generalist predators such as coccinellids have also been shown to use resources from weeds and other non-crop habitats, especially early in the growing season (Banks, 1955; Perrin, 1975; Benton and Crump, 1981; Honek, 1982; Hodek and Honek, 1996). For example, in Central Bohemia, populations of the predator *Coccinella septempunctata* L. were found to colonize habitats sequentially, starting with overwintering sites, then alfalfa and clover in early spring, followed by spring cereals later in the year (Honek, 1982). Other species use field boundaries and edges at different times throughout the season for reproduction and possible recolonization of adjacent fields (Boller et al., 1988; Wallin, 1985). Trap crops such as alfalfa interplanted with cotton may also provide a source of predators that can colonize adjacent fields and attack pest species (Corbett et al., 1991). Trap crops allow for the build up of pest and enemy populations in areas adjacent to crops being targeted for pest control. Few studies, though, have shown more than changes in the relative abundance of insects in the trap crops and other added habitats. Future studies are needed to understand the mechanisms of increased abundance and how to use this information for more effective cultural control.

5.2.2   Hypothesis 2 — Natural Enemy Abundance is Increased Because the Previously Occupied Habitat is no Longer Suitable, which Results in Higher Colonization

Unlike natural systems that typically have one disturbance over multiple years, agricultural systems are subject to multiple disturbances within and between growing seasons. Preparing the ground, planting seed, applying nutrients and pesticides, cultivation, and harvest can all act as significant disturbances to the crop ecosystem. Ecologists have begun to recognize that such disturbances can play a key role in structuring ecological communities and population dynamics (Pickett and White, 1985). Harvesting, for example, can have a tremendous detrimental effect on natural enemy populations. Honek (1982) estimated that alfalfa harvesting destroyed 90% of the recently immigrated *Coccinella septempunctata* population. Carillo (1985) demonstrated that cutting ryegrass for forage caused the European earwig, *Forficula auricularia* to immigrate to field margins. Therefore, it is important to find ways to encourage frequent colonization and recolonization of natural enemies to maintain high population densities of natural enemies.

Refuges can be created in and around crop fields to reduce the effects of disturbance on natural enemies and increase the likelihood of their recolonization. This has been examined by comparing the effects of block versus strip harvesting of alfalfa on the population dynamics of a parasitoid *Aphidius smithi* and its aphid host *Acyrthosiphon pisum* (van den Bosch et al., 1966; van den Bosch et al., 1967). Forage crops like alfalfa are cut and harvested two to four times a year. Each time, the fields are left devoid of vegetation for several days, creating a harsh microclimate where both parasitoid and host are exposed to direct solar radiation. Furthermore, they suggested that the lack of vegetation causes a decline in aphid parasitoids because of a radical reduction in their obligatory host. Altering planting and cutting dates can ameliorate these disturbances. By leaving strips of unmowed alfalfa, aphids and parasitoids are given a temporal refuge from cutting disturbances. These refuges allow *A. smithi* to retain a population in the fields so they can respond to aphid outbreaks as they occur. Additionally, it appears that *A. smithi* females gradually move from the taller, older alfalfa into the younger strips between cuttings. This increased immigration into young alfalfa puts the parasitoids in contact with young aphid colonies, where they can have the greatest suppressive effect on aphid populations. Gradual movement of parasitoids away from older plants to younger ones also means there will be fewer parasitoids at risk of being killed when the older plants are harvested. These temporal refuges reduce the effect of cutting on the parasitoid population and increase the parasitoid's overall ability to control aphid pests. Similarly, Mullens et al. (1996) found that alternating the removal of manure from poultry facilities created temporal refuges that helped increase densities of predatory mites, *Macrocheles* spp., that helped control fly pest populations.

Using a metapopulation model, Ives and Settle (1997) suggested a theoretical basis for the phenomena observed by van den Bosch (van den Bosch et al., 1966; van den Bosch et al., 1967). If fields are asynchronously planted and harvested, mobile natural enemies will have time to disperse from mature fields into younger ones. Therefore, the enemies can have a larger overall effect in controlling herbivore populations (Ives and Settle, 1997). If there are few mobile enemies in asynchronous plantings, then insect pest populations increase at alarming rates. Further studies can help determine what systems have the greatest potential for using refuges to give a greater advantage to natural enemy populations.

5.2.3   Hypothesis 3 — Natural Enemy Abundance is Increased Because a Habitat is Attractive in Some Way, which Results in Higher Colonization

Some predators and parasitoids can perceive and respond to sensory information from plants. Flowers, which are important sources of nectar for parasitoids, have been found to attract the parasitoid *Microplitis croceipes* by olfactory stimuli (Takasu and Lewis, 1993), and the parasitoid *Cotesia rubecula* by both olfactory and visual stimuli (Wäckers, 1994). Flowers and flower nectar also attract parasitoids of the tarnished plant bug (Streams et al., 1968; Shahjahan, 1974). Since many parasitoids have been found to forage for nectar and other food sources, increasing the availability and physical proximity of these sources may increase the immigration of parasitoids from other sources to target fields. This, however, remains to be definitively documented.

Natural enemies can also be attracted to plants at growth stages that may be associated with prey or hosts. The parasitoid *Campoletis sonorensis* was attracted to flowers and other plant parts that are associated with the presence of its host cotton bollworm (Elzen et al., 1983). The polyphagous heteropteran predator, *Orius insidiosus* is attracted to volatile chemicals from maize silk, which may help it feed on prey (Reid and Lapman, 1989). Other plants and volatile plant chemicals are detected by and attractive to the parasitoids *Peristenus pseudopallipes* (Monteith, 1960), *Diaeretiella rapae* (Read et al., 1970), *Heydenia unica* (Camors and Payne, 1972), *Eucelatoria* spp. (Nettles, 1979), and the chrysopid predator *Chrysoperla carnea* (Flint et al., 1979).

The reaction of insects to plant stimuli often depends on the physiological state of the insect. For example, it has been found that hungry female parasitoids responded to food-associated odors, while well-fed females responded to the host-associated odors (Takasu and Lewis, 1993; Wäckers, 1994). The ability of an insect to respond to an odor may also depend on previous experience. *Microplitis croceipes* is able to learn different odors and associate them with either host or food resources (Lewis and Takasu, 1990). Parasitic flies have also been found to be attracted to or repelled by plant odors from different trees, depending on the flies' age in relation to reproductive maturity (Monteith, 1960).

Natural enemies are also capable of perceiving and responding to other plant cues. In studying the mechanistic response of the predator *Orius tristicolor* White to a corn-bean-squash polyculture, five possible cues were described that could influence insect immigration: plant density, plant architecture, visual cues, volatile chemicals, and microclimate such as relative humidity (Letourneau, 1990). The results suggest that plant architecture and density increased colonization of the predator, regardless of prey density or plant diversity. Others have noted differences in predator abundance associated with variation in plant architecture and density, perhaps caused by microclimatic differences (Honek, 1982).

Many species have the ability to detect their prey or hosts from a distance. Frass from the larvae and scales from adult of the corn earworm *Helicoverpa zea* (Boddie) contain chemical stimuli that invoke higher activity rates and oriented host-seeking behavior in the larval parasitoid *M. croceipes*, and egg parasitoids *Trichogramma* spp. (Jones et al., 1971; Jones et al., 1973; Gross et al., 1975). However, it remains uncertain how these attractants influence the population dynamics of the parasitoids and on what spatial scales these attractants can cause increases in colonization.

Some natural enemies have also found to be attracted to volatile plant chemicals that are induced by insect herbivory. These compounds might be important in host habitat location and have been shown to be involved in the host location process. Attraction has been observed for the parasitoids *Cotesia marginiventris* (Turlings et al., 1990; Alborn et al., 1997), *Microplitis croceipes* (McCall et al., 1993), *Cortesia glomerata* (Mattiacci et al., 1994), and *Cardiochiles nigriceps* (De Moraes et al., 1998); the predaceous mites *Metaseiulus occidentalis, Phytoseiulus persimilis* (Sabelis and van de Baan, 1983), and *Amblyseius potentillae* (Dicke et al., 1990); and anthocorid predators (Drukker et al., 1995). Natural enemies have been shown to respond to plants that are typical food sources for their hosts or prey. It has been recently shown that plants give off different amounts of volatile compounds in response to different species of herbivores, and distinct parasitoid species can differentiate these chemical signals and may respond only to those compounds associated with their preferred hosts (De Moraes et al., 1998). Herbivore-induced plant volatiles have been shown to cause increased numbers of natural enemies in field situations (Drukker et al., 1995), but it is unclear at what distance natural enemies are attracted from and at what spatial scale they can be attracted. These results, however, demonstrate an enormous potential for using trap crops, intercropping, variation in planting pattern, or artificial chemicals to increase the attractiveness and colonization of species-specific natural enemies to target fields.

Another method of increasing colonization is to use artificial sprays applied to target fields. The abundance of the generalist predator, *Coleomegilla maculata* can be increased with sprays of sugar plus wheast, an artificial food source that is a mixture of a yeast, *Saccharomyces fragilis*, plus its whey substrate (Nichols and Neal, 1977). A similar result has been found for coccinellid populations using sugar solutions (Ewert and Chiang, 1966; Schiefelbein and Chiang, 1966). Adult *Chrysoperla carnea* Stephens are attracted to sucrose sprays mixed with the amino acid tryptophan, and by mixing tryptophan with artificial honeydews, greater numbers of adults will colonize sprayed fields (Hagen et al., 1976). Adding tryptophan to an artificial food source may not only increase immigration, but promote egg laying so that the predaceous larvae can have a significant effect on pest numbers in key areas. The effectiveness of both of these methods is severely reduced if other honeydew and food sources are readily available. Whether these other sources interfere with the olfactory cues or whether their presence alters predator searching behavior remains to be clarified.

5.3 NATURAL ENEMY REPRODUCTION AND LONGEVITY

Natural enemy abundance is increased if reproduction is increased by greater fecundity and longevity of adults or higher survival of offspring. Therefore, cultural practices that directly affect these processes can reduce insect pest damage. Greater reproduction and adult and larval survival often depend on the quality of the habitat. Habitat quality probably has several components, including food availability throughout the season, food abundance, microclimatic suitability, disturbance regime, and presence of natural enemies of the natural enemies. Of these factors, food availability has received the vast majority of research attention, and our analysis also concentrates on this factor.

5.3.1 Hypothesis 1 — Natural Enemy Abundance is Increased Because Food is More Abundant, which Results in Higher Reproduction, Longevity, and/or Survival

The diet of adult parasitoids and predators can have important effects on their lifetime reproductive success (Hagan, 1986; Bugg, 1987; Osakabe, 1988; Heimpel et al., 1997). Adult female parasitoids of many species feed on host insects (host feeding) or a variety of sugar sources, and both can improve egg maturation, adult maintenance, and survival (Jervis and Kidd, 1986, 1996; Heimpel and Collier, 1996; Heimpel et al., 1997; Olson and Andow, 1998). Furthermore, for those species that host-feed, the combination of host feeding and access to honey meals can significantly increase parasitoid lifetime reproductive success (Heimpel et al., 1997). Extremely low lifetime reproductive success and survival were found for individuals that did not have access to honey (Leius, 1961; Syme, 1975; Idris and Grafius, 1995; Heimpel et al., 1997; Olson and Andow, 1998). Although fewer studies have addressed how diet influences lifetime reproductive success of predators and mites, some studies have shown that sugar sources and pollen can increase their fecundity and longevity (McMurtry and Scriven, 1964; Bugg et al., 1987). Abundant sugar and pollen in the field can greatly increase lifetime reproductive success of parasitoids and predators by enhancing fecundity and longevity.

One of the most common sources of sugar and pollen in agricultural systems is from the non-crop plants that border or grow within the agricultural field. These plants (often referred to as weeds) provide sugar sources as floral and extrafloral nectar, which are visited by natural enemies (Rogers, 1985; Pemberton and Lee, 1996). Flower nectar, extrafloral nectaries, and pollen have all been shown to increase fecundity of parasitoids, predators, and mites (De Lima and Leigh, 1984; Heimpel et al., 1997). Numerous species of parasitoids are frequently observed to feed on floral (Leius, 1961, 1967; Elliott et al., 1987; Jervis et al., 1993) and extrafloral nectar (Rogers, 1985; Bugg et al., 1989, Pemberton and Lee, 1996) as well as honeydew excreted by homopteran insects (Elliot et al., 1987; Evans, 1993). Nectar is an important food source for adult parasitoids and like honeydew, the consumption of nectar can result in increased fecundity and longevity (Syme, 1975; Idris and Grafius, 1995; Jervis et al., 1996). Work by Idris and Grafius (1995) found that parasitoid fecundity was higher when *Barbarea vulgaris*, *Brassica kabar*, or *Daucus carota* flowers or honey-water was used as food, compared with no food. Hagley and Barber (1992) found that the fecundity of adult *Pholetesor ornigis* (Braconidae) increased when individuals were confined with flowers of creeping charlie (*Glenchoma hederacea* L.), dandelion (*Taraxacum officinale* Weber), and apple (*Malus domesticus* L.), but not with flowers of chickweed (*Stellaria media* L.) or Shepherd's purse (*Capsella bursapastoris* L.). Others have also reported that not all nectar sources provided benefits to parasitoids (Elliot et al., 1987), and Idris and Grafius (1995) suggested that the accessibility of nectar is related to floral characters, particularly the width of the corolla opening in relation to the size of the forager.

Honeydew from aphids has also been reported to increase fecundity in natural enemies. The fecundity of *P. ornigis* increased when individuals were confined with terminal leaves of apple with honeydew of the aphid *Aphis pomi* DeGeer, but not when confined with terminal leaves of apple without honeydew or with flowers of round-leaved mallow (*Malva neglecta* Waller) or red clover (*Trifolium pratense* L.). Parasitoids given aphid honeydew oviposited a greater proportion of their eggs than those confined with apple leaves without honeydew (Hagley and Barber, 1992)

Sugar sources have a great effect on parasitoid longevity (Leius, 1967; Syme, 1975; Heimpel et al., 1997; Olson and Andow, 1998), and are known to expand life 8 to 20 times that without a sugar source (Collier, 1995; Heimpel and Rosenheim, 1995; Heimpel et al., 1997). The life span of sugar-fed *Aphytis* spp. females varies between 2 and 6 weeks when a sugar source is provided, whereas that of sugar-deprived females rarely exceeds three days (Avidov et al., 1970; Heimpel et al., 1997). Sugar-fed *Trichogramma* females live 17 days, but only 2 to 3 days without sugar (Olson and Andow, 1998), and aphid honeydew did not extend life as long as sugar (McDougal and Mills, 1997). The longevity of *Diadegma insulare* females was significantly greater when they fed on the wildflower *B. vulgaris* than on several other wildflowers commonly found in the surrounding area or water or without food (Idris and Grafius, 1995). In addition to honey and nectar, aphid honeydew is also known to increase longevity. Adult longevity of *Diadegma insulare* and *Pholetesor ornigis* was increased when they were provided with aphid honeydew (Hagley and Barber, 1992; Idris and Grafius, 1995). However, there was no effect of longevity for *P. ornigis* when adults where confined with flowers (Hagley and Barber, 1992).

Insect predators are also known to feed on nectar and pollen (Sundby, 1967; Yokoyama, 1978; Crocker and Whitcomb, 1980; De Lima and Leigh, 1984; Bugg,

1987; Hodek and Honek, 1996). Their fecundity and longevity was enhanced by floral resources (De Lima, 1980; Agnew et al., 1982; Bugg et al., 1987). De Lima (1980) and De Lima and Leigh (1984) showed that a bigeyed bug, *Geocoris pallens* Stål, attained maximum longevity, fecundity, and per capita prey consumption rates when cotton extrafloral nectar was available in addition to prey. Extrafloral nectar alone, however, was not sufficiently nutritious to sustain reproduction (De Lima and Leigh, 1984). *Geocoris punctipes* Say and *Collops vittatus* Say were found to live twice as long on common knotweed compared to alfalfa, although the effects of alternative prey associated with the weed were not clarified (Bugg et al., 1987). It is possible that prey associated with common knotweed may compete with the pest species so that *Geocoris* spends more time searching on the knotweed than on the crop plant.

The relative abundance of prey sources can also influence natural enemies. Populations of *Coleomegilla maculata* were studied in maize monocultures and maize-bean-squash polycultures (Andow and Risch, 1985). Populations of this coccinellid beetle were greater and predation on artificial prey was higher in monocultures, which had higher prey abundance, than in the polycultures, which had food available for a longer part of the growing season. Individuals had a higher foraging success rate under higher food densities (Risch et al., 1982) and were also found to have stayed longer (Wetzler and Risch, 1984).

5.3.2   Hypothesis 2 — Natural Enemy Abundance is Increased Because Food is Available During a Longer Period of Time, which Results in Higher Reproduction, Longevity, and/or Survival

Cropping systems that maintain weeds and flowering herbs often provide the first food resources of the season, which allows for earlier development of insect predators compared to systems without the weeds and flowers. Female carabids (*Poecilus cupreus* L.) were significantly larger and had significantly more eggs earlier in the season in a cereal crop subdivided by strips of weeds and wild flowering herbs compared to a weed-free cereal area (Zangger et al., 1994). Earlier development of carabids may result in higher predator abundance early in the season and greater potential for suppressing pest populations.

Pollen has been shown to affect predaceous mite populations similarly. Pollen can increase the fecundity of predaceous mites (McMurtry and Johnson, 1965; Osakabe, 1988), and egg production was highest when tea pollen was provided to the predaceous mite, *Amblyseius sojaensis* Ehara (McMurtry and Scriven, 1964). Furthermore, it has been demonstrated that a greater percentage of the population of predaceous mite, *Amblyseius hibisci*, reached maturity on tea pollen alone than when feeding solely on spider mites (Osakabe, 1988). The poor survival on spider mites occurred because young instars of *A. hibisci* became entangled in the webbing of the spider mites. When the spider mite *Panonychus citri* was at low densities on citrus leaves, however, *A. hibisci* controlled it when tea pollen was added (Osakabe et al., 1987). The seasonal abundance of *A. hibisci* was closely correlated with peaks in flowering intensity and pollen production, and population increases could be triggered and maintained by artificially introducing pollen at weekly intervals (McMurtry and Johnson, 1965). This addition of pollen allowed *A. hibisci* populations to reach high densities before spider mite populations began to develop and also provided food during times of prey scarcity.

Increasing the fecundity and longevity of natural enemies is predicted to increase their densities and suppression of pests. Unfortunately, very few studies show how increased fecundity and longevity relate to natural enemy abundance and pest suppression. Work in cotton systems has contributed significantly to our understanding of the relationships among natural enemies, nectar sources, and prey (Lingren and Lukefahr, 1977; Agnew et al., 1982; De Lima and Leigh, 1984; Staple et al., 1997). Females of the parasitoid *Microplitis croceipes* that fed on either extrafloral nectar or sucrose in cotton fields were retained in the field significantly longer than females without any food, and they attacked hosts at a significantly higher rate than honeydew-feeding or non-feeding females (Staple et al., 1997). Likewise, the longevity of the parasitoid *Campoletes sonerensis* was shown to increase when they foraged on cotton varieties containing extrafloral nectaries, which also resulted in slightly higher parasitism rates (Lingren and Lukefahr, 1977). The imported red fire ant, *Solenopsis invicta*, is a common predator of *Heliothis* spp. in cotton, and they have been shown to visit cotton varieties with extrafloral nectaries more than nectariless varieties (Agnew et al., 1982). This increased visitation rate, however, resulted in little or no added protection to the plant from the pest (Agnew et al., 1982). Studies from other systems have also documented the influence of food sources on parasitism and predation. Parasitism of a generalist herbivore, gypsy moth (*Lymantria dispar* L.), was higher on the four main genera of plants with extrafloral nectaries than on any of five main genera of plants without extrafloral nectaries (Pemberton and Lee, 1996). In broccoli, *Diadegma insulare* parasitism was higher in crops that were surrounded by nectar-producing plants, compared to broccoli that was not surrounded by nectar-producing plants (Zhao et al., 1992).

Maintaining nectar sources in or around a cropping system can create conflicts because the effect that nectar sources have on increasing the fecundity and longevity of natural enemies can also work to increase the fecundity and longevity of herbivores (Tingey et al., 1975; Wilson and Wilson, 1976). Alternatively, the nectar plants could be so attractive that the natural enemies concentrate their foraging on the nectar plants (Naranjo and Stimac, 1987), and have no effect on the pest on the crop plant. Moreover, changes in relative plant area, structure, or complexity have the potential to increase or decrease natural enemy efficiency. For example, the egg parasitoids *Trichogramma nubilale* Ertle and Davis and *Trichogramma pretiosum* Riley have shown an inverse relationship between their searching efficiency measured by parasitism and the area of the plant being searched (Need and Burbutis, 1979; Ables et al., 1980; Burbutis and Koepke, 1981). *T. nubilale* has also shown an inverse relationship between its searching efficiency and plant complexity (Andow and Prokrym, 1990). Cultural control tactics that manipulate or add vegetation in and around crop fields will have to ensure that any benefit of increased numbers of enemies is not negated by decreases in their overall effectiveness.

Although it is not clear how effective nectar plants are as a cultural control, sugar sources increase the fecundity and longevity of many species of natural enemies, whereas the absence of sugar sources significantly reduces fecundity and longevity. Therefore, the presence of floral and extrafloral nectar and pollen may be an essential cultural practice for enabling natural enemies to have the potential to reduce pests. Exclusion of these nectar sources may limit natural enemies as a significant control factor.

In other systems, a greater abundance of consistently available prey may enable the build-up of natural enemies, resulting in greater pest suppression. Natural farming systems in Japan involve using compost for fertility, no pesticides, and continuous irrigation for growing rice. In these systems, pests such as brown planthopper *Nilaparvata lugens* are uncommon, even when there are major population outbreaks in neighboring paddies (Andow and Hidaka, 1989; Hidaka, 1990). Pest control is caused in part by natural enemies that can persist in the natural farming systems, including the wolf spider *Lycosa pseudoannulata* and the mermethid parasitic nematode *Agamermis unka*. The wolf spider is more abundant because natural farming supports abundant alternate prey, primarily small flies that are involved in sediment decomposition during their larval stages (Andow and Hidaka, 1989). The nematode is more abundant because low densities of planthoppers maintain its density (Hidaka, 1990). In these cases, a greater abundance of consistently available prey enables the build-up of natural enemies, resulting in greater pest suppression.

5.3.3 Hypothesis 3 — Natural Enemy Abundance is Increased Because the Microclimate Allows Higher Reproduction and Longevity

Predation by carabid beetles was greater in moist, shady microclimates than drier, sunny ones (Speight and Lawton, 1976; Brust et al., 1986; Perfecto et al., 1986). The amount of time available for carabids to forage for prey was greater in these habitats, probably because they could forage during the day as well as at night (Brust et al., 1986). Unfortunately, the influence of these microclimates on reproduction and longevity has not been clarified.

5.4   NATURAL ENEMY DIVERSITY

Cultural control practices can affect natural enemy species diversity, composition, and functional relationships. Species diversity is usually richest in natural systems and poorest in conventional agricultural systems (House and Stinner, 1983). Within agriculture, a similar diversity continuum exists in comparing different systems which work to add diversity (i.e., green manure, cover crops, polycultures), minimize disturbance, (i.e., reduced use of fertilizers and pesticides and no-till cultivation), or create uniformity and simplicity (i.e., intensified monocultures). Cultural control practices that add biological diversity, reduce fragmentation, and minimize disturbance should function to increase and maintain natural enemy species diversity (Kruess and Tscharntke, 1994; House and All, 1981; Drinkwater et al., 1995).

Natural enemy species diversity can be increased by the addition of biological amendments (i.e., green manure) and by the reduction of disturbances (i.e., elimination of pesticides or use of no-till practices). These methods are commonly used in organic farming (U.S. Department of Agriculture, 1980; National Research Council, 1989). Both the addition of biological materials (i.e., green manure) and the elimination of pesticides can directly and indirectly enhance natural enemy species diversity by increasing resources and reducing disturbances (Kromp, 1989; Russel, 1989). Carabid beetle communities in organic farming systems have been found to have greater carabid species diversity and abundance than conventional farming systems (Kromp, 1989). Greater parasitoid diversity and abundance were found in tomatoes on organic farms than in the tomatoes on conventional farms (Drinkwater et al., 1995). No-till or reduced tillage, in place of conventional tillage, reduced the disturbance of soil-dwelling arthropods, in turn maintaining a greater species diversity compared to conventional tillage systems (Rabatin and Stinner, 1989). No-tillage systems appear to support a larger and more diverse natural enemy community, where ground beetle (carabids and staphylinids) and spider abundance, species diversity, and biomass were higher in no-tillage compared to a moldboard plow system (House and Stinner, 1983). Natural enemy species composition can also be altered between no-tillage and conventional tillage systems. Relatively larger carabid beetles (*Pterostichus chalcites* Say and *Amphasia sericaea* Harris) have been found in no-tillage systems (Brust et al., 1986).

Reducing habitat fragmentation can increase natural enemy species diversity (Kruess and Tscharntke, 1994). The number of parasitoid species found to attack herbivores were negatively correlated with distance between crop islands and meadows (undisturbed habitats). Eight to twelve parasitoid species were found in the meadows, but only two to four species were found in the patches 500 m from the nearest meadow (Kruess and Tscharntke, 1994). Habitat fragmentation may have reduced both parasitoid biodiversity and the rate of parasitism.

Most studies on natural enemy diversity do not address how changes in diversity affect pest suppression. Brust et al. (1986), however, showed that predation on lepidopteran larvae was higher in the no-tillage systems compared to the tilled systems, and that the difference in predation rate was attributable to the greater density of large carabids in the no-tillage system.

Increased natural enemy diversity may have other beneficial effects not directly related to pest control. In reduced tillage, no-pesticide systems, Rabatin and Stinner (1989) found that there was a greater density and diversity of macro-invertebrates, which resulted in a greater diversity and density of vesicular-arbuscular mychorrhizae (VAM). The macro-invertebrates consumed and spread the VAM fungal spores. Kuikman et al. (1989) found that under drought conditions, plant nitrogen uptake was limited for plants in soil with bacteria only, but it was enhanced in soils with both a predaceous protozoa (one that preys on soil bacteria) and bacteria.

5.5 CONCLUSIONS

The spatial and temporal arrangement of the landscape can play a significant role in the colonization of certain natural enemy species. The presence and location of overwintering refuges and early season habitats can result in increased natural enemy immigration, especially early in the season when natural enemies can have the greatest effect on growing pest populations. Increased colonization, however, does not always result when sources of natural enemies are nearby in the landscape. Information about the biology, ecology, and behavior of natural enemy species may be required before it is possible to make accurate predictions for particular crop or animal systems. Cultural control tactics that manage the temporal or spatial structure of landscapes and habitats may be able to direct the movement and colonization of natural enemy populations toward insect pests.

One of the characteristics of agroecosystems is the extent and timing of disturbances. These disturbances can radically alter the population and community dynamics of both pests and natural enemies. While there is evidence that natural enemies move away from degrading habitats, it is unclear to what extent this movement takes place and whether it is sufficient to counterbalance the detrimental effects of the many sudden agricultural disturbances. Additional understanding of how and when natural enemies and pests move between habitats may better allow us to minimize the effects of disturbances and increase the recolonization of natural enemies.

The mechanisms behind natural enemy attraction have been repeatedly demonstrated for many predators and parasites. However, there has been far less work showing how the attractiveness of a habitat at a distance may lead to higher immigration by influencing the movement, foraging, and other behaviors of natural enemies. Additionally, there is little evidence that clarifies how other pests and the natural enemy's enemies may also respond to these attractive cues and the subsequent effects on pest control.

The availability of more food increases the fecundity and longevity of many natural enemies. This appears to be a general finding and probably holds across many systems. The same generalizations cannot be made for how the temporal duration of food availability will affect natural enemy longevity and fecundity, because few studies have addressed this issue. There are several details that need to be understood before it is possible to implement "more food for natural enemies" as a cultural control practice to increase natural enemy abundance. For example, the food may be more beneficial to the pests than the natural enemies, and its effects on multiple species of natural enemies are not well understood. In addition, it is uncertain how the proximity of the food resource to the crop plant affects the potential of the natural enemies to suppress pest populations. Despite this uncertainty, a speculative conclusion can be made that increasing food and extending the duration of food availability potentially increases natural enemy populations in conventional monoculture systems that are devoid of food resources early in the season when pest populations are increasing.

Much of the published work concentrates on the response of natural enemies to cultural practices. Much more work is needed to understand how changes in natural enemy populations relate to pest suppression. These kinds of relationships are necessary to complete the links from cultural practices to cultural control, and to develop a framework and functional basis for understanding cultural control using natural enemies.

As mentioned in the introduction, there are many tactics used as cultural controls that have direct effects on pests. The effects of these cultural controls on natural enemies usually have not been evaluated. The majority of examples found have focused on vegetation or habitat diversity. A few studies have documented how enemy abundance is affected by the addition of sugar (Nichols and Neal, 1977) or fertilizers (Adkisson, 1958), and Ellis et al. (1988) suggested that outbreaks of the cereal leaf beetle were caused by intensive tillage, which reduced parasitism. The vast majority of studies on cultural control, however, have ignored natural enemies, limiting our understanding of the ecology of cultural control.

There are very few studies that address the effects of cultural control practices on natural enemy species diversity and community composition. One tentative finding is that the extent of disturbance affects natural enemy species diversity and community composition. If this is generally true, agricultural practices that either add biological materials or reduce disturbance will encourage greater diversity of natural enemy species than practices that create uniformity. Under some conditions, a greater diversity of natural enemies can have a greater effect of reducing pest abundance because different species of natural enemies will feed on different prey or feed on the same prey in different ways. These ideas remain largely untested in agricultural systems, but minimizing disturbance, either by altering the type or timing of cultural practices, could maintain natural enemy species diversity that could suppress several kinds of pests.

Cultural control offers a variety of pest management tactics that can be used to improve agroecosystems economically and ecologically. It provides the opportunity to increase the abundance and effectiveness of many natural enemies. These opportunities are largely unrealized, because we cannot yet reliably use these tactics in most agroecosystems and cannot generalize known results to different systems. By increasing our knowledge of pests and enemies in relation to habitat use, we should become better able to develop and implement safe and effective cultural techniques for the management of pest insects.

REFERENCES

Ables, J. R., D. W. McCommas, Jr., S. L. Jones, and R. K. Morrison. 1980. Effect of cotton plant size, host egg location, and location of parasite release on parasitism by *Trichogramma pretiosum*. *The Southwestern Entomologist* 5(4):261-264.

Adkisson, P. L. 1958. The influence of fertilizer application on populations of *Heliothis zea* (Boddie), and certain insect predators. *Journal of Economic Entomology* 51:757-759.

Agnew, C.W., W.L. Sterling, and D.A. Dean. 1982. Influence of cotton nectar on red imported fire ants and other predators. *Environmental Entomology* 11:629-634.

Alborn, H. T., T. C. J. Turlings, T. H. Jones, G. Stenhagen, J. H. Loughrin, and J. H. Tumlinson. 1997. An elicitor of plant volatiles from beet armyworm oral secretion. *Science* 276:945-948.

Andow, D.A. 1991. Vegetational diversity and arthropod population response. *Annual Review of Entomology* 36:561-586.

Andow, D.A. 1996. Augmenting natural enemies in maize using vegetational diversity. In *Biological pest control in systems of integrated pest management*. FFTC Book Series No. 47, pp. 137-153. Food and Fertilizer Technology Center, Taipei, Taiwan.

Andow, D.A. and K. Hidaka. 1989. Experimental natural history of sustainable agriculture: Syndromes of production. *Agriculture, Ecosystems and Environment* 27:447-462.

Andow, D. A. and D. R. Prokrym. 1990. Plant structural complexity and host-finding by a parasitoid. *Oecologia* 82:162-165.

Andow, D.A. and S.J. Risch. 1985. Predation in diversified agroecosystems: Relations between a coccinellid predator *Coleomegilla maculata* and its food. *Journal of Applied Ecology* 22:357-372.

Avidov, Z., M. Balshin, and U. Gerson. 1970. Studies on *Aphytis coheni*, a parasite of the California red scale, *Aonidiella aurantii* in Israel. *Entomophaga* 15:191-207.

Banks, C. J. 1955. An ecological study of Coccinellidae (Col.) associated with *Aphis fabae* Scop. on *Vicia faba*. *Bulletin of Entomology Research* 46:561-587.

Benton, A. H. and A. J. Crump. 1981. Observations on the spring and summer behavior of the 12-spotted ladybird beetle, *Coleomegilla maculata* (DeGeer) (Coleoptera: Coccinellidae). *Journal of New York Entomological Society* 89(2):102-108.

Boller, E. F., U. Remund, and M. P. Candolti. 1988. Hedges as potential sources of *Typhlodromus pyri*, the most important predatory mite in vineyards of northern Switzerland. *Entomophaga* 33(2):249-255.

Brust, G.E. and B.R. Stinner. 1991. Crop rotation for insect, plant pathogen and weed control. In D. Pimentel (ed.), *Handbook of pest management in agriculture, 2nd edition*. CRC Press, Boca Raton, Florida.

Brust, G.E., B.R. Stinner, and D.A. McCartney. 1986. Predation by soil inhabiting arthropods in intercropped and monoculture agroecosystems. *Agriculture, Ecosystems and Environment* 18:145-154.

Bugg, R.L. 1987. Observations on insects associated with a nectar-bearing Chilean tree, *Quillaja saponaria* Molina (Rosaceae). *Pan-Pacific Entomology* 63:60-64.

Bugg, R.L., R.T. Ellis, and R.T. Carlson. 1989. Ichneumonidae (Hymenoptera) using extrafloral nectar of faba bean (*Vicia faba* L., Fabaceae) in Massachusetts. *Biological Agriculture and Horticulture* 6:107-114.

Bugg, R.L., L.E. Ehler, and L.T. Wilson. 1987. Effect of common knotweed (*Polygonum aviculare*) on abundance and efficiency of insect predators of crop pests. *Hilgardia* 55:1-51.

Burbutis, P. P. and C. H. Koepke. 1981. European corn borer control in peppers by *Trichogramma nubilale*. *Journal of Economic Entomology* 74:246-247.

Camors, F. B., Jr. and T. L. Payne. 1972. Response of *Heydenia unica* (Hymenoptera: Pteromalidae) to *Dendroctonus frontalis* (Coleoptera: Scolytidae) pheromones and a host-tree terpene. *Annals of Entomological Society of America* 65:31-33.

Carillo, J. R. 1985. Ecology of, and aphid predation by, the European earwig, *Forficula auricularia* L. in grassland and barley. Ph.D. thesis, University of Southampton.

Collier. T.R. 1995. Host feeding, egg maturation, resorption, and longevity in the parasitoid *Aphytis melinus* (Hymenoptera: Aphelinidae). *Annals of the Entomological Society of America* 88:206-214.

Coombes, D. S. and N. W. Sotherton. 1986. The dispersal and distribution of polyphagous predatory Coleoptera in cereals. *Annals of Applied Biology* 108:461-474.

Corbett, A., T. F. Leigh, and L. T. Wilson. 1991. Interplanting alfalfa as a source of *Metaseiulus occidentalis* (Acari: Phytoseiidae) for managing spider mites in cotton. *Biological Control* 1:188-196.

Corbett, A. and J. A. Rosenheim. 1996. Impact of a natural enemy overwintering refuge and its interaction with the surrounding landscape. *Ecological Entomology* 21:155-164.

Crocker, R.L. and W.H. Whitcomb. 1980. Feeding niches of the big-eyed bugs *Geocoris bullatus, G. punctipes,* and *G. uliginosus* (Hemiptera: Lygaeidae: Geocorinae). *Environmental Entomology* 9:508-513.

De Moraes, C. M., W. J. Lewis, P. W. Paré, H. T. Alborn, and J. H. Tumlinson. 1998. Herbivore-infested plants selectively attract parasitoids. *Nature* 393:570-573.

De Lima, J.O.G. 1980. Biology of *Geocris pallens* Stål on selected cotton genotypes, Ph.D. dissertation, Davis, University of California, 57 pp.

De Lima, J.O.G. and T.F. Leigh. 1984. Effect of cotton genotypes on the western bigeyed bug (Heteroptera: Miridae). *Journal of Economic Entomology.* 77:898-902.

Deng, X., N. X. Zheng, and X. F. Jia. 1988. Methods of increasing the winter survival of *Metaseiulus occidentalis* (Acari: Phytoseiidae) in northwest China. *Chinese Journal of Biological Control* 4:97-101.

Dicke, M., T. A. Van Beek, M. A. Posthumus, N. Ben Dom, H. Van Bokhoven, and A. E. De Groot. 1990. Isolation and identification of volatile kairomone that affects acarine predator-prey interactions: involvement of host plant in its protection. *Journal of Chemical Ecology* 16(2):381-396.

Dingle, H. 1972. Migration strategies of insects. *Science* 175:1327-1335.

Doutt, R. L. and J. Nakata. 1973. The *Rubus* leafhopper and its egg parasitoid: An endemic biotic system useful in grape-pest management. *Environmental Entomology* 2(3):381-386.

Drinkwater, L.E., D.K. Letourneau, F. Workneh, A.H.C. van Bruggen, and C. Shennan. 1995. Fundamental differences between conventional and organic tomato agroecosystems in California. *Ecological Applications* 5:1098-1112.

Drukker, B., P. Scutareanu, and M. W. Sabelis. 1995. Do anthocorid predators respond to synomones from *Psylla*-infested pear trees under field conditions? *Entomologia Experimentalis et Applicata* 77:193-203.

Elliot, N.C., G.A. Simmons, and F.J. Sapio. 1987. Honeydew and wildflowers as food for the parasites *Glypta fumiferanae* (Hymenoptera: Icheumonidae) and *Apanteles fumiferanae* (Hymenoptera: Braconidae). *Journal of the Kansas Entomological Society.* 60:25-29.

Ellis, C. R., B. Kormos, and J. C. Guppy. 1988. Absence of parasitism in an outbreak of the cereal leaf beetle, *Oulema melanopus* (Coleoptera: Chrysomelidae), in the central tobacco growing area of Ontario. *Proceedings of Entomological Society of Ontario* 119:43-46.

Elzen, G. W., H. J. Williams, and S. B. Vinson. 1983. Response of the parasitoid *Campoletis sonorensis* (Hymenoptera: Ichneumonidae) to chemicals (synomones) in plants: implications for host habitat location. *Environmental Entomology* 12:1873-1877.

Evans, E.W. 1993. Indirect interactions among phytophagous insects: aphids, honeydew and natural enemies. In: A.D. Watt, S.R. Leather, L.E.F. Walters, and N.J. Mills (eds.), *Individuals, Populations and Patterns in Ecology.* Intercept Press, Andover, U.K. pp. 287-298.

Ewert, M. A. and H. C. Chiang. 1966. Dispersal of three species of coccinellids in corn fields. *Canadian Entomologist* 98:999-1003.

Fell, C. M. and L. A. Hull. 1996. Overwintering of *Stethorus punctum punctum* (Coleoptera: Coccinellidae) in apple orchard ground cover. *Environmental Entomology* 25(5):972-976.

Flint, H. M., S. S. Salter, and W. S. Walters. 1979. Caryophyllene: an attractant for the green lacewing. *Environmental Entomology* 8:1123-1125.

Gillaspy, J. E. 1971. Papernest wasps (*Polistes*): observations and study methods. *Annals Entomological Society of America* 64(6):1357-1361.

Gross, H. R., Jr., W. J. Lewis, R. L. Jones, and D. A. Nordlund. 1975. Kairomones and their use for management of entomophagous insects: III. Stimulation of *Trichogramma achaeae*, *T. pretiosum*, and *Microplitis croceipes* with host-seeking stimuli at time of release to improve their efficiency. *Journal of Chemical Ecology* 1(4):431-438.

Hagan, K.S. 1986. Ecosystem analysis: Plant cultivars (HPR), entomophagous species and food supplements. In: D.J. Boethel and R.D. Eikenbary (eds.), Interactions of plant resistance and parasitoids and predators of insects. John Wiley and Sons, West Sussex, U.K., pp. 151-197.

Hagen, K. S., P. Greany, E. F. Sawall, Jr., and R. L. Tassa. 1976. Tryptophan in artificial honeydews as a source of an attractant for adult *Chrysopa carnea*. *Environmental Entomology* 5(3):458-468.

Hagley, E.A.C. and D.R. Barber. 1992. Effect of food sources on the longevity and fecundity of *Pholetesor ornigis* (Weed) (Hymenoptera: Braconidae). *Canadian Entomologist* 124:341-346.

Heimpel, G.E. and T.R. Collier. 1996. The evolution of host-feeding behavior in insect parasitoids. *Biological Reviews* 71:373-400.

Heimpel, G.E. and J.A. Rosenheim. 1995. Dynamic host feeding by the parasitoid *Aphytis melinus*: the balance between current and future reproduction. *Journal of Animal Ecology* 64:153-167.

Heimpel, G.E., J.A. Rosenheim, and D. Kattari. 1997. Adult feeding and lifetime reproductive success in the parasitoid *Aphytis melinus*. *Entomologia Experimentalis et Applicata* 83:305-315.

Hidaka, K. 1990. Natural and organic farming and insect pests. Toujusha, Tokyo (in Japanese).

Hodek, I. and A. Honek. 1996. *Ecology of Coccinellidae*. Dordrecht, The Netherlands: Kluwer Academic Press.

Hokkanen, H.M.T. 1991. Trap cropping in pest management. *Annual Review of Entomology* 36: 119-138.

Honek, A. 1982. The distribution of overwintered *Coccinella septempunctata* L. (Col., Coccinellidae) adults in agricultural crops. *Z. Angew. Ent.* 94:311-319.

House, G.J. and J.N. All. 1981. Carabid beetles in soybean agroecosystems. *Environmental Entomology* 10:194-196.

House, G.J. and B.R. Stinner. 1983. Arthropods in no-tillage soybean agroecosystems: community composition and ecosystem interactions. *Environmental Management* 7:23-28

Idris, A.B. and E. Grafius. 1995. Wildflowers as nectar sources for *Diadegma insulare* (Hymenoptera: Ichneumonidae), a parasitoid of diamondback moth (Lepidoptera: Yponomeutidae). *Environmental Entomology* 24:1726-1735.

Ives, A. R. and W. H. Settle. 1997. Metapopulation dynamics and pest control in agricultural systems. *American Naturalist* 149(2):220-246.

James, D. G. 1989. Overwintering of *Amblyseius victoriensis* Womersley (Acarina: Phytoseiidae) in southern New South Wales. *Gen. Appl. Ent.* 21:51-55.

Jervis, M.A. and N.A.C. Kidd. 1986. Host-feeding strategies of Hymenopteran parasitoids. *Biological Reviews* 61:395-434.

Jervis, M.A. and N.A.C. Kidd. 1996. Phytophagy. In: M.A. Jervis and N.A.C. Kidd (eds.). *Insect Natural Enemies*. Chapman and Hall, London, U.K, pp. 375-394.

Jervis M.A., N.A.C. Kidd, M.G. Fitton, T. Huddleston, and H. Dawah. 1993. Flower-visiting by hymenopteran parasitoids. *Journal of Natural History* 27:-67-105.

Jervis, M.A., N.A.C. Kidd, and G.E. Heimpel. 1996. Parasitoid adult feeding behavior and biocontrol — a review. *Biocontrol News and Information* 17:11-22.

Johnson, C.G. 1960. A basis for a general system of insect migration and dispersal by flight. *Nature* 186:348-350.

Jones, R. L., W. J. Lewis, M. Beroza, B. A. Bierl, and A. N. Sparks. 1973. Host-seeking stimulants (kairomones) for the egg parasite, *Trichogramma evanescens*. *Environmental Entomology* 2(4):593-596.

Jones, R. L., W. J. Lewis, M. C. Bowman, M. Beroza, and B. A. Bierl. 1971. Host-seeking stimulant for parasite of corn earworm: isolation, identification and synthesis. *Science* 173:842-3.

Kamil, A.C., J.R. Krebs, and H.R. Pulliam (eds.). 1987. *Foraging behavior.* Plenum Press, New York.

Kido, H., D. L. Flaherty, D. F. Bosch, and K. A. Valero. 1984. French prune trees as overwintering sites for the grape leafhopper egg parasite. *Am. J. Enol. Vitic.* 35(3):156-160.

Kromp, B. 1989. Carabid beetle communities (Carabidae, Coleoptera) in biologically and conventionally farmed agroecosystems. *Agriculture, Ecosystems and Environment* 27:241-251.

Kruess, A. and T. Tscharntke. 1994. Habitat fragmentation, species loss, and biological control. *Science* 264:1581-1584.

Kuikman, P.J., M.M.I. van Vuuren, and J.A. van Veen. 1989. Effect of soil moisture regime on predation by protozoa of bacterial biomass and the release of bacterial nitrogen. *Agriculture, Ecosystems and Environment* 27:271-279.

Leius, K. 1961. Attractiveness of different foods and flowers to the adults of some hymenopterous parasites. *Canadian Entomologist* 90:369-376.

Leius, K. 1967. Food sources and preferences of adults of a parasite, *Scambus buolianae* (Hymenoptera: Ichneumonidae), and their consequences. *Canadian Entomologist* 99:865-871.

Letourneau, D. K. 1990. Mechanisms of predator accumulation in a mixed crop system. *Ecological Entomology* 15:63-69.

Lewis, T. and J. W. Stephenson. 1966. The permeability of artificial windbreaks and the distribution of flying insects in the leeward sheltered zone. *Annals of Applied Biology* 58:355-363.

Lewis, W. J. and K. Takasu. 1990. Use of learned odours by a parasitic wasp in accordance with host and food needs. *Nature* 348:635-636.

Lingren, P.D. and M.J. Lukefahr. 1977. Effects of nectariless cotton on caged populations of *Campoletus sonerensis*. *Environmental Entomology* 6:586-588.

Mattiacci, L., M. Dicke, and M. A. Posthumus. 1994. Induction of parasitoid attracting synomone in brussels sprouts plants by feeding of *Pieris brassicae* larvae: role of mechanical damage and herbivore elicitor. *Journal of Chemical Ecology* 20(9):2229-2247.

Mattson, W.J., Jr. 1980. Herbivory in relation to plant nitrogen content. *Annual Review of Ecology and Systematics* 11:119-161.

McCall, P. J., T. C. J. Turlings, W. J. Lewis, and J. H. Tumlinson. 1993. Role of plant volatiles in host location by the specialist parasitoid *Microplitis croceipes* Cresson (Braconidae: Hymenoptera). *Journal of Insect Behavior* 6(5):625-639.

McDougal, S.J. and N.J. Mills. 1997. The influence of hosts, temperature and food sources on the longevity of *Trichogramma platneri*. *Entomologia Experimentalis et Applicata* 83:195-203.

McMurtry, J.A. and G.T. Scriven. 1964. Studies on the feeding, reproduction, and development of *Amblyseius hibisci* (Acarina: Phytoseiidae) on various food substances. University of California Citrus Research Center and Agriculture Experiment Station, Riverside, paper no. 1531.

McMurtry, J.A. and H.G. Johnson. 1965. Some factors influencing the abundance of the predaceous mite *Amblyseius hibisci* in sourthern California (Acarina: Phytoseiidae). *Annals of the Entomological Society of America* 58:49-56.

Monteith, L. G. 1960. Influence of plants other than the food plants of their host on finding by tachinid parasites. *Canadian Entomologist* 92:641-652.

Mullens, B. A., N. C. Hinkle, and C. E. Szijj. 1996. Impact of alternating manure removal schedules on pest flies (Diptera: Muscidae) and associated predators (Coleoptera: Histeridae, Staphylinidae; Acarina: Macrochelidae) in caged-layer poultry manure in Southern California. *Journal of Economic Entomology* 89(6):1406-1417.

Müller, C. B. and H. C. J. Godfrey. 1997. Apparent competition between two aphid species. *Journal of Animal Ecology* 66:57-64.

Murphy, B. C., J. A. Rosenheim, R. V. Dowell, and J. Granett. 1998. Habitat diversification tactic for improving biological control: parasitism of the western grape leafhopper. *Entomologia Experimentalis et Applicata* 87:225-235.

Murphy, B. C., J. A. Rosenheim, and J. Granett. 1996. Habitat diversification for improving biological control: abundance of *Anagrus epos* (Hymenoptera: Mymaridae) in grape vineyards. *Environmental Entomology* 25:495-504.

Naranjo, S.E. and J.L. Stimac. 1987. Plant influences on predation and oviposition by *Geocoris punctipes* (Hemiptera: Lygaeidae) in soybeans. *Environmental Entomology* 16:182-89.

National Research Council. 1989. Alternative agriculture. National Academy Press, Washington, D.C.

Nichols, P. R. and W. W. Neal. 1977. The use of food wheast as a supplemental food for *Coleomegilla maculata* (DeGeer) (Coleoptera: Coccinellidae) in the field. *Southwestern Entomologist* 2(3):102-105.

Need, J. T. and P. P. Burbutis. 1979. Searching efficiency of *Trichogramma nubilale*. *Environmental Entomology* 8:224-227.

Nettles, W. C. 1979. *Eucelatoria* sp. females: Factors influencing response to cotton and okra plants. *Environmental Entomology* 8:619-623.

Olson, D.M. and D.A. Andow. 1998. Larval crowding and adult nutrition effects on longevity and fecundity of female *Trichogramma nubilale* Ertle & Davis (Hymenoptera: Trichogrammatidae). *Environmental Entomology* 27: 508-514.

Osakabe, M. 1988. Relationships between food substances and developmental success in *Amblyseius sojaensis* Ehara (Acarina: Phytoseiidae). *Applied Entomology and Zoology* 23:45-51.

Osakabe, M., K. Inoue, W. Ashihara. 1987. Effect of *Amblyseius sojaensis* Ehara (Acarina: Phytoseiidae) as a predator of *Panonychus citri* (McGregor) and *Tetranychus kanzawai* (Acarina: Tetranychidae). *Applied Entomology and Zoology* 22:594-599.

Pedigo, L.P. 1996. *Entomology and pest management, 2nd edition*. Prentice Hall, Upper Saddle River, New Jersey.

Pemberton, R.W. and J. Lee. 1996. The influence of extrafloral nectaries on parasitism of an insect herbivore. *American Journal of Botany* 83:1187-1194.

Perfecto, I., B. Horwith, J. Vandermeer, B. Schultz, H. McGuinness, and A. Dos Santos. 1986. Effects of plant diversity and density on the emigration rate of two ground beetles, *Harpalus pennsylvanicus* and *Evarthus sodalis* (Coleoptera: Carabidae), in a system of tomatoes and beans. *Environmental Entomology* 15:1028-1031.

Perrin, R. M. 1975. The role of the perennial stinging nettle, *Urtica dioica*, as a reservoir of beneficial natural enemies. *Annals of Applied Biology* 81:289-297.

Pickett, S. T. A. and P. S. White, eds. 1985. *The ecology of natural disturbance and patch dynamics*. San Diego: Academic Press.

Rabatin, S.C. and B.R. Stinner. 1989. The significance of vesicular-arbuscular mycorrhizal fungal-soil macroinvertebrate interactions in agroecosystems. *Agriculture Ecosystems and Environment* 27:195-204.

Read, D. P., P. P. Feeny, and R. B. Root. 1970. Habitat selection by the aphid parasite *Diaretiella rapae* (Hymenoptera: Braconidae) and hyperparisite *Charips brassicae* (Hymenoptera: Cynipidae). *Canadian Entomologist* 102:1567-1578.

Reid, C. D. and R. L. Lapman. 1989. Olfactory response of *Orius insidiosus* (Hemiptera: Anthocoridae) to volatiles of corn silk. *Journal of Chemical Ecology* 15:1109-1115.

Riechert, S.E., L. Provencher, and K. Lawrence. 1999. The potential of spiders to exhibit stable equilibrium point control of prey: Tests of two criteria. *Ecological Applications* 9:365-377.

Risch, S.J., R. Wrubel, and D.A. Andow. 1982. Foraging by a predaceous beetle, *Coleomegilla maculata* (Coleoptera: Coccinellidae), in a polyculture: Effects of plant density and diversity. *Environmental Entomology* 11:949-950.

Rogers, C.E. 1985. Extrafloral nectar: entomological implications. *Bulletin of the Entomological Society of America* 31:15-20.

Rosenheim, J.A. 1998. Higher-order predators and the regulation of insect herbivore populations. *Annual Review of Entomology* 43:421-447.

Russel, E.P. 1989. Enemies hypothesis: a review of the effect of vegetational diversity on predatory insects and parasitoids. *Environmental Entomology* 18:590-599.

Sabelis, M. W. and H. E. van de Baan. 1983. Location of distant spider mite colonies by phytoseiid predators: demonstration of specific kairomones emitted by *Tetranychus urticae* and *Panonychus ulmi*. *Entomologia Experimentalis et Applicata* 33:303-314.

Schellhorn, N.A. and D.A. Andow. 1999. Cannibalism and interspecific predation: Roles for oviposition and foraging behavior. *Ecological Applications* 9:418-428.

Schiefelbein, J. W. and H. C. Chiang. 1966. Effects of spray of sucrose solution in a corn field on the populations of predatory insects and their prey, *Entomophaga* 11(4):333-339.

Sengonca, C. and B. Frings. 1989. Enhancement of the green lacewing *Chrysoperla carnea* (Stephens), by providing artificial facilities for hibernation. *Turkiye Entomoloji Dergisi* 13(4):245-250.

Shahjahan, M. 1974. *Erigeron* flowers as food and attractive odor source for *Peristenus pseudopallipes*, a braconid parasitoid of the tarnished plant bug. *Environmental Entomology* 3:69-72.

Sotherton, N. W. 1984. The distribution and abundance of predatory arthropods overwintering on farmland. *Annuals of Applied Biology* 105:423-429.

Southwood, T.R.E. 1962. Migration of terrestrial arthropods in relation to habitat. *Biological Review* 37:171-214.

Speight, M.R. and J.H. Lawton. 1976. The influence of weed-cover on the mortality imposed on artificial prey by predatory ground beetles in cereal fields. *Oecologia* 23: 211-223.

Staple, J. O., A. M. Cortesero, C. M. De Moraes, J. H. Tumlinson, and W. J. Lewis. 1997. Extrafloral nectar, honeydew, and sucrose effects on searching behavior and efficiency of *Microplitis croceipes* (Hymenoptera: Braconidae) in cotton. *Environmental Entomology* 26:617-623.

Stary, P. and J. P. Lyon. 1980. *Acyrthosiphon pisum ononis* (Hom., Aphididae) and *Ononis* species as reservoirs of aphid parasitoids (Hym., Aphididae). *Acta Entomol. Bohemsolov.* 77:65-75.

Stern, V. 1991. Environmental control of insects using trap crops, sanitation, prevention, and harvesting. In D. Pimentel (ed.), *Handbook of pest management in agriculture, 2nd edition*. CRC Press, Boca Raton, Florida.

Stinner, B.R. and G.J. House. 1990. Arthropods and other invertebrates in conservation-tillage agriculture. *Annual Review of Entomology* 35:299-318.

Streams, F. A., M. Shahjahan, and H. G. LeMasurier. 1968. Influence of plants on the parasitization of the tarnished plant bug by *Leiophron pallipes*. *Journal of Economic Entomology* 61:996-999.

Sundby, R.A. 1967. Influence of food on the fecundity of *Chrysopa carnea* Stephens (Neuroptera: Chrysopidae). *Entomophaga* 12:475-79.

Syme, P.D. 1975. The effects of flowers on the longevity and fecundity of two native parasites of the European pine shoot moth in Ontario. *Environmental Entomology* 4:337-346.

Takasu, K. and W. J. Lewis. 1993. Host- and food-foraging of the parasitoid *Microplitis croceipes*: learning and physiological effects. *Biological Control* 3:70-74.

Teetes, G.L. 1991. Environmental control of insects using planting time and plant spacing. In D. Pimentel (ed.), *Handbook of pest management in agriculture, 2nd edition*. CRC Press, Boca Raton, Florida.

Thomas, M. B., S. D. Wratten, and N. W. Sotherton. 1991. Creation of "island" habitats in farmland to manipulate populations of beneficial arthropods: predator densities and emigration. *Journal of Applied Ecology* 28:906-917.

Tingey, W.M., T.F. Leigh, and A.H. Hyer. 1975. *Lygus hesperus*: Growth, survival, and egg laying resistance of cotton genotypes. *Journal of Economic Entomology* 68:28-30.

Turlings, T. C. J., J. H. Tumlinson, and W. J. Lewis. 1990. Exploitation of herbivore-induced plant odors by host-seeking parasitic wasps. *Science* 250:1251-1253.

United States Department of Agriculture. 1980. Report and Recommendations on Organic Farming. U.S. Government Printing Office, Washington, D.C.

van den Bosch, R., C. F. Lagace, and V. M. Stern. 1967. The interrelationship of the aphid, *Acyrthosiphon pisum* and its parasite, *Aphidius smithi*, in a stable environment. *Ecology* 48(6):993-1000.

van den Bosch, R., E. I. Schlinger, C. F. Lagace, and J. C. Hall. 1966. Parasitization of *Acyrthosiphon pisum* (Harris) by *Aphidius smithi* Sharma & Subba Rao a density dependent process in nature (Homoptera: Aphididae) (Hymenoptera: Aphididae). *Ecology* 47:1049-1055.

Vorley, V. T. and S. D. Wratten. 1987. Migration of parasitoids (Hymenoptera: Braconidae) of cereal aphids (Hemiptera: Aphididae) between grassland, early-sown cereals and late-sown cereals in southern England. *Bulletin of Entomological Research* 77:555-568.

Wäckers, F. L. 1994. The effect of food deprivation on the innate visual and olfactory preferences in the parasitoid *Cotesia rubecula*. *Journal of Insect Physiology* 40(8):641-649.

Wallin, H. 1985. Spatial and temporal distribution of some abundant carabid beetles (Coleoptera: Carabidae) in cereal fields and adjacent habitats. *Pedobiologia* 28:19-34.

Wetzler, R.A. and S.J. Risch. 1984. Experimental studies of beetle diffusion in simple and complex crop habitats. *Journal of Animal Ecology* 53:1-19.

Williams, D. W. 1984. Ecology of the blackberry-leafhopper-parasite system and its relevance to California grape agroecosystems. *Hilgardia* 52:1-33.

Wilson, R.L. and F.D. Wilson. 1976. Nectariless and glaborous cottons: Effect on pink bollworm in Arizona. *Journal of Economic Entomology* 69:623-4.

Wratten, S. D. and C. F. G. Thomas. 1990. Farm-scale spatial dynamics of predators and parasitoids in agricultural landscapes. In R. G. H. Bunce and D. C. Howard (eds.) *Species Dispersal in Agricultural Habitats*. London: Belhaven Press. pp. 219-237.

Yokoyama, V.Y. 1978. Relation of seasonal changes in extrafloral nectar and foliar protein and arthropod populations in cotton. *Environmental Entomology* 7:799-802.

Zangger, A., J-A. Lys, and W. Nentwig. 1994. Increasing the availability of food and the reproduction of *Poecilus cupreus* in a cereal field by strip-management. *Entomologia Experimentalis et Applicata*. 71:111-120.

Zhao, J.Z., G.S. Ayers, E.J. Grafius, and F.W. Stehr. 1992. Effects of neighboring nectar-producing plants on populations of pest Lepidoptera and their parasitoids in broccoli plantings. *Great Lakes Entomologist* 25:253-258.

CHAPTER 6

Implementation of Ecologically-Based IPM

G. W. Cuperus, P. G. Mulder, and T. A. Royer

CONTENTS

1-56670-478-2/00/$0.00+$.50
© 2000 by CRC Press LLC

6.1   INTRODUCTION

Integrated pest management (IPM) is a systems approach that provides an eco-logically-based solution to pest control problems. IPM is defined here as **a sustainable approach to managing pests that combines biological, cultural, physical, and chemical tools in a way that minimizes economic, health, and environmental risks.** It is a proven approach that balances economic, environmental, and societal (health) objectives. Previous authors had a vision for IPM that focused on "integrated control," a strategy involving primarily chemical and biological control (Chant 1964, Pickett 1961, Pickett et al. 1958, Collyer 1953, Grison 1962, Stern et al. 1959, van den Bosch and Stern 1962, Franz 1961). While the latter two groups of authors broadened their scope of integrated control to include cultural and other means of physical pest control, their discussions still centered around classical chemical and biological controls. (van den Bosch and Stern 1962, Franz 1961). IPM has expanded its scope over the past 40 years to encompass a variety of applications in rural and urban settings. This expansion has resulted in a scientific exploration to discover new tools for maintaining pest populations at acceptable levels while sustaining an ecological balance. In addition to this expansion, IPM has become a target for change. IPM practitioners first recognized the need for this change as public concern over pesticide issues came to the foreground (Carson 1962). This concern has blossomed with the advent of additional pest control and regulatory issues. Resistance management, worker protection standards, water quality concerns, and food quality protection represent only a portion of the issues confronting IPM implementation today.

6.1.1   What Does IPM Entail?

In order to implement an effective IPM program today, basic changes in current decision-making processes may be required. Such programs must merge ecology, economics, and environmental concerns with practical management concepts. Growers must recognize that their decisions have consequences that reach far beyond the immediate time and location of their operation. They need to incorporate information gained from the use of key tools such as long-term planning, crop monitoring, and good recordkeeping, to make sound management decisions.

An integrated program implies the merging of disciplines, resources, and management strategies into a multifaceted system. In contrast to unilateral control strategies, such as simply applying a pesticide, IPM typically involves implementation of several strategies in an integrated system that optimizes pest control (Zalom et al. 1992).

The term **pest** refers to insects, weeds, diseases, rodents, and other organisms that compete with humans for food and shelter, or affect human health. The paradigm for managing pests has broadened significantly in definition and implementation over the last 30 years, evolving from single-tactic pest management practices, largely with chemical pesticides, to multiple pest management (insects, weeds, diseases) systems, integrated crop management, and finally to integrated resource management. The refinement, implementation, and practice of IPM must fulfill broadened expectations. We should ask if our present infrastructure supports the adoption and implementation of IPM programs that can fulfill these expectations. Does IPM have the flexibility to grow in an age of increasing emphasis on biotechnology? Should the use of transgenic crops be considered part of an IPM program? Answers to each of these challenging questions will ultimately determine the future direction of IPM implementation. Our greatest challenges, as scientists who develop and refine new IPM programs, will be to decipher the practical aspects that the end-user is willing to adopt, then integrate these aspects into an easily managed system (Doutt and Smith 1971). Regardless of the emergence of new tactics and technologies, IPM programs must ultimately be considered in the context of basic ecological principles within the surrounding environment (Cate and Hinkle 1994).

Management is a decision-making process required to produce a commodity in a planned, systematic way. An IPM program strives to keep pest problems from reaching economically damaging levels, while maintaining consistent profits and simultaneously limiting adverse environmental and social concerns.

IPM programs often substitute management time and expertise for off-farm inputs such as pesticides, fertilizers, and fuel. Significant time is required for training, monitoring, quality control, and maintenance of the program. There is often a trade-off between management time and capital input such as pesticides. Osteen and Szmedra (1989) document some trade-offs between pesticide use and a decrease in labor requirements (Figure 6.1). These trade-offs, coupled with a trend towards increasing farm size, create a new challenge of implementing ecologically-based IPM programs. Large farm size, limited personnel, and access to inexpensive yet effective pesticides have increased grower reliance on these materials (Benbrook

Index 1965 = 100

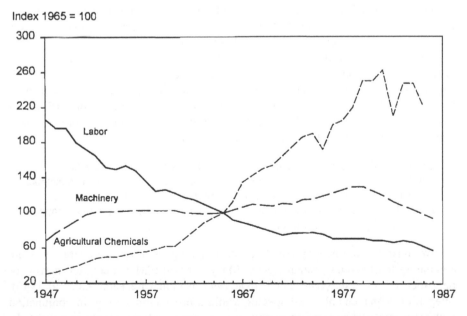

FIGURE 6.1 Pesticide, labor, and machinery use in agriculture over time (Osteen and Szmedra, 1989).

et al. 1996). In fact, Benbrook et al. (1996) suggests that decisions about pesticide application (particularly herbicides) have fallen victim to the marketing wizardry of the chemical industry. As a consequence, many growers make these treatment decisions based purely on economic rather than ecological principles. A sound IPM program should be flexible enough to account for balancing the impact of each factor (thresholds, ecology, natural control, biocontrol, sustainability, cultural control, chemical control, environment, economics, physical control) within the system (National Research Council 1989, USDA Agricultural Research Service 1993, National Coalition on Integrated Pest Management 1994).

6.1.2   IPM: A Focus On Ecology

A comprehensive IPM program requires fundamental understanding of the ecological relationships among crop plants, herbivores or competitors, and the environment. In the 1950s, the original concepts of IPM were developed by practitioners who focused on ecological approaches. Recently, an extensive dialog has surfaced on the need to return IPM to its ecological roots (National Research Council 1996, Benbrook et al. 1996). Pest management approaches must be understood in terms of their interactions with other aspects of each ecosystem. IPM does not stand as an entity onto itself. Cate and Hinkle (1994) tracked the evolution of the ecological paradigm from which IPM was conceived. The original philosophy attempted to develop a system based upon a fundamental understanding of plant/pest interactions that maintained pest populations at sub-economic levels. Kogan (1998) discussed the complex agricultural and socioeconomic interactions that extend far beyond plant/pest interactions and more realistically characterize implementation of IPM in cropping systems. It is now recognized that IPM programs must consider ecological and sociological forces while meeting the goal of maintaining pest populations below economically damaging numbers through an integrated approach.

Integrating several disciplines into commodity management programs has progressed well in some production settings (Cuperus et al. 1990, Cuperus et al. 1992, Collins et al. 1992). Caddel et al. (1995, 1996) applied an interdisciplinary approach to single production issues, which was well received by growers and/or IPM consultants. Comprehensive IPM programs, however, are substantially more complex. Kogan (1998) suggested three levels of integration:

Level I — Single pest management approach (species/population integration).
Level II — Integration of multiple species (insects, weeds, and diseases) and methods for their management in a crop (community level integration).
Level III — Integration of multiple species within the context of the total cropping system (Agroecosystem level).

We must develop new systems that reach the highest level of integration while continuing to effectively manage pests. Many criticize IPM as being too narrow or only focused on pests (Cate and Hinkle 1994, National Research Council 1996). Integrating management of all species within a production setting, in conjunction with environmental and socioeconomic aspects should alleviate these criticisms. As we develop IPM systems, we must consider those involved in the business of production agriculture and the constraints they operate under while producing our food and fiber. These constraints include, but are not limited to, economic investment, time, personnel, marketing, environment, worker protection, and food safety. These aspects are constantly being juggled by growers as they seek to comply with regulations and remain financially solvent.

6.2 PROCESSES THAT DEFINE IPM

The three basic "forces" that shape the design of IPM programs are economics, the environment, and sociology.

6.2.1 Economics

On the surface, chemical pest control appears to be a break-even proposition. Pest control represents 13 to 34% of a farmer's variable crop production costs, yet pests are responsible for crop losses of 10 to 30% (Cuperus et al. 1997). Producers spend over $10.5 billion to purchase 1.25 billion pounds of active ingredients for pest control (Aspelin 1997, Benbrook et al. 1996). Various measures of pesticide use (kg active ingredient used, A.I./acre, number of applications/acre, or cumulative environmental impact) cloud the discussions and often allow people to draw differing conclusions.

IPM has successfully led the way in educating the public about issues associated with the use and/or misuse of insecticides. Since 1964, quantities of insecticides applied have declined by over 40% (Anderson 1994). In addition, use of fungicides has also declined slightly (from 9 to 7%). However, there is room for improvement. In reality, herbicide use patterns have not changed dramatically over the past 10 years. The amount of herbicide applied/acre has increased from 24% of the total pesticide quantity applied in 1964 to 76% in 1982 (Benbrook et al. 1996, Lin et al. 1995, Anderson 1994, Mayerfeld et al. 1996). In addition, the percentage of acres cultivated for weed control and banded application of herbicides (IPM suggested practices) has decreased (Buhler et al. 1992, Liebman et al. 1996, Mayerfeld et al. 1996). If IPM is to become widely accepted, we must educate the end-user, balance the "forces" that shape its design, and convince the grower of its value for providing increased economic return (Hutchins 1995).

Economics represent the cornerstone of a rational approach to pest management. IPM has consistently focused on balancing inputs with returns to maximize profits. Early IPM efforts focused on development of criteria for economic decisions such as economic injury levels and economic thresholds. While many of these were based largely on field observations without experimentation, they provided guidelines for more rational pesticide use (Pedigo and Higley 1992). Future comprehensive economic thresholds will integrate dynamic marketing strategies and economic values with variable control costs to make better choices for long-term economic return. The power of computers, coupled with specialized decision support software that is available from both public and private sectors, lets users estimate cost:benefit relationships for most management inputs. It also allows for examination of whole-farm utilization of economic resources and estimated payback over time (Stark et al. 1992, Stritzke et al. 1996, Flinn et al. 1995).

Economic risk is often identified as the major factor that prevents adoption of IPM approaches (Cuperus and Berberet 1994, Sorensen 1994). All stages of commodity production, including processing and distribution, have risks. Inputs such as pesticides are often viewed as a means to help reduce the perceived risks associated with the enterprise. Economic decision support tools, such as economic injury levels, were developed to reduce risks and establish a rational approach to using pesticides (Pedigo and Higley 1992). The development of economic injury levels depends upon a thorough understanding of plant:pest relationships within a cropping system. Pedigo and Higley (1992) suggest that most available decision support information are best guesses. Of particular concern are type II errors where treatment is needed and not made, thereby threatening the crop and enterprise.

Successful IPM systems must reduce the risk of making a type II error, through a quantitative understanding of pest/injury relationships. Promotion of IPM often suggests that "risk" of losses may be experienced from implementation so growers often view IPM adoption as an additional source of "risk" to their enterprise. The promotion of IPM must emphasize management of the surrounding ecological system at a cost that does not adversely affect quality. Growers and IPM practitioners must be persuaded to accept "management," not "control," of pest problems. Once accepted, this concept usually leads to thorough economic analyses of several scenarios of management. If growers cannot be persuaded to adopt some level of IPM through an economic analysis of the options, then regulatory or environmental constraints may ultimately dictate their management levels. Fortunately, the literature is replete with evidence on the short-term economic benefits from IPM and how it has reduced pesticide use (Beingolea 1981, French 1982, Hussey 1985, Frisbie and Adkisson 1985, Readshaw 1984, Strayer 1971, Whalon and Croft 1983, Wiley 1978).

Over the past 50 years, agricultural enterprises have evolved from essentially a "way of life" into business enterprises that are influenced by societal, political and economic forces on a global scale. To remain competitive, producers and processors are expected to make rapid changes in production practices and switch to alternative enterprises that may require significantly different equipment and IPM systems. IPM practitioners need detailed information on partial budgets, production costs, and potential profits to provide the services that producers will demand.

In many cropping systems, potential losses from pests represent a relatively small portion of the production budget, yet they drive management decisions because of producers' concern over risk (Cuperus and Berberet 1994, Stark et al. 1990). As an example, consider the seasonal impact of defoliating caterpillars on crops such as soybean or peanut. Because of production of excess foliage, both of these leguminous plants can compensate for and/or tolerate appreciable injury without significant yield loss (Higley 1994, Pedigo et al. 1986, Higley 1992, Smith and Barfield 1982). Over the course of research on this complex relationship between defoliation and plant yield, two major factors were identified that ultimately determined insect economic thresholds: time and environment. Smith and Barfield (1982) coined the term "temporal tolerance" to describe the relationship between plant age and the degree of reaction to injury in peanut. Likewise, Higley (1994), Pedigo et al. (1986), and others demonstrated the importance of protecting yield-producing structures (flowers, pods, and seeds) from insect injury during the critical reproductive period of soybean. Environmental factors (biotic and abiotic) mediate the final effects of defoliant injury (Higley 1994). Biotic factors include other pests (e.g., diseases and weeds) and natural enemies (e.g., predators and parasites), while abiotic factors include the effects of climate (e.g., rainfall/irrigation, temperature, etc.). The point of this discussion is that while little effect is seen from loss of seedling plants and/or severe defoliation near harvest in leguminous crops such as peanut or soybean, growers continue to apply pesticides prophylactically or preventatively for insect pests that occur on these plants during non-critical periods (Criswell et al., unpublished data).

Marketing is a critical but often neglected component of IPM decision support systems. Market price and demand may fluctuate 50 to 100%, depending on conflicting customer demands such as the quality of the product, presence of pesticide residues, or the product availability during peak market demand (Stritzke et al. 1996, Stark et al. 1990). While moving toward identity preservation, and quality control, IPM practitioners must understand these market forces and subsequently develop specific pest management programs to meet customer preferences.

An example of a pest management system that reflects the needs of a changing marketplace was reported by Owen (1996) and Suter (1995), in the areas of fresh fruit and vegetable production where:

  i. Consumers demand produce that is cosmetically perfect and without pesticide residues.
 ii. Food processors who purchase fresh produce are concerned about pesticide residues (over 50% of processors have changed purchasing contracts to minimize residues in their products).
iii. Surveys indicate that more than 50% of grocers test fresh produce for pesticide residues, yet 98% indicate that they have not heard of IPM (Owen 1996, Suter 1995).

What is suggested by this production system and its market characteristics is that the desire for residue-free produce competes directly with the demand for "perfect" (blemish-free) produce. Which voice does the grower heed? Only through effective IPM and public education can he meet the demands of both.

Over the past 25 years, losses caused by pests have not changed substantially (National Research Council 1989). Producers who are averse to risk respond by making pesticide applications that are preventive (Cuperus and Berberet 1994) and not necessarily economically pragmatic. In addition, farmers are continually challenged by the introduction of exotic pests such as tropical soda apple, leafy spurge, and Karnal bunt. Once established, the first response is typically to apply pesticides for control, with limited success. This only emphasizes the importance of gathering detailed information on the distribution, phenology, control, and management of new risks before responding with area-wide, blanket treatments. Information gained about pest distribution, pesticide use patterns, and pest management practices can affect international market demand. For example, information on the distribution of

Karnal bunt is mandatory before export can occur. As new international trade agreements evolve, innovative and comprehensive pest management practices will become increasingly important.

6.2.2 Environmental

A basic principle of IPM emphasizes protection of land, water, and wildlife species. Environmental risks associated with pesticides include detrimental effects to beneficial and non-target organisms, aquatic toxicity in surface and groundwater, and avian toxicity (Benbrook et al. 1996). Not only are scientists concerned, but surveys also indicate that the general public often links environmental degradation with pesticides (Tables 6.1 and 6.2). Environmental concerns are often important in determining the direction for IPM programs, in part, through influencing resource allocations (Kogan 1998, Benbrook et al. 1996, Cuperus et al. 1997). The use of DDT and other chlorinated hydrocarbons caused pesticide poisonings and reduced numbers of certain wildlife species (birds), which spawned public awareness and created a need for programs like IPM (Kogan 1998). Early implementation was also stimulated by public concern over the impact of using these materials in agriculture, as highlighted in *Silent Spring* (Carson 1962).

The future of IPM will likely seek a balance between the economics of production and environmental stewardship. Future IPM programs will extend well beyond agriculture into the agriculture:urban interface. Societal, political, and economic pressures are forcing production systems to integrate Best Management Practices (BMPs) that target protection of water quality (Jacobsen 1997). Producers are starting to understand the importance and difficulty of balancing environmental and economic aspects of pest management decisions.

Table 6.1 Oklahoma City Respondents' Indications of Environmental Concerns. Oklahoma, 1994 (Shelton et al., 1997)

Concern	Respondents	Percent
Polluted Air	107	27
Waste Management	50	13
Pesticides	49	12
Contaminated Water	43	11
Oil Fields	30	8
Chemical Waste	10	2
Other	10	2
No Answer	99	25

Table 6.2 Respondents' Perceptions on Sources of Contamination in Oklahoma Surface and Ground Waters. Oklahoma, 1991 (Shelton et al., 1997)

	Hazardous Wastes		Pesticides		Petroleum		Nitrates		Bacteria		Heavy Metals	
	N	%	N	%	N	%	N	%	N	%	N	%
Surface	95	25	101	27	43	11	49	13	46	12	13	4
Ground	104	28	71	19	66	18	46	12	35	9	23	6

IPM consultants must consider the environmental costs of pest management. Such integration is already available in some decision support tools. Higley and Wintersteen (1992) outlined a method for estimating the economic costs of environmental impact when selecting a pesticide. Economic costs focus on producer perceptions of pesticide efficacy and potential costs to the local environment. Numerous decision support systems exist that are capable of integrating pesticide efficacy and fate with economic costs (Pratt et al. 1993). Kovach et al. (1992) developed an index that considers the economic cost incurred through pesticide use on human health, the environment, and other related components, and used this index in estimating a relative value for these components to aid in choosing cost-effective pest management approaches. Riha et al. (1996) examined this perspective from a much broader scale, asking consultants to integrate planning and implementation over a longer time frame, on both micro and macro scales, and more from an ecological perspective. The greatest challenge that IPM consultants face is the sustainable management of agricultural production systems in conjunction with preservation of natural resources.

6.2.3 Sociology

In agricultural and urban settings, social issues affect the acceptability of pesticides as a management tool. Some central concerns include endangered species, safety of farm workers, and food quality protection. Society benefits from the appropriate implementation of IPM programs. IPM is a socially acceptable approach for regulating pest species because it is comprehensive and flexible, enabling farmers, urban dwellers, school systems, and municipalities to develop more sustainable, environmentally sound, and economically viable systems (Buttel et al. 1990).

In light of these criteria, several resounding questions surface. Is IPM to the point of meeting the expectations of reduced pesticide inputs promised in the late 1960s and early 1970s? Probably not, particularly as it relates to herbicide use. Virtually 100% of the corn and soybean acres throughout the Corn Belt are treated with herbicides. While insecticide use declined on land where corn was rotated with other crops, the number of acres treated is surging due to the failure of crop rotation as a rootworm management tool in the eastern Corn Belt. Insecticide use on continuous corn exceeds 90% with little to no input from scouting. Has IPM succeeded despite nearly three decades of support from the Federal government? From a sociological vantage point, perhaps not. Have we approached a time when pest management decisions will be made in the best interests of society versus the most expedient and profitable vantage point of the producers? Possibly. Coble et al. (1998) recently proposed the use of pesticide prescriptions to salvage valuable uses of high-risk pesticides. This proposal may represent a move toward regaining some of the ardent sociological supporters of IPM while saving effective tools that might otherwise be eliminated because of standards of the Food Quality Protection Act (FQPA 1996).

6.2.4 Food Safety

Because personal health and well-being are highly valued in today's society, consumers have great concern about the safety of the food supply. The choices that

Table 6.3   Percent of Public Expressing Trust at
Different Stages of Food Production and
Distribution Systems (Sachs et al., 1987)

	1965	1984
Farmer care	81.5	61.6
Production regulations	97.7	45.8
Retail inspections	94.0	48.9

consumers make through purchasing food items dictate which foods are produced, processed, and distributed. It is increasingly evident that consumers desire a safe, wholesome food supply, produced without harm to the environment. Although scientists generally agree that microbial contamination poses the biggest threat to food safety, the general public believes that pesticides and their residues are the most critical food safety issue (Cuperus et al. 1991). National surveys show that over 65% of consumers are concerned with the safety of the food supply because of pesticide residues (Pomerantz 1995).

Over the past 25 years, consumer attitudes have reflected increasing concern about the safety issues in food production systems (Table 6.3) (Sachs et al. 1987). These concerns are believed to be a result of a limited understanding of science and agriculture, and dread of cancer. The public wants assurance that food they eat is produced and processed with the safest possible IPM systems (Thompson and Kelvin 1996). Shelton et al. (1997) suggested that the public is more concerned about pesticide exposure as it relates to the health of farm workers and maintaining a clean environment than they are about their own exposure to pesticides (Table 6.4). In one study, consumers changed their buying patterns because of safety concerns regarding pesticide residues in produce (Pomerantz 1995). Other studies indicate that consumers are willing to pay a small premium for food commodities having less exposure to pesticides (Collins et al. 1992). Farmers and processors must be educated as to the most effective ways to respond to these public concerns.

The most critical determinants for adoption of IPM practices, at least in the short term, are demographic and sociological in nature. Less than 2% of the U.S. population is directly involved in farming; therefore, tremendous challenges in education and communication must be met if a public consensus is to occur regarding the importance of IPM in sustaining a wholesome, abundant food supply, and maintaining safe environments for humans and other species.

Table 6.4   Oklahoma Survey Response to Perceived Hazard to Wildlife,
the General Public, and Farm Workers from Exposure to
Pesticides. Oklahoma, 1991 (Shelton et al., 1997)

Target Group	Level of Hazard							
	No Hazard		Very Little		Some		Great Deal	
	N	%	N	%	N	%	N	%
Wildlife	15	4	38	10	136	36	165	44
Farm Workers	8	2	42	11	174	46	133	36
Farmer/Owner	7	2	58	16	174	46	118	31
General Public	15	4	78	21	173	46	91	24

Table 6.5 Major Wheat Pests and Pest Management Strategies Utilized (Cuperus et al., 1992, Stuckey et al., 1990)

Pest	Resistant Varieties	Crop Rotation	Planting Date	Tillage	Soil Fertility	Pesticides	Biological Control
Diseases[1]							
Leaf Rust	1	—	3	—	—	1	—
Tan Spot	—	2	3	2	—	2	—
Wheat Streak							
Mosaic Virus	3	3	2	1	—	—	—
Take-all	2	1	2	3	2	—	—
Insects[2]							
Greenbug	—	—	—	2	—	1	2
Hessian Fly	1	—	1	1	—	3	—
White Grub	—	2	2	3	—	2	—
Weeds[2]							
Mustards	—	3	3	3	—	1	—
Bromus spp.	—	1	2	2	3	2	—

[1] Key to effectiveness of practice: 1 = highly effective control measure, 2 = moderately effective, 3 = slightly effective.
[2] Key to impact of practice: 1 = excellent management tool, 2 = some management impact, 3 = small impact, no impact.

6.3 DEVELOPING AN IPM SYSTEM

The basic tools for developing IPM systems include: cultural control, mechanical control, host plant resistance, biological control, reproductive manipulation, chemical control and regulatory control. For most cropping systems, an ecologically-based IPM approach will emphasize those types of controls that are critical to environmental stewardship: cultural practices, host plant resistance, and biological control.

I. Cultural practices alter the ecology of the cropping system and must play a fundamental role in developing an IPM approach. Pedigo (1995) called this approach "preventative pest management." Some cultural methods purposefully alter habitats to limit pest species, including crop rotation, tillage, fertility management, and adjustment of planting dates (Table 6.5) (Cuperus et al. 1992, Stuckey et al. 1990).

- Crop rotation — Many pest populations can be reduced through crop rotation. Classic examples include Mexican and northern corn rootworms, weeds such as cheat in wheat, and many phytophagous nematodes and diseases (Stuckey et al. 1990).
- Modifying planting date is a long-known technique for altering the synchrony of hosts and pests, to allow the crop to avoid infestation. Examples include targeting planting dates to reduce boll weevils in cotton, wheat streak mosaic virus (Willis 1984) (Figure 6.2A), and Hessian fly in wheat (Stuckey et al. 1990) (Figure 6.2B).
- Grazing alfalfa hay stands to reduce abundance of insects and weeds (Dowdy et al. 1992) (Table 6.6).

A

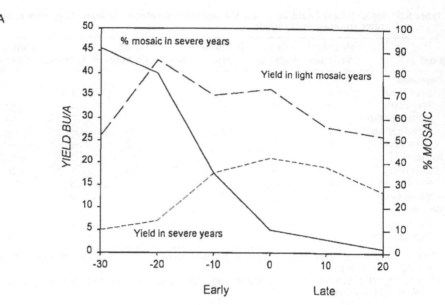

B

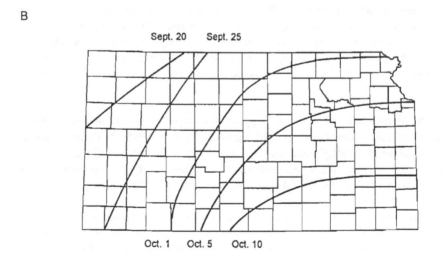

FIGURE 6.2 (A) Effects of modifying planting date on wheat streak mosaic virus and wheat yield. (B) Suggested planting dates for Kansas wheat, based on approximate Hessian fly-free dates (Willis 1984, Stuckey et al., 1990).

Table 6.6 Reducing Populations of
Alfalfa Weevil, *Hypera postica*
(Gyllenhal) by Fall-cutting or
Grazing (Dowdy et al., 1992)

Treatment	Eggs/0.1 m²	
	1984	1985
Fall-cut	12.6 ± 1.6	29.3 ± 2.7
Grazed	9.9 ± 2.5	34.9 ± 3.1
Control	45.7 ± 4.5	67.2 ± 4.5

II. Mechanical and physical methods include:

- Tillage to physically bury or injure pests. It is a particularly effective tool for managing numerous weed species and some pathogens and insects.
- Screening/cleaning produce to remove weed seeds, insects, and seed-borne pathogens. These sanitation processes help remove residual infestation sources. This approach represents a core management program for grain storage, food processing, and food distribution systems.
- Pest populations can be managed by adjusting temperature or moisture/humidity. Approaches such as aeration of stored grain have great value in controlling insects and microorganisms.

III. Host Plant Resistance is defined as the heritable qualities possessed by the plant which influence the ultimate degree of damage caused by the pest (Painter 1951). Host plant resistance is a foundation of ecologically-based IPM systems. Host plants and animals have different capacities for resisting diseases, insects, or nematodes. As an example, plant breeders working on alfalfa have incorporated resistance to many insects and diseases in multiple-pest resistant cultivars. Through the use of host plant resistance, producers realized increased stand persistence and net returns to investment by over $100/acre over a 5-year stand life (Stark et al. 1992).

A continuing challenge to agriculture is to preserve resistance once developed. Consider the following:

Newly released resistant cultivars of wheat are challenged continuously by greenbug *Schizaphis graminum* (Rondami). Greenbug populations are genetically diverse; as resistant cultivars become widely deployed, they create a strong selection pressure for those individuals (biotypes) that can survive on resistant plants. The selection process allows a biotype that is virulent on newly released resistant varieties to become more common and to perhaps someday dominate the population (Porter et al. 1997). This shift in the genetic makeup of the greenbug population leads to the loss of plant resistance because it becomes ineffective.

Genetic engineering is rapidly providing new avenues for development and deployment of pest-resistant cultivars. Through this process, corn, cotton, and other crops are modified to produce biological products such as the endotoxin associated with *Bacillus thuringiensis* (Berliner). This technology presents new opportunities and challenges for the development and implementation of comprehensive IPM programs. Programs that utilize these transgenic lines must include strategies on managing resistant cultivars in conjunction with a refuge of susceptible plants. These strategies will help combat the development of virulent insects capable of survival and production on resistant cultivars.

IV. *Biological control.* Natural control includes the combined effects of biotic and abiotic agents that serve to maintain a more or less fluctuating pest population density within definable upper and lower limits over time. Biological control, a component of natural control, is defined as the use of biological control agents including predators, parasites, pathogens, antagonists and competitors to reduce losses from pests (Cate and Hinkle 1994). IPM practitioners are being challenged to fully utilize native beneficial species in future IPM systems and to incorporate their impact into future programs employing the use of biocontrol organisms.

Biological control, cultural control, and host plant resistance are the three key regulatory factors that maintain pest populations at levels below economic thresholds.

V. *Reproductive manipulation (Genetic control) methods* include the use of organisms that have been altered reproductively to manage pests. An example is the release of sterilized males against the screwworm, thereby delivering lethal genes into the population. Reproductive manipulation methods also include use and manipulation of pest mating strategies to reduce or eliminate their populations. On a simple scale, this involves monitoring pest populations with pheromone-baited traps and attempting to use capture data to aid in scouting and/or treatment decisions. More complex means of reproductive manipulation involve mating disruption, which uses pheromone dispensers to mimic, compete with, or mask the natural hormone (Beers et al. 1993), thereby reducing successful mating and reproductive capacity of the pest population.

VI. *Legislative and regulatory control methods* include such tools as inspections and quarantines to prevent the introduction and spread of pests into new habitats. These methods are particularly important in limiting introductions of weed, insect, and disease organisms into new territories as global trade increases.

VII. *Chemical control methods.* Pesticide use in IPM programs has become a political football; passed, punted, and even fumbled by those brave enough to challenge an established, effective industry to concentrate their efforts into safer, more "IPM-friendly" compounds (Royer et al. unpublished). Throughout the literature, several constraints have been identified as reasons for not adopting IPM practices (Sorensen 1993). One of the top three constraints was listed as "an EPA pesticide regulatory process that is burdensome, expensive, time-consuming and unclear" (Sorensen 1993). In addition, supporters of IPM have been unwilling to take a united stand on the controversial question, "Should the primary goal of IPM be to reduce pesticide use?" (Gray 1995a). A slim majority (about 51%) of Cooperative Extension Service IPM Coordinators felt that the chief goal of IPM was not to reduce pesticide use. This attitude may reflect a remnant of the prevailing opinion among extension entomologists in the early 1970s who felt the primary goal was to use all pest regulating factors, including chemicals to decrease pest populations and thereby avoid or reduce preventive insecticide programs that are based on a calendar schedule of application (Petty 1973). Burns et al. (1987) provided similar suggestions, but also called for reduced pesticide inputs and increased emphasis on natural control.

In light of the controversy surrounding the issue of pesticide reduction in IPM programs and the challenge by the Clinton Administration that IPM be implemented on 75% of the nation's managed acres by the year 2000, one major question surfaces: "Would we gain political support for IPM programs if our primary goal was a reduction in pesticide use?" In our reluctance to take a stand on this issue, IPM supporters may only be avoiding the inevitable loss of these tools brought on by the Food Quality Protection Act (FQPA 1996). With increased emphasis on biointensive pest management (Benbrook et al. 1996), the recent suggestion of prescription pesticide use for high-risk materials (Coble et al. 1998) and the EPA's initiative of giving the "fast-track" to many of the safer, environment-friendly pesticides, we may want to reflect on where our past experiences might be directing future decisions on pesticide issues. Chemical control is not limited to the use of pesticides. Semiochemicals, attractants and repellents are also being developed, either to enhance pesticide efficacy or directly control pests. As an example, the compound cucurbitacin is an obligatory feeding stimulant that affects the northern corn rootworm (NCR) *Diabrotica barberi* Smith & Lawrence and western corn rootworm (WCR) *Diabrotica virgifera virgifera* LeConte. It is mixed with the conventional insecticide carbaryl and allows for better, safer control of adult beetles at lower dosages.

Chemical control tools over the past 25 years have changed dramatically. The chemical nature and delivery of pesticides has experienced a significant shift to newer, reduced-rate technologies and selective chemistry that is tailored to individual pest problems. Both of these approaches will have less direct environmental impact. Decision support systems for pesticide applications have improved greatly because forecasting systems that are based on a biological understanding of pest populations have been developed for diseases (Jacobsen 1997, Wu et al. 1996), weeds (Stritzke et al. 1996), and insects (Flinn et al. 1995, Harris et al. 1997, Sparks 1995, Ring and Harris 1983). In addition, many of these systems consider environmental and economic impacts of chemical pest control decisions.

Scouting/monitoring is fundamental to decision-making in IPM systems. It does not operate, however, exclusive of other tools available in IPM programs. Many proactive, long-range tools for management (e.g., variety selection or crop rotation) can provide overall reductions in pest populations and cause scouting/monitoring to be less time-consuming. By definition, IPM is information intensive and site specific. Even within a state or region, variables such as soil type, location, market opportunities, or climatic differences will affect the design of an IPM program for a specific crop. The following are considerations for implementing an effective IPM program.

1. Identify and define the roles of all persons involved, including farm-ranch managers, crop consultants, pesticide applicators, warehouse managers, distributors, and other personnel. This is particularly important when using contractors or private pesticide applicators.

2. Set realistic management objectives for individual components, including yield goals, quality controls, and potential economic returns. This is critical to a partial budgeting operation.

3. Develop conceptual short- and long-range management alternatives to help manage the commodity and pests within that commodity. Many of the most effective management tools such as crop rotation and variety selection are long-range in nature.

4. Establish an understanding of marketing alternatives and consumer attitudes. Marketing is a critical component that directly influences the contribution of pest management decisions; of critical importance is the development and use of dynamic economic thresholds that are sensitive to fluctuations in costs, market prices, consumer attitudes, and management restrictions. The changing demographics of today's society and consumer concerns about production practices will alter marketing challenges and opportunities.

5. Establish quality-control monitoring programs for components of IPM systems including production, storage, and processing; this should include monitoring the quality of the commodity, extent of pest infestation, environmental parameters, and biological control agents within the system.

6. If pesticide applications are judged to be profitable according to decision guidelines, choose the most selective pesticide and apply it at proper dosages with appropriate safety equipment to reduce hazards to workers. Provide worker safety training to increase efficiency and reduce potential liability.

7. Maintain complete records to document profitability, implement quality control of pesticides, maintain food safety records, and fulfill legal requirements related to pesticide applications. Accurate records allow the managers to chart pest population densities and develop an understanding of historical trends related to abiotic and/or biotic factors in that given environment.

The implementation of IPM must be flexible, so that it can be tailored to individual situations. Managers can thereby develop appropriate packages that fit the specific needs of their operation in light of economics, availability, environment, socioeconomics, and other important factors within that location.

6.3.1   The Adoption of IPM

Education, economics, and societal perceptions influence the rate of adoption of any new technology (Buttel et al. 1990). In order to increase the rate of adoption for IPM, farmers must be satisfied that, when implemented properly, IPM does not increase risk in their operation. Like other groups of people, farmers make decisions and act on them basically to satisfy their own preferences (Coleman 1994). Unless farmers perceive a need to change from a system with preventative applications of pesticides being a frequent occurrence, adoption of IPM technology will be slow to occur.

In this next section, we will identify several key attributes that must be satisfied if IPM adoption is to proceed. These include: relative advantage of the technology, compatibility with current technologies, perceived complexity, ease of application, and observability.

Relative advantage — When attempts are made to promote implementation of new IPM systems or techniques, the relative advantage of the program should be readily documented and demonstrated to farmers. Especially important is demonstrating the relative advantage to limit the risks of erroneous decisions. Hutchins (1995) discussed IPM implementation and suggested that, if the technology has a relative advantage such as profit, it will achieve rapid, widespread adoption. The relative advantage, however, must be developed in context of farming decisions and local production practices. Hutchins described IPM as a business opportunity within the context of farming operations and stated that:

i. Value is the key concept of IPM, but it can only be defined by the user.
ii. If true value is present, it will be exploited in a business enterprise.
iii. IPM is input-dependent and all control tactics need to be considered based on their ability to add value to the pest management decision.
iv. Free enterprise must apply to agriculture as it does with any other business. In other words, there must be a competitive benefit for early adopters and a penalty for those who fail to adopt.

Compatibility — New pest management approaches must be compatible with existing production systems. Producers should be able to integrate their pest management program into their machinery management, harvest management, tillage, and other existing operations. Adoption will be minimal if the technology cannot be melded into these existing operations.

Perceived complexity — If producers perceive the technology and its integration as being too complex, they will not adopt. Scientists often present detailed information to potential adopters about the biology of the pest, the ecology of the system, and monitoring techniques used in developing the IPM system which makes it seem complex. The final product placed in the hands of producers must be designed as user-friendly as possible. IPM is replete with computer-assisted decision support systems that were not adopted extensively because they were too difficult to use. This attribute is negatively related to the rate of adoption; however, complex technologies can be "reinvented" and presented differently (i.e., as a series of steps) emphasizing the ease of application and observability. In addition, crop consultants can serve as liaisons for many new technologies, thereby increasing adoption of methods that farmers may be hesitant to incorporate into their present operation.

Ease of application and observability — Herbert (1995) and Steffey (1995) discussed how producers may often select different components that best fit their production systems. The components must be "displayed" in a way that convinces growers that: 1) the tactic works, and 2) it can be incorporated into their present system. An IPM system can be implemented over time, through increased exposure to new tools and familiarity with their practical use.

The Cooperative Extension Service has a rich tradition of enhancing adoption through demonstration. Observation is a key reason demonstrations are established and field days organized. To improve adoption, clientele gain a better appreciation and familiarity by physically observing the systems or approaches. Because of the

Table 6.7   Point System Assignment for IPM Implementation (Adapted in part from Jacobsen, 1997)

Value of Approaches for IPM		
0 Low	0.5 Medium	1.0 High
Preventative treatments Limited use of ecologically based approaches	Resistant varieties Scouting	Long-term planning Focus on ecologically-based approaches
	Uses consultant or CCA	Uses advanced disease Advisories
Uses CCA or applicator Cultural tactics of pest management used	Some integration of biological control Cultural and mechanical tactics of pest management used	Full integration of biological control Using cultural, mechanical and biological tactics of pest management where applicable

importance of observation, most agricultural and urban service companies use extensive demonstrations and clearly identify the activity/product by placing signs on variety trials, pesticide evaluations and machinery/tillage displays.

The typical pattern of adoption begins slowly. During initial attempts, early adopters experiment, then implementation increases through personal transfer of information (Table 6.7). Rapid transfer of technology occurs when most of the five key requirements are met and a group of key community leaders is willing to implement a new technique (Seyum 1997, Hamilton et al. 1997).

6.3.2   Measuring Integrated Pest Management

Several efforts have characterized or measured IPM adoption ranging from simple to complex. Some key reasons for measuring IPM adoption include:

* Better targeting of research needs and educational delivery.
* Documenting progress for commodity, state, and federal funding support.
* Data support for quantifying and qualifying federal and state cost-share programs.
* Quantifying economic return for use in marketing efforts.

Many authors have examined adoption of IPM on a continuum from low to high levels (Kogan 1998, Benbrook et al. 1996, Sinclair et al. 1997, Hamilton et al. 1997, Lambur et al. 1985) (Figure 6.3) with different opinions on where a farmer's program should be placed within that continuum. Vandeman et al. (1994) showed that adoption of IPM is occurring with over 10% of producers in the U.S. using at least one IPM practice in their production system. This percentage was much greater in highly managed crops (e.g., grapes, oranges and almonds). Other authors using more rigorous criteria for defining IPM adoption, showed only 4 to 8% of farmers were using IPM at a high level of implementation in their production systems (Benbrook et al. 1996).

While the previous authors suggest that adoption of IPM, at some level, is occurring in U.S. agriculture, others contend that long-term solutions to pest problems must involve more than single-tactic management schemes (Cate and Hinkle

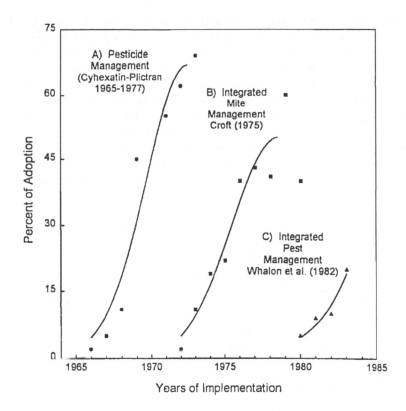

FIGURE 6.3 Progression of typical IPM adoption in an apple system, as integration intensity increases, Michigan 1965–1984 (Lambur et al., 1993). (A) Single-facet pest management of mites using Plictran from 1965–1977; (B) Multifaceted management of mites using a few IPM strategies; (C) Labor-intensive integrated pest management of the apple production system for mite control, using several IPM strategies.

1994). This holistic approach to IPM emphasizes management of the ecological unit and enhancement of natural and biological control. The ecological unit may encompass a large geographic area, including the non-crop ecosystem and surrounding landscape. Area-wide suppression and/or eradication programs have attempted this approach; however, politics and ideologies often run the program, rather than ecological, biological, and production characteristics (Cate and Hinkle 1994). Implementation of area-wide IPM programs would allow large-scale use of *enhancement practices* (resistant plants, biological control, cultural practices, genetic controls, mating disruption, trapping, etc.) that would accentuate natural control of pests, and more prudent use of *intervention practices* (pesticides, early harvest, tillage, etc.) when pest populations exceed economic thresholds.

Paramount to any discussion on implementing IPM is recognition that its complexity can delay adoption (Lincoln and Parencia 1977, Pimentel 1982, Reichelderfer et al. 1984). Therefore, as IPM programs are developed, they should be constructed within the framework of practical farm operations (Doutt and Smith 1971). Recently, scientists in Georgia have developed such a program that utilizes several aspects of a peanut farmer's operation (planting date, soil type, variety selection, etc.) in developing a risk index for tomato spotted wilt virus in a given area (Brown et al. 1998). Similarly, Damicone et al. (1994) developed an advisory for management of early leaf spot in peanut that utilizes weather-based indices in triggering treatment suggestions. Use of these indices may ultimately provide consultants and farmers with useful thresholds for plant pathogens, an area that has historically been lacking in IPM programs. In addition, development of these thresholds could cause modifications of arthropod thresholds where the animal vectors the disease.

Regardless of our location on the continuum of IPM evolution, nearly every supporter of such programs agrees that specific criteria should be met and questions answered before widespread use of IPM practices can be implemented. These criteria include:

1) A clear definition of IPM by the administration, within the context of the ecological unit described by Cate & Hinkle (1994).
2) Incentives to use IPM or the perception that economic benefits of the program justify the increased demands on management (Sorenson 1993).
3) Sustained funding and resources for IPM programs at the implementation level with teams of scientists, growers, consultants, and industry representatives.
4) Increased knowledge of practical means for using IPM alternatives and support for the growing industry of consultants that will deliver the services.
5) Decreased impact of an EPA regulatory process that is burdensome, time-consuming, expensive, and unclear (Sorenson 1993) while educating the agency about the need for a full toolbox of chemicals to choose from in developing and refining IPM programs (Schrimpf 1998).

The simplest and most direct measurements of adoption focus on individual tactics such as the use of resistant varieties or scouting. Several authors characterize this as level I or low-level IPM (Benbrook et al. 1996, Kogan 1998, Zalom et al. 1990). While adoption of single tactics is easily measured, the results may not give a true representation of the extent of IPM programs being employed. Herbert (1995) indicated producers readily adopt single tactics and over years may build an IPM system tailored to their local needs and production practices. Unfortunately, they may not actually identify themselves as practicing IPM because a clear, concise definition of IPM adoption levels by the government is lacking.

Several authors examined, through multiple factor analysis, the frequency with which IPM programs include features such as scouting, host plant resistance, and some integration of biological control (Teetes et al. 1997, Benbrook et al. 1996). IPM unites numerous cultural, biological and management factors into a system that can weigh management practices into an index of adoption. This approach allows farmers to rate the extent of their IPM adoption on a continuum from low to high (Hollingsworth et al. 1996, Benbrook et al. 1996). A system for rating IPM adoption in commercial wheat elevators is depicted in Table 6.8. This approach more appropriately defines IPM as a system and allows improved flexibility in targeting research and education programs. If such a system is to become widely utilized, each component must be clearly defined and scaled, and end-users educated about the interactions between each facet of implementation.

Table 6.8 Integrated Pest Management Criteria for Commercial Wheat Elevators in the United States (Cuperus et al., 1997)

Factor	Points
Sanitation	120
Weekly sanitation schedule	10
Grass mowed	10
Rodent traps in place	10
Facility bird- and rodent-proofed	20
Loading	100
Check loads received for moisture	10
Check loads received for dockage	10
Check loads received for insects	10
Segregate based on moisture	10
Aeration	130
Aeration fans present	20
Thermocouples present and maintained	20
Air flow rate known for all fans	10
Monitoring	90
Check grain temperatures once/week	20
Check insects at least monthly	20
Check for roof leaks	10
Use insect traps	10
Pesticides/safety	120
Respiratory equipment present and used	20
Bonus points	110
IPM training at the elevator	20
Emergency Action Plan in place and used	10
Written preventative maintenance plan in place	10
Total possible points	670
Basic IPM elevator requires 70% of total points	472
Advanced IPM systems 80% of total points	536

6.3.3 Implementation of IPM — Case Histories

6.3.3.1 Alfalfa Integrated Management (AIM)

Alfalfa is grown on 140,000 to 240,000 hectares in Oklahoma and most is sold for dairy, horse, and beef production (Ward et al. 1994). Climatic factors, pests including insects (alfalfa weevil, aphids, cutworms), diseases (especially soil-borne pathogens), and weeds make alfalfa an expensive and risky crop to grow. Oklahoma State University has an excellent interdisciplinary research program, but the interface between research, extension, and farmers was not effective in promoting adoption of IPM (Stark et al. 1990). Farmers were not growing improved pest-resistant varieties, nor were they using improved monitoring systems for optimal timing of insecticide applications. In 1983, an initiative began with the Oklahoma Alfalfa Hay and Seed Association to develop an implementation program that focused on linking research, extension, and farmers to current, research-based information. The first

Table 6.9   Annual Operating Budget for Established Alfalfa in Oklahoma (Enterprise Budgets 1994)

Operating Inputs	Units	Single Unit Price (U.S. Dollars)	Quantity	Total Value – Single Unit Price × Quantity (U.S. Dollars)
Phosphorous (P_2O_5)*	kg/ha	0.15	71.4	10.71
Potassium (K_2O)	kg/ha	0.12	35.7	4.28
Insecticide — Parathion	ha	3.03	3 (# of applications per year)	9.10
Swathe and Bale	Mg	22.78	4.08	92.94
Custom hauling	Bales.	0.40	128	51.20
Implement Rental — Fertilizer Spreader	ha	0.91	1 trip	0.91
Establishment prorated**	ha	36.30	.2	7.26
Annual operating capital	DOLLARS/ha	8.5%	5.48	0.47
Machinery labor	HRS/ha	5.50/hr	.140	0.77
Machinery fuel, repairs		1.10		DOLLARS/ha
Total operating costs				$178.74/ha

* Phosphorous and Potassium amounts are actual pounds of material.
** Seed allocated over 5 years.

approach was to examine a typical operating budget for alfalfa (Table 6.9) that showed significant costs in insect management (10%). More importantly it showed:

• Over 50% of the budget was related to machinery.
• Stand longevity was critical to extended profitability (Stark et al. 1992, Stark et al. 1990).
• Nearly 50% of the alfalfa is sold for dairy and horse feed; therefore, marketing was a significant factor in pest management decisions.

The Alfalfa Integrated Management (AIM) team has organized and developed programs jointly with producers over the past 15 years, focusing on field demonstrations, helping producers set priorities of management, and balancing inputs with profitability. The results were impressive, with rapid adoption of improved varieties from 5 to nearly 75% (Stark et al. 1990) and a great increase in adoption of scouting and decision-making protocols for insecticide use against key insect pests (Mulder et al. unpublished). The apparent reasons for rapid adoption included:

• Farmers had ownership and investment of the educational program.
• Farmers could see the long-term advantage of adoption. Planting improved varieties and increased stand longevity by an average of one year (Stark et al. 1990). Adoption of improved scouting and decision-making protocols for insecticide use reduced inputs and increased net return.
• Most IPM measures adopted were readily compatible with ongoing farm activities. Improved varieties required no changes. Improved monitoring increased time and energy devoted to management, but compensated for the extra effort by reducing risk of economic damage by proper timing of insecticide applications.

6.3.3.2 *Stored Product Research and Education*

In Oklahoma, when stored-product IPM research and extension work began in the mid-1980s, the state was losing millions of dollars from insect damage, grain shrinkage, excessive nonproductive pesticide inputs, and wasted electrical costs (Cuperus et al. 1990, Reed et al. 1993). Pesticide inputs were extensive, control failures were common, and worker safety concerns were high. Surveys completed in 1990 showed that 75% of personnel using fumigants were not observing proper safety procedures. In addition, insect populations with resistance were causing control failures, and illegal residues in stored grain were common (Zettler and Cuperus 1990). Farm storage and elevator management costs were high, which limited potential profit. The financial viability of the entire grain-trade storage and distribution system in Oklahoma was threatened by these factors.

In response to this crisis, research and extension teams developed management strategies based on the ecology of the system, which is regulated primarily by temperature. Manipulating grain temperature using aeration in conjunction with improved monitoring, sanitation, and grain management, has been effective in significantly reducing or eliminating use of pesticides entirely. Information leading to adoption of IPM was delivered through grain management workshops in which over 2500 registrants attended demonstrations, short safety courses, and hands-on fumigation workshops.

The impact of these programs has resulted in a significant change in manager philosophy from preventative treatments to predictive managed systems. Results of research and education programs lead to a reduction of fumigant use, with a decrease from 2.6 fumigations/yr in 1986 to less that 1.2 applications currently (Cuperus et al. 1990, Kenkel et al. 1994).

Why were these technologies adopted?

* The program was designed and funded by farmers' associations and elevator representatives to make sure the approach was targeted and relevant.
* Relative advantage was demonstrated with a reduction of inputs and risk.
* Worker safety improved and profitability increased. Because this program focused on elevator facilities and ongoing management, the approach was practical for those in the business.
* Because the approach was similar to ongoing practices, it could be easily repackaged and not perceived to be complex.
* Elevator operators and producers were encouraged to participate in hands-on workshops.
* Impacts were readily observed in reduced pest levels, increased sanitation levels and decreased reliance on insecticides.

6.3.4 Marketing an IPM System

One question that is paramount in discussions with farmers and processors: how can they recoup investments in IPM? They must clearly see an economic incentive for adoption to occur. Several publications have addressed key incentives to adoption (Benbrook et al. 1996, Hutchins 1995, Herbert 1995) and usually focus on a few key elements to implementation. Key elements to improve marketing of IPM concepts should include:

    i. Focusing on economics (Hutchins 1995).
    ii. Emphasizing incentives, including monetary and other incentives (e.g., IPM producer of the year).
    iii. Improving communication to clearly define IPM programs.
    iv. Characterizing intended user groups.
    v. Dispelling the idea that IPM adoption is going to increase risk of significant losses due to pest damage.
    vi. Maintaining funding for research and education.
    vii. Understanding producer expectations.
    viii. Promoting privatization of consultant enterprises that utilize an IPM approach and strengthening their partnership with academia. Most producers and urbanites already obtain the majority of their information from secondary providers as opposed to academia (Cuperus & Berberet 1994).
    ix. Allowing retail growers to promote IPM-grown produce.
    x. Minimizing the influence of regulatory policies on IPM implementation, development and refinement. This would allow for a complete set of tools to choose from in constructing IPM programs.
    (Herbert 1995, Steffey 1995, Benbrook et al. 1996, Schrimpf 1998)

6.4   IPM AND THE FUTURE

6.4.1   Case History — The Corn/Soybean Production System

Corn and soybean are grown on over 65 million acres in the U.S. Both crops have proven to be very compatible for developing a crop rotation system that effectively reduces several insect pests, plant pathogens, and weeds. Key pests of corn that are targets of IPM implementation include the European corn borer (ECB), *Ostrinia nubilalis* Hübner, and the corn rootworm complex consisting of the NCR, and WCR. These pests have traditionally been managed using strategies that include host plant resistance, crop rotation, tillage, and chemical control. In recent years, there have been increasing challenges to the effectiveness of this system. The development of new pest problems and the introduction of new technologies for managing these pests has contributed to these challenges. The movement of these species into soybeans presents new challenges for developing comprehensive pest management strategies.

The ECB was introduced into the U.S. in 1917. By the 1950s, it had become established as a major pest in all areas of the corn belt. Damage on corn by ECB tunneling into stalks results in weakened plants and allows secondary fungal infections to further degrade stalks, resulting in lodged plants and loss of ears at harvest. Yield is also reduced because tunneling disrupts nutrient flow through the plant. In general, ECB can cause yield reductions of 5 to 15%, depending upon the infestation level.

A number of strategies exist for managing ECB, but the major ones are focused in two areas: host plant resistance and chemical control. Through the years, corn varieties have been developed that resist infestation or tolerate physical injury from ECB tunneling. Some varieties have been selected that contain levels of a plant hormone called DIMBOA, which is toxic to early instars when they attempt to feed on resistant plants (Scott et al. 1964). This hormone is present in high levels in plants <40 cm in height, but as the plant grows, DIMBOA becomes diffused within the plant tissues, and is no longer present at high enough levels to kill ECB larvae. Corn varieties with tolerance to ECB were also developed to sustain high yields by withstanding tunneling long enough to be harvested before lodging occurs.

There are numerous chemical insecticides available for control of ECB, including various formulations of the microbial agent *Bacillus thuringiensis* (*B.t.*). Although not commonly used as an insecticide preparation, *B.t.* has become a powerful tool for managing ECB through the development of bioengineered corn varieties that have the capability to produce the *B.t.* toxin within their tissues. Concern is now focused on how to manage this new tool to reduce selection for strains of ECB that are resistant to the *B.t.* toxin. Several strategies have been proposed in managing this technology to limit resistance of ECB; however, the majority of such plans are voluntary (EPA 1998). A failure to adopt resistance management plans could allow resistance to develop and create more pressure from regulatory agencies to limit the use of transgenic *B.t.* in other high-value crops (Ostlie et al. 1997). The North Central Regional technical committee on "Ecology and Management of the ECB and other Stalk-Boring Lepidoptera" (NC-205) have outlined a 5-step approach in implementing a resistance management plan for *B.t.* corn hybrids (Ostlie et al. 1997). This plan calls for monitoring and scouting of both B.t. and non-*B.t.* corn, planting 20 to 40% non-*B.t.* hybrids as a primary refuge for ECB larval populations and continuing to use an IPM approach for all pests within the local agro-ecosystem. If this tool is to remain effective for the long term, comprehensive management strategies focused on using *B.t.* corn in a sustainable way must be developed and used.

The corn rootworm complex, consisting of the NCR, WCR, and the spotted cucumber beetle (SCB), *Diabrotica undecimpunctata howardi* Barber, have been and continue to be the recipient of intense management. The NCR and WCR overwinter as eggs in the soil. They hatch the following spring, and feed on corn roots. They feed on a very limited range of host plants and basically depend upon corn for survival. For over 50 years, both species could be managed by crop rotation with soybeans, which was not a host for NCR and WCR larvae. In recent years, both species have developed a means of surviving in a corn/soybean rotation. Eggs of the NCR are capable of surviving for 2 years in the soil (Krysan et al. 1986) and then hatch predominantly in those years when corn is planted in the rotation. The WCR adult females have begun to lay eggs in soybean fields, which overwinter and hatch the following year where the same field is planted to corn (Levine, 1995). While these biological traits have not yet spread through the entire population, they are causing serious problems in parts of the corn belt.

In Nebraska, where continuous corn is grown over large geographic areas, the WCR can be managed through insecticide applications directed at control of the larvae, or adults. Chemical control has met with some environmental and social problems throughout the years. Problems associated with control of larvae using granular insecticide applications include issues of health hazards to the handlers and applicators, killing of non-target animals, as well as environmental and groundwater contamination. Insecticide control directed at adult beetles has resulted in the selection of adult populations that are resistant to several insecticides (Meinke et al. 1998). Despite the occurrence of the abberations, there are current attempts to evaluate the methodology needed to manage corn rootworms on an area-wide basis (Gray 1995b). This presents an additional set of challenges. The future of corn rootworm management will require more sophisticated strategies that combine several tactics that are compatible with the total production system.

6.4.2   New Tools — New Challenges

The future of IPM will be an exciting and challenging era with tremendous changes in technology and the potential of decreased resources for research and education. This technological era with Internet capabilities provides tremendous opportunities to access information quickly, but it also presents challenges in reaching audiences appropriately.

- Precision agriculture is defined as the utilization of key technologies, including global positioning systems (GPS), geographic information systems (GIS), soil mapping and yield monitoring, into a system that applies inputs only when necessary and in the appropriate amount (MacRae 1997). Precision farming packages are rapidly being developed and implemented. This approach allows farmers to optimize economic returns and minimize environmental impacts by maintaining better control of input costs. Present precision agriculture systems include components that manage nutrients, map weeds, and provide decision support for insect management (MacRae 1997, Fitzner 1996). Other researchers have integrated GIS/GPS technologies with economic and environmental models to better target IPM programs and to elucidate the tradeoffs that could occur (Tian et al. 1996). By integrating models and GIS technologies, Tian et al. (1996) found that approximately 10% of the cotton acreage that could be involved in a boll weevil eradication program would pose a problem with environmental standards for management of water quality and other issues.
- Area-wide approaches to IPM have been advocated for nearly 30 years. Early thoughts focused on area-wide applications of insecticides, but recent applications are attempting to utilize ecologically-based techniques such as mating disruption to control oriental fruit moth, grape berry moth, and peachtree borer in fruit trees, or sanitation and aeration to control insects in stored grain systems. The concept of an area-wide approach in suppressing pests through a unified community effort over large geographic areas, coordinated by organizations, and focused on population suppression, was traditionally developed as a mandatory participation program (Kogan 1998). Efforts to date include: the codling moth (CM), *Cydia pomonella* L., in the northwestern United States; the boll weevil, *Anthonomus grandis grandis* Boheman, in the cotton belt; leafy spurge in the northwest; corn rootworm in the midwest; and stored grain pests throughout the plains states.

- Biotechnology has made an impact on agricultural production systems and will make even greater contributions in the future. Transgenic plants with resistance to insects and diseases are becoming widely available. In 1998, up to 20 million acres were planted to genetically modified corn that contains the gene for producing the *B.t.* endotoxin. Some genetically engineered crops are designed specifically to encourage pesticide use such as herbicide-tolerant soybeans that can withstand broadcast application of Roundup®.* The greatest challenges to biotechnology include: further understanding of cost:benefit relationships; consumer acceptance issues of resistance management; and compatibility of these technologies with comprehensive IPM programs. All of these challenges will eventually be integrated into new pest management systems that may someday emphasize prescription production and/or application.
- Food safety — The Food Quality Protection Act (FQPA 1996) will force farmers and IPM consultants to examine differing pesticide toxicities, re-educate themselves and clientele on use of pesticides, and refocus on the total food production, processing, and distribution systems from a food safety perspective. Producers, processors, grocers, and consumers are all involved in the food distribution system, and each has a vital role in understanding perceptions and practices regarding food safety. IPM users must link key food safety components and integrate IPM into the food distribution system. This new legislation (FQPA) will help link food systems together promoting the concept of IPM, provided that it does not encourage the elimination of several useful tools (chemicals) for developing such programs.
- The concept of using IPM to sell produce and vice versa has been explored by many authors (Bruhn et al. 1992, Anderson 1993, Burgess 1989, Burgess et al. 1990, Collins et al. 1992, Hollingsworth 1994, Hollingsworth et al. 1993, Anderson et al. 1996). By using labels and other active marketing strategies (newspapers, brochures, etc.), awareness and acceptance of IPM by consumers is enhanced because they believe it will have positive consequences for human health and environmental stewardship (Anderson et al. 1996). Similarly, IPM certification can help farmers build consumer demand for products grown with these methods and ultimately can allow growers the opportunity to obtain a premium economic return for investing in IPM methodologies (Burgess 1989, Hollingsworth et al. 1993). The marketing of IPM-grown products to consumers has begun in the U.S. and Europe. Throughout supermarkets in the northeastern U.S., a grocery chain, Wegman's, offers fresh produce that has an IPM label. The labeling criteria were developed and licensed through Cornell University (Benbrook et al. 1996). Presently, the economic return from these attempts is under scrutiny. In Europe, "eco" labels have captured over 4% of the market share (Matteson 1995, 1996). While these recent efforts have captured national and international attention, markets continue to change significantly both in field crops (Ward et al. 1994) and in horticultural crops (Suter 1995).
- Biotechnological advances will play a major role in future IPM systems and allow the development of host plants containing high levels of resistance to many pests. Stewardship of these techniques will be a priority because of the large investment made in developing the technology.
- Social issues will continue to be important to the future of IPM. Society's concern for the structure of agriculture, the safety of the food supply, and the preservation

* Registered trademark of Monsanto Co., Life Sciences/Ag Sector, St. Louis, MO.

of endangered wildlife will direct major changes in the future of IPM. IPM must be viewed as a sound investment for preserving or protecting natural resources. Traditionally, IPM represented a reaction to crises often created by traditional pest control approaches. This thought process must change if IPM is to move forward into public awareness and acceptance. With less than 2% of the U.S. population directly involved with commercial agriculture, IPM's future will depend upon how well communication flows between the scientific community, consultants, producers, the public and decision-makers. If IPM remains focused narrowly in agriculture, two situations could develop. First, this focus could create an atmosphere for increased governmental regulation of production and farming practices. Second, the entire concept of IPM will become totally privatized with little to no public investment or support. In contrast, if IPM is developed with a common understanding as an investment, then asking for implementation investment and adoption by farmers and subsequent use by consumers is likely to occur.

REFERENCES

Anderson, M.D. 1993. Pesticides and their alternatives. Perspectives of New England vegetable grower. Summary report. School of Nutrition, Tufts Univ. Medford, MA.

Anderson, M. 1994. Agricultural Resources and Environmental Indicators. Agricultural Handbook Number 705. Economic Research Service, U.S. Department of Agriculture, Washington, D.C.

Anderson, M.D., C.S. Hollingsworth, V. VanZee, W.M. Coli, and M. Rhodes. 1996. Consumer response to integrated pest management and certification. *Agriculture, Ecosystems and Environment* 60:97-106. Elsevier Science.

Aspelin, A.L. 1997. Pesticide industry sales and usage. 733-R-97-002. U.S. Envir. Prot. Agency. Washington, D.C. 20460.

Beers, E.H., J.F. Brunner, M.J. Willett, and G.M. Warner. 1993. Orchard pest management. A resource book for the Pacific Northwest. Good Fruit Grower, Washington State Fruit Commission. Yakima, WA. 276 pp.

Beingolea, O. 1981. Economic constraints to the development and implementation of integrated pest control. In *Conf. Future trends integrated pest management, Belagio, Italy.* London Cent. Overseas Pest Res. pp. 50-54.

Benbrook, C.M., E. Groth III, J.M. Harroran, M.K. Hansen, and S. Marquardt. 1996. *Pest Management at the Crossroads.* Consumers Union. Yonkers, NY.

Brown, S.L., J.W. Todd, A.K. Culbreath, H. Pappu, J. Baldwin, and J. Beasley. 1998. Tomato spotted wilt of peanut: Identifying and avoiding high-risk situations. University of Georgia, Cooperative Extension Service Bulletin 1165. 10 pp.

Bruhn, C., S. Peterson, P. Phillips, and N. Sakovidh. 1992. Consumer response to information in integrated pest management. *J. Food Safety* 12:315-326.

Buhler, D.D., J.L. Gunsolus, and D.F. Ralson. 1992. Integrated weed management techniques to reduce herbicide inputs in soybean. *Agron. J.* 84:973-978.

Burgess, R.B. 1989. IPM market survey final report. Finger Lakes Research. Victor, N.Y.

Burgess, R., J. Kovach, C. Petzoldt, A. Shelton, and J. Tette. 1990. Results of IPM marketing survey. In *Reports from the 1989 IPM Research, development and implementation projects in vegetables.* IPM publication 109. N.Y. State IPM program. N.Y. State Ag. Exp. Sta. Geneva, N.Y.

Burns, A.J., T.H. Coaker, and P.C. Jepson. 1987. *Integrated pest management.* Academic Press. London. 304 pp.

Buttel, F.H., G.W. Gillespie, Jr., and A. Power. 1990. Sociological aspects of agricultural sustainability in the United States: a New York case study. In *Sustainable Agricultural Systems*. Soil and Water Conservation Society. Ankeny, IA. pp. 515-532.

Caddel, J.L., J.F. Stritzke, P.G. Mulder, R.L. Huhnke, R.C. Berberet, and C.E. Ward. 1995. Alfalfa harvest management — discussions with cost-benefit analysis. Oklahoma State Univ. Oklahoma Coop. Ext. Serv. Circ. E-943.

Caddel, J.L., J.F. Stritzke, P.G. Mulder, G.V. Johnson, C.E. Ward, R.L. Huhnke, and R.C.Berberet. 1996. Alfalfa stand establishment, questions and answers. Oklahoma State Univ. Oklahoma Coop. Ext. Serv. Circ. E-949.

Carson, R. 1962. *Silent Spring*. Fawcett Crest, New York.

Cate, J.R. and M.K. Hinkle. 1994. *Integrated pest management: the path of a paradigm*. National Audubon Society. Washington, D.C.

Chant, D.A. 1964. Strategy and tactics of insect control. *Canadian Entomol*. 96:182-201.

Coble, H.D., A.R. Bonanno, B. McGaughey, G.A. Purvis, and F.G. Zalom. 1998. Feasibility of prescription pesticide use in the United States. *Pesticide & Tox. Chem. News*. Vol 26 No 51. Oct. 1998.

Coleman, J.F. 1994. A rational choice perspective on economic sociology. In *The Handbook of Economic Sociology*. Smola, N.J. and R. Swedberg, Eds., Princeton Univ. Press, Princeton.

Collins, J.K., G.W. Cuperus, B.O. Cartwright, J.A. Stark, and L.L. Ebro. 1992. Consumer attitudes on production systems for fresh produce. *J. of Sust. Agric*. 3:456-466.

Collyer, E. 1953. Insect population balance and chemical control of pests. In *Chem & Ind*. pp. 1044-1046.

Croft, B.A. 1975. Integrated control of apple mites. Mich. State Univ. Ext. Serv. Bull. E-825.

Cuperus, G.W. and R.C. Berberet. 1994. Training specialists in sampling procedures. In *Handbook of sampling methods for arthropods in agriculture*. Pedigo, L.P. and G.D. Bunting, Eds. CRC Press, Boca Raton.

Cuperus, G.W., G. Johnson, and W.P. Morrison. 1992. IPM in wheat and stored grain. in *Successful Implementation of Integrated Pest Management for Agricultural Crops*. Leslie, A. and G.W. Cuperus, Eds. Lewis Publishers, Boca Raton. pp. 33-55.

Cuperus, G., R. Berberet, and P. Kenkel. 1997. The Future of integrated pest management. In *Integrated Pest Management*. University of Minnesota, St. Paul, MN.

Cuperus, G.W., R.T. Noyes, W.S. Fargo, B.L. Clary, D.C. Arnold, and K. Anderson. 1990. Successful management of high-risk stored wheat system in Oklahoma. *Amer. Entomol*. 36:129-134.

Cuperus, G.W., P. Kendall, S. Rehe, S. Sachs, R. Frisbie, K. Hall, C. Bruhn, H. Deer, F. Woods, B. Branthaver, G. Weber, B. Poli, D. Buege, M. Linker, E. Andress, W. Wintersteen, F. Dost, J. Damicone, D. Herzfeld, J. Collins, B. Cartwright, and C.C McNeal. 1991. Integration of food safety and water quality concepts throughout the food production, processing, and distribution educational programs: using Hazard Analysis Critical Control Point (HACCP) philosophies. Okla. Coop. Ext. Serv. Circ. E-903.

Damicone, J.P., K.E. Jackson, J.R. Sholar, and M.S. Gregory. 1994. Evaluation of a weather-based spray advisory for management of early leaf spot of peanut in Oklahoma. *Peanut Sci*. 21:115-121.

Doutt, R.L. and R.F. Smith. 1971. The pesticide syndrome-diagnosis and suggested prophylaxis. In *Biological control*. Huffaker, C.B., Ed. Plenum Press, NY. 511 pp.

Dowdy, A.K., R.C. Berberet, J.F. Stritzke, J.L. Caddel, and R.W. McNew. 1992. Late fall harvest, winter grazing, and weed control for reduction of alfalfa weevil (Coleoptera:Curculionidae) populations. *J. Econ. Entomol*. 85: 1946-1953.

Enterprise budgets. 1994. Dept. of Agric. Econ. Okla. Coop. Ext. Serv. Stillwater, OK.

Environmental Protection Agency. 1998. Final Report of the FIFRA Scientific Advisory Panel. Subpanel on *Bacillus thuringiensis* Plant-Pesticides and Resistance Management.

Fitzner, M. 1996. Emerging issues influencing integrated pest management. In *Proceedings of the Third National IPM Symposium/Workshop*. United States Dept. of Agric. Misc. Publ. No. 1542. pp. 169-180.

Flinn, P., D. Hagstrum, and L. Song. 1995. Stored grain advisor. Coop. Ext. Serv., Kansas State Univ. S-86.

Food Quality Protection Act. 1996. 104th Congress. House of Representatives. Report 104.

Franz, J. 1961. The ecological effect of the control of insects by means of viruses and/or bacteria as compared with chemical control. *I.U.C.N. Symposium,* Warsaw, July 1960. pp. 93-105.

French, R.A. 1982. Diffusion of integrated pest management (IPM) innovation in New Zealand from researcher to farmer. In *Proc. Australos. workshop dev. Implementation IPM, Auckland.* Auckland, New Zealand. Gov. Print. pp. 52-66.

Frisbie, R.E. and P.L. Adkisson 1985. IPM: Definitions and current status in U.S. agriculture, In *Biological control in Agricultural IPM Systems.* Hoy, M.A. and D.C. Herzog, Eds. Academic. New York. pp. 41-51.

Gray, M.E. 1995a. Status of CES — IPM Programs: Results of a National IPM Coordinators Survey. *Amer. Entomol.* 41:136-138.

Gray, M.E. 1995b. Area wide pest management for corn rootworms: fantasy or realistic expectations? In *Proceedings Ill. Agric. Pest. Conf.* Coop. Ext. Serv., Univ. of Ill. at Urbana-Champaign. pp. 101-106.

Grison, P. 1962. Development et perspectives de la lutte biologique. *Entomophaga* 7: 325-335.

Hamilton, G.C., M.G. Robson, G.M. Ghidiu, R. Samulis, and E. Prostko. 1997. 75% adoption of integrated pest management by 2000? A case study from New Jersey. *Amer. Entomol.* 43:74-78.

Harris, M.K., Millar, J.G., and A.E. Knutson. 1997. Pecan nut casebearer (Lepidoptera: Pyralidae) sex pheromone used to monitor phenology and estimate effective range of traps. *J. Econ. Entomol.* 90:983-987.

Herbert, D.A., Jr. 1995. Integrated pest management systems: Back to basics to overcome adoption obstacles. *J. Agric. Entomol.* 12:203-210.

Higley, L.G. 1992. New understandings of soybean defoliation and their implications for pest management. In *Pest management in soybean.* Copping, L.G., M.B. Green, and R.T. Rees, Eds. Elsevier, London.

Higley, L.G. 1994. Insect injury to soybean. In *Handbook of Soybean Insect Pests.* Higley, L.G. and J. Boethel, Eds. Entomol. Soc. of Amer. Pub. pp. 11-13.

Higley, L.G. and W. Wintersteen. 1992. A new approach to environmental risk assessment of pesticides as a basis for incorporating environmental costs into economic injury levels. *Amer. Entomol.* 38:34-39.

Hollingsworth, C.S. 1994. Integrated pest management certification: A sign by the road. *Am. Entomol.* 40:74-75.

Hollingsworth, C.S., W.M. Coli, and R. Hazzard. 1996. Integrated pest management Massachusetts Guidelines: Commodity specific definitions. Univ. of Mass. AG-1206.

Hussey, N.W. 1985. Biological control — A commercial evaluation. *Biocontrol News Inf.* 6:93-99.

Hutchins, S. 1995. Free enterprise: the only sustainable solution to IPM implementation. *J. Agric. Entomol.* 12:211-217.

Jacobsen, B. 1997. Role of plant pathology in integrated pest management. *Ann. Rev. of Phytopath.* 35:373-391.

Kenkel, P., J.T. Criswell, G. Cuperus, R. Noyes, K. Anderson, and W. Fargo. 1994. Stored product integrated pest management. *Food Reviews International* 10:177-193.

Kogan, M. 1998. Integrated pest management: Historical perspectives and contemporary developments. *Ann. Rev. Entomol.* 43:243-70.

Kovach, J., C. Petzoldt, J. Degni, and J. Tette. 1992. A method to measure the environmental impact of pesticides. New York Food and Life Sciences Bulletin No. 139. New York Agric. Exp. Sta., Geneva, NY.

Krysan, J.L., D.E. Foster, T.F. Branson, K.R. Ostlie, and W.S. Cranshaw. 1986. Two years before the hatch: rootworms adapt to crop rotation. *Bull. Entomol. Soc. Am.* 32:250-253.

Lambur, M.T., M. Whalon, and F.E. Fear. 1985. Diffusion theory and integrated pest management: illustrations from Michigan fruit IPM program. *Bull. Entomol. Soc. Am.* 31(3):40-46.

Levine, E. 1995. Rootworm problems in first year corn: an increasing problem? In *Proc. Ill. Agric. Pest. Conf.* Coop. Ext. Serv., Univ. of Illinois at Urbana-Champaign. pp. 133-135.

Liebman, M., C.L. Mohler, and C.P. Staver. 1996. *Ecological management of agricultural weeds.* Cambridge University Press.

Lin, B.H., M. Padgitt, L. Bull, H. Delvo, D. Shank, and H. Taylor. 1995. Pesticide and fertilizer use and trends in U.S. Agriculture. Agricultural Economics report No. 717. Economic Research Service, U.S. Department of Agriculture, Washington, D.C.

Lincoln, C. and C.R. Parencia. 1977. Insect pest management in perspective. *Bull. Entomol. Soc. Am.* 23:9-14.

MacRae, I.V. 1997. Site Specific IPM. In *Integrated Pest Management.* University of Minnesota, St. Paul, MN.

Matteson, P.C. 1995. The 50% pesticide cuts in Europe: A glimpse of our future. *Amer. Entomol.* 41:210-220.

Matteson, P.C. 1996. Green labels in the Netherlands: careful negotiations and clearer choices *Pest Camp.* 6:3-6.

Mayerfeld, D.B., G.R. Hallberg, G.A. Millar, W.K. Wintersteen, R.G. Hartzler, S.S. Brown, M.D. Duffy, and J.R. DeWitt. 1996. Pest management in Iowa: Planning for the future. Integrated farm management report number 17, Iowa State University Cooperative Extension. June 1996.

Meinke, L.J., B.D. Siegfried, R.J. Wright, and L.D. Chandler. 1998. Adult susceptibility of Nebraska western corn rootworm (Coleoptera:Chrysomelidae) populations to selected insecticides. *J. Econ. Entomol.* 91:594-600.

National Coalition on Integrated Pest Management. 1994. Toward a goal of 75 percent cropland under IPM by 2000. Jan., Austin, TX.

National Research Council, Board on Agriculture. 1989. Alternative agriculture. Nat. Acad. Press, Washington, D.C.

National Research Council. 1996. *Ecologically-based Pest Management: New Solutions for a New Century.* Nat. Acad. Press, Washington, D.C.

Osteen, C. and P. Szmedra. 1989. Agricultural pesticide use trends and policy issues. United States Dept. of Agric. Econ. Res. Serv. Rept. No. 622.

Ostlie, K.R., W.D. Hutchison, and R.L. Hellmich. 1997. *B.t.* corn & European corn borer — long-term success through resistance management. NCR Pub. 602. Univ. of Minn. St. Paul, MN.

Owen, G. 1996. The grocer and the potential industry development. In *Proc. 15th Annual Horticultural Industry Show.* Okla. Coop. Ext. Serv., Stillwater, OK. pp. 15-16.

Painter, R.H. 1951. *Insect resistance in crop plants.* Macmillan, New York. 520 pp.

Pedigo, L. and L. Higley. 1992. The economic injury level concept and environmental quality: A new perspective. *Amer. Entomol.* 38: 12-21.

Pedigo, L.P. 1995. Closing the gap between IPM theory and practice. *J. Agric. Entomol.* 12:171-181.

Pedigo, L.P., S.H. Hutchins, and L.G. Higley. 1986. Economic injury levels in theory and practice. *Annu. Rev. Entomol.* 31:341-368.

Petty, H.B. 1973. An applied insect pest management program and population threshold. In *25th Illinois Operator Spray School Summary of Presentations, 24-25 January, 1973.* Univ. of Ill. Coop. Ext. Serv., Urbana-Champaign. pp. 66-68.

Pickett, A.D., W.L. Putman, and E.J. LeRoux. 1958. Progress in harmonizing biological and chemical control of orchard pests in eastern Canada. In *Proc. 10th Int. Congr. Entomol.* Montreal. pp. 169-174.

Pickett, A.D. 1961. The ecological effects of chemical control practices on arthropod populations in apple orchards in Nova Scotia, Canada. *I.U.C.N. Symposium* Warsaw, July 1960. pp. 19-24.

Pimental, D. 1982. Perspectives of integrated pest management. *Crop Prot.* 1:5-26.

Pomerantz, M.L. 1995. A profile of the fresh produce consumer. *The Packer.* 54:30-39.

Porter, D.R., J.D. Burd, K.A. Shufran, J.A. Webster, and G.L. Teetes. 1997. Greenbug (Homoptera: Aphidae) biotypes: selected by resistant cultivars or preadapted opportunists. *J. Econ. Entomol.* 90:1055-1065.

Pratt, P., E. Williams, J. Limsupavanich, and G.W. Cuperus. 1993. WEA$CON. Okla. State Univ. Coop. Ext. Serv. Software Serv. CSS-59.

Readshaw, J.L. 1984. Application of pest management programmas. In *Proc. Aust. Appl. Entomol. Res. Conf., 4th Adelaide.* Gov. Print. Adelaide, Australia. pp. 150-162.

Reed, C., J. Pederson, and G.W. Cuperus. 1993. Efficacy of grain protectants against stored grain insects in wheat stored on farms. *J. Econ. Entomol.* 86:1590-1598.

Reichelderfer, K.H., G.A. Carlson, and G.A. Norton. 1984. Economic Guidelines for crop pest control. *Food Agric. Org.*, Rome. 93 pp.

Riha, S., L. Levitan, and J. Hutson. 1996. Environmental-impact assessment — the quest for a holistic picture. In *Proceedings of the Third National IPM Symposium/Workshop.* United States Dept. of Agric. Misc. Publ. No. 1542. pp. 40-58.

Ring, D.R. and M.K. Harris. 1983. Predicting pecan nut casebearer (Lepidoptera:Pyralidae) activity at College Station, Texas. *Environ. Entomol.* 12:482-486.

Sachs, C., D. Blair, and C. Richter. 1987. Consumer pesticide concerns: a 1965 and 1984 comparison. *J. Consumer Affairs* 231:96-107.

Schrimpf, P.J. 1998. Prevent the squeeze on our food and fiber. *Ag Consultant.* Supplement. 54(5):1-16.

Scott, G.E., A.R. Hallauer, and F.F. Dicke, 1964. Types of gene action conditioning resistance to European corn borer leaf feeding. *Crop Sci.* 4: 603-606.

Seyum, D. 1997. Adoption of peanut integrated management practices in Atoka/Bryan and Caddo Counties of Oklahoma. Ph.D. Dissertation. Okla. State Univ. Stillwater, OK.

Shelton, K., G. Cuperus, M. Smolen, J. Criswell, C. Koelsch, and K. Pinkston. 1997. The urban environment: Oklahoma attitudes and practices. Okla. Coop. Ext. Serv. Circ. IPM-6.

Sinclair, J.B., M. Kogan, and M.D. McGlamery. 1997. Guidelines for the Integrated pest management of soybean pests. Nat. Soybean Res. Lab. Pub. 2. Coll. of Agric., Cons. and Env. Sciences, Univ. of Illinois at Urbana-Champaign, Urbana, IL.

Smith, J.W., Jr. and C.S. Barfield. 1982. Management of preharvest insects. In *Peanut Science and Technology.* Pattee, H. and C. Young, Eds. Amer. Peanut Res. and Edu. Soc., Inc. Yoakum, TX.

Sorensen, A. 1994. Constraints to the adoption of integrated pest management in stored grain., Nat. Found. for Int. Pest Manag., Austin, TX.

Sorensen, A. 1993. Regional producer workshops: constraints to the adoption of integrated pest management. Center for Agric. in the Environ. Amer. Farmland Trust. 60 pp.

Sparks, D. 1995. A budbreak-based chilling and heating model for predicting first entry of pecan nut casebearer. *Hort. Science* 30: 366-368.

Stark, J.A., J. Limsupavanich, G. Cuperus, C. Ward, and R. Huhnke. 1992. Profalf: An interdisciplinary expert system for alfalfa management. Okla. State Univ. Coop. Ext. Serv. Software Series CSS-66.

Stark, J.A., G. Cuperus, C. Ward, R. Huhnke, L. Rommann, P. Mulder, J. Stritzke, G. Johnson, J.T. Criswell, and R. Berberet. 1990. Integrated practices: a case study of Oklahoma alfalfa management. Okla. Coop. Ext. Serv. Circ. E-899.

Steffey, K. 1995. IPM today: are we fulfilling expectations? *J. Agric. Entomol.* 12:183-190.

Stern, V.M., R.F. Smith, R. Van den Bosch, and K.S. Hagan. 1959. The integrated control concept. *Hilgardia* 29:81-101.

Strayer, J. 1971. The pest management concept: The extension entomologist's view. Proc. Tall Timbers Conf. Ecol. Anim. Control Habitat Manag., 3rd Tallahassee. Tall Timbers Res. Sta. Tallahassee. pp. 21-28.

Stritzke, J.F., G.W. Cuperus and S. Zhang. 1996. WEEDALF$: Integrated expert system for weed management in alfalfa. In *Proc. N. Amer. Alfalfa Imp. Conf.* p. 30.

Stuckey, R., J. Nelson, G. Cuperus, E. Oelke, and H.M. Bahn. 1990. *Wheat Pest Management*, Ext. Serv./USDA, Washington, D.C.

Suter, D. 1995. Mid-south region's value-added food industry: an analysis of fruit and vegetable processors. M.S. Thesis. Okla. State Univ. Stillwater, OK.

Teetes, G.L., B.B. Pendleton, and R.D. Parker. 1997. Quantifying the use of IPM by sorghum growers. TPMA project termination report, Dept. of Entomol. College Station, TX 50 pp.

Thompson, J.S. and R.E. Kelvin. 1996. Suburbanites' perceptions about agriculture: the challenge for the media. *J. Applied Comm.* 80:11-20.

Tian, X., G. Sabbath, G. Cuperus, and M. Gregory. 1996. Environmental impact of the Oklahoma boll weevil program. *J. Water Res.* 32:1027-1037.

USDA, Agricultural Research Service. 1993. USDA programs related to integrated pest management. USDA Program Aid 1506.

Vandeman, A., J. Fernandaz-Cornejo, and B.L. Lin. 1994. Adoption of integrated pest management in the U.S. agriculture. *Agric. Infor. Bull.* No. 707. Econ. Res. Serv., Washington, D.C.

Van den Bosch, R. and V.M. Stern. 1962. The integration of chemical and biological control of arthropod pests. *Annu. Rev. Entomol.* 7:367-386.

Ward, C., R. Huhnke, and G. Cuperus. 1994. Alfalfa buyer preferences by Oklahoma and Texas dairy producers. Okla. Coop. Ext. Circ. E-936.

Whalon, M.E., M.F. Berney & S.L. Battenfield. 1982. *Biological monitoring in apple orchards: an instruction manual.* Mich. State Univ. Manual No. 3.

Whalon, M.E. and B.A. Croft. 1983. Implementation of apple IPM. In *Integrated management of insect pests of pome and stone fruits.* Croft, B.A. and S.C. Hoyt, Eds. Wiley-Interscience. New York. pp. 411-449.

Wiley, W.R.Z. 1978. Barriers to the diffusion of IPM programs in commercial agriculture. In *Pest Control Strategies.* Smith, B.H. and D. Pimentel, Eds. Academic. New York. pp. 285-308.

Willis, W. 1984. Wheat diseases. S-23, Coop. Ext. Serv. Kansas State Univ., Manhattan, KS. 31 pp.

Wu, L., J.P. Damicone, and K.E. Jackson. 1996. Comparison of weather-based advisory programs for managing early leaf spot on runner and spanish peanut cultivars. *Plant Disease* 80(6) 640-645.

Zalom, F.G., C.V. Weakly, M.P. Hogffman, L.T. Wilson, J.L. Grieshop, and G. Miyao. 1990. Monitoring tomato fruitworm eggs in processing tomatoes. *Calif. Agric.* 44:12-14.

Zalom, F.G., R.E. Ford, R.E. Frisbie, C.R. Edwards, and J.P. Tette. 1992. Integrated pest management: Addressing the economic and environmental issues of contemporary agriculture. In *Food, Crop Pests, and the environment: The need and potential for biologically intensive integrated pest management.* Zalom, F.G. and W.E. Fry, Eds. APS Press. St. Paul, MN.

Zettler, L. and G.W. Cuperus. 1990. Pesticide resistance in *Tribolium* and *Rhyzopertha* in Oklahoma. *J. Econ. Entomol.* 83:1677-1681.

Biological Control

CHAPTER 7

Biological Control of Insects

James Robert Hagler

CONTENTS

1-56670-478-2/00/$0.00+$.50
© 2000 by CRC Press LLC

7.1   INTRODUCTION

Throughout history, a relatively small number of insect species have threatened human welfare by transmitting disease, reducing agricultural productivity, damaging forests and urban landscapes, or acting as general nuisances. Humans have attempted to eradicate, control, or manage these pests using a wide variety of methods including chemical, biological, cultural, and mechanical control (National Academy of Sciences, 1969). The main strategy used in the second half of the 20th century for controlling pests has been the use of chemical pesticides (van den Bosch, 1978; Casida and Quistad, 1998).

The pesticide revolution began in the early 1940s with the development of synthetic pesticides. These pesticides showed a remarkable ability to kill pests without any apparent side-effects. The early success of synthetic pesticides led many experts to believe that they had discovered the "silver bullet" for pest control. As a result, biological, cultural, and mechanical controls were often underutilized or disregarded as viable pest management strategies. Although pesticides provided a short-term solution for many pest problems, the long-term negative effects of using pesticides did not begin to surface until the late 1950s. In 1962, Rachel Carson's book *Silent Spring* provided the general public with the first warning that many pesticides produced undesirable side-effects on our environment (Carson, 1962). Further consequences of overreliance on pesticides became apparent over the next few decades. For example, prior to the 1940s, it was estimated that insects destroyed 7% of the world's crops. By the late 1980s, crop destruction due to pests had risen to 13% (Wilson, 1990). This doubling of crop damage since the pesticide revolution occurred despite a 12-fold increase in pesticide use (Poppy, 1997). The increase in crop destruction is due, in part, to increased incidence of pesticide resistance, secondary pest outbreaks, and natural enemy destruction. These problems, coupled with increasing environmental concerns and pesticide costs, have forced growers to seek more environmentally safe and cost-effective pest control strategies. One of the most promising, yet underused, pest control strategies is biological control.

This chapter will provide readers with a general review of the fundamental principles of biological control, including the history, the methods, and the agents used for biological control. Central to this review is discussion of the key issues surrounding implementation of biological control in the new millenium.

7.2   DEFINITION OF BIOLOGICAL CONTROL

Entomologists have struggled with a definition for biological control for almost a half-century. In 1919, the eminent biological control researcher H. S. Smith defined biological control simply as "the control or regulation of pest populations by natural enemies" (Debach and Rosen, 1991). He defined a natural enemy as any biological organism that exerts the control. His definition only included the use of predators, parasitoids, and pathogens as biological control agents.

Biological control is the deliberate exploitation of a natural enemy for pest control. In other words, biological control is an activity of man. This differs from natural control, which is unassisted pest regulation due to biotic (e.g., predators, parasites, and pathogens) and abiotic (e.g., weather) forces (Debach and Rosen, 1991). Recently, a working group from the National Academy of Sciences broadened the definition of biological control beyond living organisms to include the use of genes or gene products to reduce pest populations (National Research Council, 1987). In 1995, the U.S. Congress, Office of Technology Assessment defined "biologically based technologies for pest control" (BBTs). BBTs included the use of predators, parasitoids, pathogens, pheromones, natural plant derivatives (e.g., pyrethrums, nicotine, etc.), insect growth regulators, and sterile insect releases as biological control agents (U.S. Congress, 1995).

Variations in the definition of biological control might seem trivial, yet those who prefer the more narrow definition are concerned that these other pest management approaches might garner most of the research dollars at the expense of the traditional biological control approaches. For this review, I will use the strictest definition of biological control and consider only predators, parasitoids, and pathogens as biological control agents. Pheromones, natural plant compounds, insect growth regulators, sterile insect releases, and genetic manipulations will be regarded here as parabiological control agents (Sailer, 1991). Although I do make a distinction between biological control and parabiological control, it is important to understand that parabiological control tactics will be of the utmost importance to enhancing the future success of the traditional biological control approaches. It is likely that parabiological control tactics will be included in the definition of biological control more frequently in the years to come because they are usually selective and environmentally benign.

7.3 HISTORY OF BIOLOGICAL CONTROL

One of the oldest-known methods used to control pests is the deliberate exploitation of their natural enemies. The first documented evidence of the use of natural enemies to control pest populations came from China and Yemen. Hundreds of years ago, ant colonies were moved between fields for controlling pests in tree crops (Coulson et al., 1982). Linnaeus made written reports of the use of predators to control pests in 1752 (Van Driesche and Bellows, 1996). In 1762, the first planned successful international movement of a natural enemy was undertaken. The mynah bird was introduced from India to control the red locust, *Nomadacris septemfasciata* in Mauritius. By 1772, this bird was credited for successfully controlling a locust pest (Debach & Rosen, 1991).

The so-called "modern age of biological control" began in 1888 when natural enemies were collected in Australia and imported to California to control the cottony-cushion scale, *Icerya purchasi* Maskell. This project is considered one of the major milestones in the history of entomology. The cottony-cushion scale was discovered in Menlo Park, California in 1868. This scale was not native to California, therefore it lacked any co-evolved natural enemies. The scale population exploded and within 20 years it had destroyed the citrus industry in California. In 1886, C.V. Riley (Chief of the Division of Entomology of the USDA), Albert Koebele, and D.W. Coquillett (and many others), initiated a classical biological control program targeted at the cottony-cushion scale. It was believed that this scale originated in Australia, so that is where the researchers searched for its natural enemies. The cottony-cushion scale was difficult to locate in Australia because the native natural enemy complex there was very effective at suppressing the pest population. However, a few scales were discovered that were either parasitized by a fly, *Cryptochetum iceryae* (Williston) or being eaten by a lady bird beetle, *Vedalia cardinalis* (later named *Rodolia cardinalis* [Mulsant]). These two natural enemies were shipped from Australia to California and placed into screened cages in citrus orchards for further evaluation. The lady beetle had a voracious appetite specifically for cottony cushion scale and within a couple of months had completely devoured all of the scales within the cages. The beetles were then distributed to a few growers in California and released into open citrus orchards for their establishment. By the end of the decade, the cottony cushion scale was fully controlled by the lady beetle. To date, this is perhaps the greatest example of a successful biological control program (Caltagirone and Doutt, 1989). Ironically, the overwhelming success of this effort proved to be a problem for subsequent biological control programs, because every subsequent research program was expected to yield equally impressive results.

Over the past 110 years there have been dozens of successful biological control programs initiated. Unfortunately, there have also been many failures. A database has been developed by the International Institute of Biological Control (IIBC), called BIOCAT, that is accessible on the World Wide Web. This database summarizes both successful and unsuccessful classical biological control programs (Greathead and Greathead, 1989). It also provides interesting insights into the patterns that exist between successful and unsuccessful programs.

7.4   BIOLOGICAL CONTROL — ITS ROLE IN IPM

Integrated pest management, or IPM, is a pest management approach that incorporates several different management strategies into one overall program (Stern et al., 1959). Ideally, IPM programs are designed to provide environmentally friendly and sustainable pest control. Ironically, before the insecticide revolution, the fundamental principles of IPM were being readily used for pest control. There was an enormous amount of effort dedicated to studying insect pest biology and nonchemical pest control strategies (Kogan, 1998). During this time, there were no "silver bullets" for pest control, so entomologists were forced to "integrate" biological, cultural, physical, and mechanical controls. Biological control is only one of the components of IPM. Biological control was a popular pest management strategy because it complemented many of the other IPM tactics. However, in the late 1940s, synthetic pesticides became the dominant method for pest control. Pesticides were not only incompatible with most other IPM tactics, but they were used without any regard to those alternate approaches.

The "re-invention" of IPM originated in the late 1950s when researchers began to realize that chemical pest control was not an effective strategy. The development of resistance to pesticides, the occurrence of secondary pest outbreaks, along with the harmful effects of pesticides on natural enemies and on the environment forced us to reexamine the fundamental concepts of IPM. Today, the frequency that IPM is being used as it was originally defined is rising (Kogan, 1998).

The future of IPM relies on our ability to get back to the basics of pest management. Emphasis needs to be replaced on studying the ecology of pests and their natural enemies and using IPM tactics that are compatible with biological control. In order for biological control to achieve wide-scale success, it is critical that environmentally benign, area-wide IPM tactics are used in concert with biological control. The principles of the IPM approach to pest management are discussed in greater detail elsewhere in this edition.

7.5   TYPES OF BIOLOGICAL CONTROL

The three basic types of biological control are conservation, introduction, and augmentation (Waage and Mills, 1992). Conservation involves preserving and/or enhancing natural enemies that are already present in the environment. Introduction involves importing and releasing exotic (non-indigenous) natural enemies against foreign and indigenous pests. Augmentation involves mass-rearing natural enemies in the laboratory and releasing them into the environment. These strategies are not mutually exclusive. For example, conservation should also be practiced when augmentation and introduction are employed.

7.5.1   Conservation of Natural Enemies

Conservation of natural enemies means enhancing or protecting the environment for natural enemies. It differs from natural control in that it is a conscious management decision. Conservation is achieved by using pest control tactics that preserve or enhance natural enemies (e.g., planting refuge crops) or by avoiding pest control tactics that are harmful to them (e.g., broad-spectrum pesticides). Conservation of natural enemies is a biological control tactic that should be a component of every pest management program, but, unfortunately, is underutilized due to the planning and effort required. Some of the methods used for conserving natural enemy populations include: avoiding the use of broad-spectrum insecticides; planting cover crops or refuge crops; and providing food supplements for natural enemies (see Van Driesche and Bellows, 1996 for more detail).

The use of broad-spectrum chemical insecticides is the major reason that the potential for conservation has not been reached. Most predators and parasitoids are vulnerable to insecticides. Unfortunately, the application of broad-spectrum insecticides is far too often the first and only method used for pest management (van den Bosch, 1978). Recently, more selective insecticides have been developed that are more compatible with conservation. Some examples of selective insecticides include the use of genetically engineered crops (e.g., Bt cotton), insect pathogens, and chemical formulations that contain pest-specific substances that interfere with the pest's endocrine system (i.e., insect growth regulators) (U.S. Congress, 1995). The use of pest-specific insecticides should decrease pest populations while conserving natural enemy populations. Before applying any insecticides, the applicator should be aware of the chemical's effect on non-target natural enemies (Jones et al., 1998).

Another tactic for conservation is to provide cover crops or refuge crops for predators and parasitoids. Cover and refuge crops, planted within and adjacent to high cash crops, serve to help attract, maintain, or increase predator and parasitoid populations by providing them with a more suitable habitat to survive. Growers can conserve predators and parasitoids in their orchards (e.g., pecans and apples) by planting leguminous cover crops (e.g., clover), which attract numerous natural enemy species and sometimes replenish the soil with nutrients (e.g., nitrogen) (Bugg et al., 1991). However, some cover crops may increase the cost of production because they require extra maintenance, water, or fertilizer beyond that required for the cash crop.

Refuge crops can also be planted adjacent to other crops in order to provide predators and parasitoids with a supplemental food source. For example, many parasitoid species rely on nectar-producing plants for energy. Sometimes, plants that are known to yield a high volume of nectar are planted near other crops to serve as an "energy source" for foraging parasitoids. Similarly, pollen is an excellent food supplement for many predator species. Sometimes pollen-rich plants (e.g., sunflowers) are planted near crops to enhance predator populations. Additionally, refuge crops can provide natural enemies with an insecticide-free habitat when adjacent fields are being treated with insecticides. Insecticide-free areas can serve as an invaluable refuge for natural enemies that might be otherwise exposed to harmful insecticides (Van Driesche and Bellows, 1996).

7.5.2 Introduction of Natural Enemies (Classical Biological Control)

Insects are often introduced into new areas either accidentally or purposefully. Sometimes these introduced insects (also known as exotic or non-indigenous insects) find a suitable host plant(s) in the new habitat in which they can survive and reproduce. When an exotic insect is introduced into a new area, it often does not have any co-evolved natural enemies to suppress its population. As a result, the exotic insect soon becomes a pest. The cottony-cushion scale scenario described above is a perfect example of an insect that was accidentally introduced into an area in which it did not have any co-evolved natural enemies. As a consequence, the cottony-cushion scale, which is not a pest in its native land of Australia, became a destructive pest in California (Caltagirone and Doutt, 1989).

The gypsy moth is another example of an introduced insect becoming a significant pest. In 1869, a scientist attempting to develop the silk industry in America purposefully brought gypsy moths into the U.S. from Europe (Debach and Rosen, 1991). Unfortunately, a few of the captive moths escaped and reproduced. In a very short period of time, with no native natural enemies to control them, the gypsy moth became (and continues to be) the major forest pest in the United States (Elkinton and Liebhold, 1990).

When an exotic insect establishes itself in a new area as a pest, the first place to search for potential biological control agents is in the pest's native habitat. Often, an introduced insect has co-evolved natural enemies in its native habitat that kept it from becoming a pest. If the origin of the pest is known, then natural enemies can be imported from its homeland and introduced into the new habitat. Importing and introducing an exotic natural enemy is also known as classical biological control. Classical biological control is probably the most successful, yet controversial type of biological control (U.S. Congress, 1995; Waage, 1996). Classical biological control requires more forethought and research than conservation or augmentation. Great care must be taken when attempting to establish non-indigenous natural enemies into a new region in order to minimize the chance of creating further unforeseen ecological problems (Waage and Mills, 1992).

Classical biological control is researched and implemented by scientists and is usually funded by federal or state governments. It is not unusual for a classical biological control program to take five to ten years to complete. However, the economic benefits derived from a successful classical biological control program are usually impressive. The benefit-to-cost ratio can range from 10:1 to 100:1 (Tisdell, 1990).

Several basic principles should be followed when selecting a classical biological control agent. The single greatest characteristic is that the agent must have a narrow host range, both to increase the effect on the target pest and to minimize any possible effects on non-target organisms (Debach and Rosen, 1991; Waage and Mills, 1992). It is for this reason that specialist parasitoids are generally regarded as better candidates for classical biological control than generalist predators. The natural enemy should also originate from a region with a climate similar to the one in which it is being introduced. Obviously, if the exotic natural enemy cannot survive and reproduce, it will not be an effective biological control agent. Additionally, the exotic organism should be (although not always) easy to capture in large numbers in its native habitat or be easy to rear (Debach and Rosen, 1991). The chances of establishing an exotic natural enemy are greatly increased if thousands or even millions of individuals can be released over a period of several years. Finally, every precaution needs to be taken to ensure that the exotic natural enemy itself does not become a pest. Before any classical biological control agent is introduced into a new area it must be extensively studied as an individual and as part of its new environment (see Waage and Mills [1992] and Van Driesche and Bellows [1993] for thorough reviews of the scientific protocols used for classical biological control).

7.5.3 Augmentation of Natural Enemies

Another type of biological control is augmentation, which consists of augmenting existing populations by producing natural enemies in the laboratory and releasing them into the field. The augmentation of natural enemy populations is the biological control equivalent to insecticide applications (Table 7.1). Unlike conserved or introduced natural enemies, augmented natural enemies are not necessarily expected to

Table 7.1   A Generalized Comparison of the Attributes of Augmented Natural Enemies and Conventional Pesticides

Attribute	Predators	Parasitoids	Pathogens	Conventional Pesticides
Host Range	Moderate/Wide	Narrow	Narrow	Wide
Commercial Availability	Low	Low	Medium	High
Shelf Life	Short (days)	Short (days)	Short/Moderate (weeks-months)	Long (years)
Cost	High	High	Moderate	Low
Ease of Application	Difficult	Difficult	Easy	Easy
Effectiveness	Low	Low/Moderate	Low/Moderate	High
Compatibility with Pesticides	Low	Low	High	High
Environmental Impact	Low	Low	Low	High
Occurrence of Resistance	None	None	Low	High

survive into the next year. However, when augmentation is combined with effective conservation, natural enemy populations may increase over time.

The most widely used augmentative biological control agents are insect pathogens. Currently, several pathogens are commercially available for controlling a wide variety of pests. In many cases, predators and parasitoids are not viable augmentative biological control agents because they are not practical or economically feasible to mass-produce (Grenier et al., 1994). There are several logistical difficulties that must be overcome before predators and parasitoids become widely used for augmentative biological control. Currently, most predator and parasitoid species are being reared on their prey (host) at high cost. Inexpensive artificial diets might make the mass production of predators and parasitoids economically feasible (Grenier et al., 1994).

Once the difficulties of developing artificial diets are overcome, then quality control studies are needed to test the efficacy of the biological control agents in the field (Hoy et al., 1991). Predators and parasitoids reared for successive generations on artificial diet in the laboratory might not perform as well as their native counterparts (i.e., they might become domesticated) (Hagler and Cohen, 1991; van Lenteren et al., 1997). Additionally, the production, distribution, and application of augmented biological control agents needs to be standardized so that their full potential is realized (Hoy et al., 1991; Smith, 1996; Obrycki et al., 1997; O'Neil et al., 1998; Ridgway et al., 1998). Augmentative biological control is not just a matter of ordering a package of natural enemies, releasing them into the field, and waiting for the control to happen. Both the suppliers and users of natural enemies need to have an understanding of how to apply the agent properly and of its limitations. End-users need to apply the agent in sufficient quantities to ensure effective pest management when the target pest is most vulnerable (Smith, 1996). For example, it would not be practical to release an egg parasitoid when there were no pest eggs present in the field. Also, it is important that the biological control agent is applied in a manner to minimize its mortality. For example, most parasitoids should be released during the cool part of the day and away from direct sunlight.

Whereas predators and parasitoids have been used sparingly for augmentative biological control, there are circumstances where they have been used successfully (Hoffmann et al., 1998). They are often used for controlling pests on high cash crops that are grown in small fields (e.g., strawberries) (Hoffmann et al., 1998). Additionally, predators and parasitoids are often released into barnyards, interior landscapes, greenhouses, and home gardens where insecticide applications are impractical because of the proximity to large numbers of humans and livestock.

The concept of augmentative biological control has generated an enormous amount of public interest over the past decade. Many small businesses have begun to market predators, parasitoids, and pathogens as "environmentally friendly" and "natural" alternatives for pest control. Currently, there are over 100 companies in North America that are dedicated to selling beneficial organisms (i.e., predators and parasites) for augmentative biological control use (Hunter, 1994). Although probably environmentally safe, these biological control agents might be serving only as a placebo to the end-user (Harris, 1990). More thorough field studies are needed to evaluate the efficacy of augmentative biological control agents before they are sold to consumers (Hagler and Naranjo, 1996). Additionally, the quality of predators and parasitoids reared for successive generations in captivity need further examination (Hopper et al., 1993).

7.6 GROUPS OF NATURAL ENEMIES

Natural enemies are classified into three major groups; predators, parasitoids, or pathogens. Predators and parasitoids are often collectively referred to as macrobiological control agents and pathogens are often called microbiological control agents, or simply microbials. A fourth classification of natural enemies, that of parabiological control agents (Sailer, 1991), is often included when the broadest definition of biological control is used (U.S. Congress, 1995).

Natural enemy communities are often large and complex, with a wide array of interactions occurring at any given time (e.g., predator-prey interactions, hyperpredation, competition, etc.). An excellent review of the types of natural enemy interactions that can occur is provided by Sunderland et al. (1997).

7.6.1 Predators

Insect predators, including representatives from most of the major orders in the class Insecta, are abundant in agroecosystems, urban environments, and aquatic habitats (Table 7.2). Most insect predators feed on a wide variety of prey, consume many prey throughout their immature and adult life stages, rapidly devour all or most of their prey, and prey on insects and mites smaller than themselves (Sabelis, 1992; Lucas et al., 1998). Although predators are regarded as a major biological control force, remarkably little is known about their prey choices in the field. Complex interactions among predators and prey make each predator assessment unique and difficult to describe (Hagler and Naranjo, 1996; Sunderland, 1996; Naranjo and Hagler, 1998).

Table 7.2   A Listing of Some of the Common Predators Found in Agroecosystems

Order	Family	Predator	Prey*	Reference
Orthoptera	Mantidae	Praying mantids	Large and small insects	Van Driesche and Bellows, 1996
Dermaptera	Labiduridae	Earwigs	Caterpillars, many others	Knutson and Ruberson, 1996
Thysanoptera	Aleolothripidae	Predaceous thrips	Spider mite eggs	Knutson and Ruberson, 1996
Heteroptera	Anthocoridae	Minute pirate bugs	Insect eggs, soft-bodied insects, small insects	Hagler and Naranjo, 1996
	Lygaeidae	Big-eyed bugs	Insect eggs, soft-bodied insects, small insects	Knutson and Ruberson, 1996
	Miridae	Plant bugs	Insect eggs, soft-bodied insects, small insects	Hagler and Naranjo, 1994
	Nabidae	Damsel bugs	Insect eggs, small insects	Knutson and Ruberson, 1996
	Reduviidae	Assassin bugs	Small insects, caterpillars	Knutson and Ruberson, 1996
	Pentatomidae	Predaceous stink bugs	Small caterpillars	Knutson and Ruberson, 1996
Neuroptera	Chrysopidae	Lacewings	Aphids, soft-bodied insects	Flint and Driestadt, 1998
Coleoptera	Coccinellidae	Lady beetles	Aphids, soft-bodied insects, insect eggs	Flint and Driestadt, 1998
	Carabidae	Ground beetles	Insect eggs, soft-bodied insects, caterpillars	Knutson and Ruberson, 1996
	Staphylinidae	Rove beetles	Small insects	Knutson and Ruberson, 1996
	Melyridae	Soft-winged flower beetles	Insect Eggs, soft-bodied insects, small caterpillars	Knutson and Ruberson, 1996
Diptera	Cecidomyiidae	Predaceous midges	Aphids	Flint and Driestadt, 1998
Hymenoptera	Formicidae	Ants	Insect eggs, soft-bodied insects, small insects	Knutson and Ruberson, 1996
	Vespidae	Hornets, yellow jackets	Caterpillars, small insects	Flint and Driestadt, 1998
	Sphecidae	Digger wasps, mud daubers	Caterpillars, small insects	Flint and Driestadt, 1998

* Virually all of the predators listed here are generalist predators and feed on many types of prey.

Most predators are generalist feeders that can and will feed on a wide variety of insect species and life stages (Whitcomb and Godfray, 1991). Some predators, such as lady beetles and lacewings may prefer certain prey (e.g., aphids) (Obrycki and Kring, 1998), but they will attack many other prey that they encounter. Unfortunately, many important predator species are cannibalistic and/or feed on other beneficial insects (Sabelis, 1992). For example, green lacewings and praying mantids are notorious for preying on younger and weaker members of their own species. Most predators have a host range that also includes other beneficial insects. It is not uncommon for higher-order predators to feed on other predators or parasitoids (Polis, 1994; Sunderland et al., 1997; Rosenheim, 1998). Additionally, some predator species can be pests. Perhaps the best example of an insect possessing the characteristics of both a pest and a predator is the fire ant, *Solenopsis* spp. The fire ant is a voracious predator on the eggs of many lepidopteran pests. However, the fire ant is also a major pest because it inflicts painful stings to animals and constructs nests that are detrimental to landscapes (Lofgren et al., 1975; Way and Khoo, 1992).

Predators must eat many prey items during their immature and/or adult stages in order to survive. The number of prey needed for a given predator species to complete its development varies among species. Some predators, such as some lacewing species, are only predaceous during their immature stages. The adults of these lacewings only feed on nectar or water.

The time spent handling prey varies by predator species and life stage. Handling times can vary from a few seconds to several hours (Cloarec, 1991; Wiedenmann and O'Neil, 1991). Most predators are highly mobile, and are only briefly associated with their prey. The predator quickly devours a single prey item and then moves on to feed again. The relatively short period of time that predators are associated with their prey, coupled with the lack of evidence of feeding (i.e., they often totally devour their prey) are two of the many reasons that make it difficult to quantify predation in the field (Hagler et al., 1991).

Generally, predators attack and feed on arthropods that are smaller and weaker than themselves (Whitcomb and Godfrey, 1991; Sabelis, 1992; Lucas et al., 1998). Preying on smaller animals allows them to use brute force to capture and kill prey. Some predators, however, are able to kill and consume prey many times their size by using artifacts such as venoms, traps (pitfall traps, webs, etc.), and modified body structures (raptorial forelegs, body spines, modified mouthparts).

Predators consume their prey in one of two different ways. Some predators (e.g., beetles, dragonflies, praying mantids) use biting or chewing mouthparts for consuming their prey. Chewing predators usually capture smaller prey using their powerful mandibles, and totally devour prey (Figure 7.1). Others (e.g., true bugs) use piercing and sucking mouthparts for consuming prey. Piercing and sucking predators quickly pierce their prey with needle-like mouthparts, inject potent digestive enzymes, and suck up the internal liquefied nutrients from their victims (Figure 7.2). Typically, piercing and sucking predators do not totally devour their prey (Cohen, 1998).

Predators search for their prey using one of two strategies. Some groups of predators actively stalk their prey. Stalking predators are usually very quick and

Figure 7.1   A lady beetle devouring an aphid with its powerful chewing mandibles.

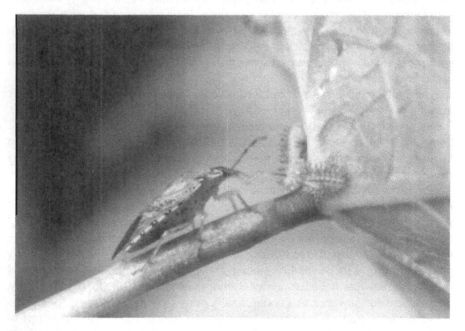

Figure 7.2   A spined soldier bug piercing and sucking nutrients from a Mexican bean beetle larva.

mobile (e.g., lady beetles). Other groups of predators patiently sit and wait for mobile prey to walk into an ambush. Ambush predators are usually well camouflaged (e.g., praying mantids) and use the element of surprise for attacking unsuspecting prey (Cloarec, 1991; Sabelis, 1992).

Predators are important natural agents, and as a group, are usually best suited for conservation because of their generalist feeding habits. Every effort should be made to conserve or enhance indigenous predator populations using one or more of the conservation tactics described previously. If a given predator species is to be considered for classical biological control, extensive research will be needed to ensure that non-target organisms will not be impacted.

The potential for using predators for augmentative biological control has not been fully realized. Mass-producing predators is costly and difficult. Furthermore, research aimed at testing the efficacy of predators reared on artificial diets is lacking (Leppla and King, 1996). For instance, there is always the possibility that predators reared for successive generations on an inanimate artificial diet will become domesticated. Domesticated predators may be unable to perform as efficiently as their native counterparts (Hagler and Cohen, 1991). Hopefully, in the near future, inexpensive and effective artificial diets will be developed that will facilitate the research, mass production, and application of predators as augmentative biological control agents (Grenier et al., 1994).

7.6.2 Parasitoids

Parasitoids are often referred to in the entomology literature as parasites. Although these two terms are often used interchangeably, a distinction should be made between them. A parasitoid ultimately consumes and kills its hosts, whereas a true zoological parasite (e.g., tapeworm) does not. Virtually all arthropod "parasites" are true parasitoids (Godfray, 1994).

Parasitoids are abundant in virtually all agroecosystems and urban environments. However, they are not as widespread in the class Insecta as predators. Almost all of the major parasitoid species occur in the orders Hymenoptera (wasps) (approximately 78% according to Feener and Brown [1997]) and Diptera (flies) (Table 7.3). Almost every insect pest, predator, and parasitoid has one or more parasitoid species that attacks it. Parasitoids that attack insect pests are commonly known as primary parasitoids, while those that attack other parasitoids are known as hyperparasitoids. Obviously, hyperparasitoids are not ideal candidates for biological control (Sullivan, 1987).

Parasitoids have many characteristics that distinguish them from predators (Table 7.1). Generally, parasitoids have a narrow host range; feed on only one host throughout their life span; attack hosts larger than themselves; feed on their host only during their immature stage (although the adults of some species may feed on hosts); and are immobile as immatures and free-living as adults (Sabelis, 1992).

Usually, a parasitoid species will attack a specific life stage of its host. Thus, parasitoids are classified as egg parasitoids, larval (nymphal) parasitoids, or adult parasitoids. Some parasitoid species will oviposit in one life stage, but emerge in a later life stage. Such parasitoids are named accordingly. For example, *Chelonus* sp. nr. *curvimaculatus* is an egg-larval parasitoid of pink bollworm (Hentz et al., 1998).

Table 7.3   A Listing of Some of the Common Parasitoids Found in Agroecosystems

Order	Family	Host	Internal/ External	Reference
Diptera	Tachinidae	Beetles, butterflies, and moths	Internal	Knutson and Ruberson, 1996
	Nemestrinidae	Locusts, beetles	Internal	Flint and Dreistadt, 1998
	Phoridae	Ants, caterpillars, termites, flies, others	Internal	Flint and Dreistadt, 1998
	Cryptochaetidae	Scale insects	Internal	Flint and Dreistadt, 1998
Hymenoptera	Chalcididae	Flies and butterflies (larvae and pupae)	Internal or External	Flint and Dreistadt, 1998
	Encyrtidae	Various insects eggs, larvae or pupae	Internal	van Driesche and Bellows, 1996
	Eulophidae	Various insects eggs, larvae or pupae	Internal or External	van Driesche and Bellows, 1996
	Aphelinidae	Whiteflies, scales, mealybugs, aphids	Internal or External	van Driesche and Bellows, 1996
	Trichogrammatidae	Moth eggs	Internal	Flint and Dreistadt, 1998
	Mymaridae	True bugs, flies, beetles, leafhoppers eggs	Internal	Flint and Dreistadt, 1998
	Scelionidae	Insects eggs of true bugs and moths	Internal	Flint and Dreistadt, 1998
	Ichneumonidae	Larvae or pupae of beetles, caterpillars and wasps	Internal or External	Flint and Dreistadt, 1998
	Brachonidae	Larvae of beetles, caterpillars, flies and sawflies	Internal (Mostly)	Knutson and Ruberson, 1996

The narrow host range exhibited by parasitoids makes them ideal biological control agents. Most parasitoids only attack one species or a group of related species. Therefore, parasitoids are well suited for conservation, augmentation, and classical biological control. To date, parasitoids are the most important of the macrobiological control agents used for classical biological control programs. Because most parasitoids are species- and stage-specific, it is critical that they are present in the habitat when their host is at its vulnerable stage of development. Therefore, the timing of a parasitoid release is of utmost importance. It would not be effective to release an egg parasitoid if only the larval stage of the targeted pest was present in the field.

Parasitoids have evolved a much more intricate relationship with their hosts than predators have with their prey. Adult parasitoids are free-living and usually feed on honeydew, nectar, pollen, or water in order to survive. However, some adult species are predaceous and will prey on their hosts by piercing soft-bodied prey (i.e., whiteflies and aphids) with their ovipositor or mouthparts and eating the juices that leak out of the wounded host. This type of behavior, known as "host feeding," leads to the death of the host and usually enhances the impact of the parasitoid on the host population (Jervis and Kidd, 1986; Heimpel and Collier, 1996).

Figure 7.3 A parasitoid parasitizing a gypsy moth caterpillar.

Unlike predators, parasitoids are only (highly) mobile and able to seek out their host during the adult stage. Typically, adult females lay one or more eggs in (endoparasitoid) or on (ectoparasitoid) a host (Figure 7.3). When the egg hatches, the larva begins to feed on its host. Parasitoids do not immediately kill their hosts. The immobile larvae utilize the host as food and shelter throughout their development. The parasitoid-host relationship is more efficient than a predator-prey relationship, requiring far less food for survival (Hassell and Godfray, 1992).

The mode of life adapted by parasitoids has greatly limited their freedom of action; they have become highly adapted to certain niches. In particular, the larval stages of parasitoids have become intimately connected to and dependent on their hosts both for their shelter and food. Consequently, parasitoids are usually smaller than their host. To this end, biological control by parasitoids is subtler than a population of pests being devoured by predators. However, it is usually easy to detect parasitism. For example, immature whitefly parasitoids can be readily seen within large whitefly nymphs; many caterpillar egg parasitoids cause their host to turn black; and aphid parasitoids turn the aphids black and "mummified."

Searching is vital to the success of parasitoids. Parasitoids are much more efficient than predators at searching and locating their hosts. They have an uncanny ability to locate prey, even at very low host densities, using chemical cues (Vet and Dicke, 1992; Godfray, 1994). For example, some parasitoids locate their host by homing in on long-range chemical cues produced by undamaged plants (Udayagiri and Jones, 1993) and plants that have been damaged by caterpillars (Paré and Tumlinson, 1997). Once the plant has been located, the female wasp then begins to use short-range chemical cues produced directly by the pest (Tumlinson et al., 1993). Good searching capacity allows parasitoids to control pest populations more efficiently than predators.

Parasitoids as a group are commonly used for all types of biological control. Every effort should be made to conserve or enhance native parasitoid populations. Parasitoids are particularly susceptible to broad-spectrum pesticides, therefore applications of these materials should only be used as a last resort (Theiling and Croft, 1988; Jones et al., 1998).

The narrow host range exhibited by most parasitoid species makes them ideal candidates for classical biological control. As with predators, the full potential for using parasitoids for augmentative biological control has not been fully realized. Mass production of parasitoids is easier and less expensive than the mass production of predators, but research is still needed to further develop rearing procedures. To date, the greatest use for parasitoids as augmentative biological control agents has been in the greenhouse industry (van Lenteren and Woets, 1988). An enormous amount of progress has been made over the past decade in developing parasitoids for augmentative biological control (van Lenteren et al., 1997). In the near future, the application of parasitoids will be a common pest control tactic.

7.6.3 Pathogens

Just like vertebrates, insects are susceptible to a variety of pathogens. The pathogens used for biological control of insects include bacteria, fungi, viruses, protozoans, and nematodes. Within each of these groups, there are hundreds or thousands of species that are known to attack insects. However, only a few have been used for pest control. Naturally occurring pathogens commonly attack insects, causing illness and sometimes death (Figure 7.4). Often the sub-lethal effects of pathogens can alter insect behavior to prevent insect reproduction.

Dozens of pathogens have been mass-produced and marketed as "biological insecticides" (Cook et al., 1996). These pathogens, mainly bacteria (*Bacillus thuringiensis*), have been used for controlling a wide variety of pests. Most pathogens are applied directly to crops using standardized pesticide sprayers or dispersed through irrigation water (Chapple et al., 1996). Commercially available pathogens are attractive biological control agents because they usually have a narrow host range, are environmentally safe, and are biodegradable (Table 7.1). Unfortunately, microbials only account for about 2.0 to 5.0% of the world pesticide market (Payne, 1989; Ridgway and Inscoe, 1998). Currently, there are numerous other pathogen species that show promise as biological insecticides. However, more progress is needed toward developing better mass production systems and more stable formulations (Roberts et al., 1991).

Most types of pathogens share some of the pitfalls associated with chemical insecticides. For example, insects can develop resistance to pathogens if they are constantly exposed to them (McGaughey and Beeman; 1988, McGaughey, 1994;

Figure 7.4   A nuclear polyhedrosis virus attacking and killing a beet armyworm. An infected (top) and healthy caterpillar (bottom).

McGaughey and Whalon, 1992). Additionally, the development and registration of pathogens is often difficult and costly. Some pathogens have a short shelf life or field life. Improved formulations for pathogens may increase shelf life and field persistence, and ensure that the pathogens rapidly move from infected individuals to uninfected ones, killing the hosts.

Ironically, the narrow host range exhibited by most pathogens, which is a desirable quality for biological control, has limited commercial pathogen development. The pesticide industry is reluctant to invest in products that have a narrow host range, and thus, a narrow sales market (Waage, 1996).

The basic approaches used for exploiting pathogens are mainly by conservation and augmentation. Classical biological control of insects is only rarely attempted with pathogens, since most diseases are distributed worldwide (Milner, 1997). Naturally occurring pathogen populations are usually conserved through some form of microhabitat manipulation (Fuxa, 1987; Roberts et al., 1991) in order to create favorable conditions for pathogen reproduction. Most pathogens thrive in warm, moist habitats. Augmented pathogens, like predators or parasitoids, can be applied by either inoculation or inundation. For inoculation, the pathogen is released in low numbers where it maintains and spreads itself throughout the pest population. For inundation, the pathogen is applied in large quantities just like a chemical pesticide. In this case, the pathogen is not necessarily expected to spread throughout the pest population (Fuxa, 1987).

7.6.3.1   Bacteria

Many different insect species are infected and killed by bacteria. Bacterial pathogens are the most common type of pathogen used for biological control. Currently, there are several formulations that are registered for commercial pesticide use.

Bacillus thuringiensis or Bt, is the most widely applied biological control agent (Cook et al., 1996). Bt exerts its toxicity only after phytophagous insects have ingested it. The Bt toxin is a high molecular weight protein crystal that causes paralysis of the insect's gut, followed by a general paralysis and insect death (Gill et al., 1992). The major advantages of using Bt (and most other bacterial pathogens) for pest control are that it is specific, effective, environmentally safe, and rapidly kills its host. Additionally, Bt has a short residual period so it is an ideal candidate for pest control on fruits and vegetables, in urban areas (parks), and near streams and ponds (Pinnock et al., 1977).

Several different formulations and varieties of Bt exist. Early Bt products only controlled lepidopteran larvae. These products continue to be used successfully to control lepidopteran pests. Subsequently, Bt products have been developed specifically for controlling beetles (e.g., Colorado potato beetle and elm leaf beetle).

Recently a second generation of Bt products has been developed. Bt has been incorporated into plant tissue (e.g., cotton, potato, tomato, and corn) using genetic engineering technology. The use of crops that have been genetically modified to contain Bt have generated an enormous amount of scientific and ethical debate. On the one hand, crops that contain Bt have automatic and specific pest protection. Therefore, the labor and costs associated with applying conventional insecticides are eliminated. Additionally, natural enemies are conserved because the Bt toxin does not affect them (Meeusen and Warren, 1989). On the other hand, pests are constantly exposed to Bt, even when control is not needed. This constant exposure will undoubtedly increase the incidence of pest resistance to Bt (Meeusen and Warren, 1989; Tabashnik, 1994; Gould, 1994). Additionally, the incorporation of an "automatic" pest control tactic means that consumers must pay for the pest control even if it is not needed. The development and application of genetically engineered crops will be the focus of much more research and scientific and ethical debate in the years to come (U.S. Congress, 1995; Rice and Pilcher, 1998).

7.6.3.2   Fungi

It has been estimated that over 700 species of fungi infect insects; however relatively few (approximately 17) have been developed and used for insect control (Roberts et al., 1991; Fuxa, 1987; Jaronski, 1997). Compared to most other types of pathogens, fungi have a relatively wide host range. For example, *Beauveria bassiana* has been identified as a potential biological control agent of many different arthropod pests (e.g., beetles, ants, termites, true bugs, grasshoppers, mosquitoes, and mites). The wide insect host range of some fungi has caused concern regarding safety to non-target organisms. Honey bees are susceptible to *Beauveria* and *Metarhizium* (Roberts et al., 1991). Clearly, thorough research needs to be conducted on the host specificity of entomopathogenic fungi and methods need to be developed that will minimize adverse effects on non target organisms.

Fungi differ from most of the other types of insect pathogens in that they do not have to be ingested in order to invade their host (Hajek and St. Leger, 1994). Fungi can enter their host through natural openings in the insect cuticle and spread to the hemocoel (Ferron, 1978). Because fungi infect insects by penetrating the cuticle, direct contact between the fungi and the insect host is necessary. The time required to kill an insect by fungal infection can be from only a few days to several weeks (generally 3 to 7 days), depending on the fungus (Jaronski, 1997). The ability of fungi to infect the insect's external integument makes them good candidates for controlling piercing/sucking herbivores, which are usually immune to other pathogens due to their feeding behavior (Roberts et al., 1991).

Most entomopathogenic fungi have many biotic and abiotic limitations that limit their wide-scale development and application. The biotic limitations are poorly understood, but they are primarily associated with the penetration of the fungus into the host's integument (i.e., the degree of contact and infectivity). A better understanding of the factors that affect the ability of a given fungus to penetrate and invade its host are of paramount importance in the future development of fungi as viable biological control agents. Also, better fungal formulations are needed to improve their overall shelf life, virulence, infectivity, and persistence (Milner, 1997; Fuxa, 1987; Jaronski, 1997).

Many abiotic factors also limit the use of fungi for controlling insect pests. Most fungi require a cool and moist environment (>90% humidity) to germinate (Ferron, 1978). Once they germinate, then their efficacy can be maintained at moderate humidity (i.e., approximately 50%) and temperatures between 20 to 30°C. However, there are a few strains of fungal pathogens that are effective in arid environments (Bateman et al., 1993).

Unlike other pathogens, fungi grow well on simple and inexpensive media. This characteristic, coupled with their relatively wide host range, makes many fungal pathogens potentially good candidates for commercial production.

7.6.3.3 Viruses

For the most part, viruses have been used for classical and augmentative biological control (Roberts et al., 1991). Nuclear polyhedrosis viruses (NPVs) comprise the major group of viruses that attack insects. Most NPVs attack and infect young lepidopteran larvae that have ingested virus particles. Death by viral infection usually takes several weeks. The relatively slow speed at which viruses kill their host has hindered their acceptance as a widely used biological control tactic (Bonning and Hammock, 1996).

The efficacy of most viruses is heavily influenced by prevailing environmental factors. The transfer of a virus to a host usually requires the virus to survive in soil litter and on plant surfaces, before they are moved passively by abiotic and biotic agents. Additionally, the efficacy of many viruses is adversely affected by direct sunlight.

The major advantages of using viruses for insect control are that they are host specific and environmentally safe (Bonning and Hammock, 1996). Again, their narrow host specificity makes them desirable candidates for biological control, but limits their commercial development (Roberts et al., 1991).

7.6.3.4   Protozoa

Many indigenous protozoans infect and kill insects. The most common group of protozoans is microsporidia (Brooks, 1988; Henry, 1981). Over 250 species have been described; however, it is believed that thousands of additional species probably exist (Maddox, 1987). There is a lack of research documenting the effectiveness of protozoans as biological control agents because they are difficult to diagnose and identify (Hazard et al., 1981).

Most microsporidia are transmitted to insects by oral ingestion of spores. However, some species are transmitted transovarially via the egg or by parasitoids (Andreadis, 1987; Siegel et al., 1986). In most instances, insects infected with indigenous microsporidia go unnoticed because they kill their host so slowly that it is difficult to differentiate between disease-caused mortality and natural mortality. It is for this reason that the sub-lethal effects of microsporidia infections may cause the most significant reductions in pest populations. For example, insects that are infected with sub-lethal amounts of microsporidia may have reduced fecundity and reduced mating. This ultimately results in lower pest populations over subsequent generations (Canning, 1982; Maddox, 1986). The sub-lethal effects of microsporidia infections on pest populations is a research area that needs to be more thoroughly examined.

Protozoans have not been developed as microbial insecticides because they do not cause rapid mortality. Additionally, because they are obligate parasites and cannot be grown on artificial media (they must be produced in living host cells), commercial development of protozoans is impractical. It is probably more realistic to consider protozoans as natural control agents.

7.6.3.5   Nematodes

Entomophagous nematodes are probably among the most potentially useful and commercially attractive type of pathogen. Nematodes (the name is derived from the Greek word for thread) are slender, tubular (non-segmented) worm-like organisms that can be found throughout the world inhabiting both soil and water. Many of the species are barely visible to the naked eye.

The class Nematoda contains a wide variety of species. Most species are free-living and feed on bacteria, fungi, and algae. Many nematode species are pests that parasitize animals (including humans) and plants. Nearly 40 families of nematodes are known to exclusively parasitize and feed on arthropods. To date, the most beneficial nematodes are found in the families Heterorhabditidae and Steinernema-tidae (Georgis, 1990). Both of these families are obligate parasites that have evolved a symbiotic relationship with pathogenic bacteria (e.g., *Xenorhabdus* and *Photo-rhabdus*) (Poinar, 1990). The nematodes provide the "transportation" for the bacteria by penetrating the insect through the mouth, anus, or spiracles (heterorhabditids can also penetrate the cuticle) (Georgis, 1992). Once in the host, the nematodes release the bacteria, which quickly multiply and kill the host. In turn, the nematodes use the bacteria and the insect cadaver as a source of food and shelter (Kaya and Gaugler, 1993). The nematodes then mature, mate, and reproduce in the host tissue. Infective-stage juveniles emerge from the cadaver and search for a new host (Georgis, 1992).

Nematodes have characteristics that make them outstanding candidates for all types of biological control (e.g., conservation, augmentation, and classical) and potentially competitive with insecticides for marketability. Nematodes are highly mobile, and can find and kill a new host in just a few days (Gaugler, 1988). Several nematode species are easily mass-produced *in vitro* and applied into the field using standardized pesticide sprayers or irrigation systems (Georgis and Hague, 1991). Additionally, nematodes and their bacterial symbiots are safe to higher order animals and plants. Finally, nematodes have a relatively wide host range which makes them more likely to be developed commercially (Georgis, 1992).

One major drawback associated with nematodes is their susceptibility to desic-cation and ultraviolet light (Georgis, 1992; Gaugler et al., 1992). As with most other groups of pathogens, most nematode species prefer a cool and moist environment to survive. Additionally, their relatively wide host range suggests that non-target organisms might be impacted by nematode applications (Akhurst, 1990). However, a recent study suggests that non-target effects of nematodes on predators and para-sitoids is minimal (Georgis et al., 1991).

Nematology as a science is still in its infancy. However, discoveries over the past two decades have shown that they have enormous potential for controlling many types of insect pests under certain environmental conditions (Webster, 1998). Major barriers to overcome before they are widely accepted as viable biological control agents include storage and shipping of large-scale supplies of nematodes. Increased effort is needed to search and screen for more virulent strains of the nematode/bac-terium complex that can survive under a wider variety of environmental conditions (Georgis, 1990).

7.6.4 Parabiological Control Agents

Although parabiologicals are excluded from the traditional definition of biological control (Debach and Rosen, 1991), they provide specific pest control and work syn-ergistically with predators, parasitoids, and pathogens. Parabiologicals include sterile insect releases, pest-specific pheromones, and insect growth regulators (Sailer, 1991).

7.6.4.1 Sterile Insect Release

Sterile insect release involves exposing *in vitro* reared insects to radiation and releasing them into the field to mate with native insects. This technique is an enormously successful pest control tactic for certain pests. The doses of radiation sterilize the laboratory-reared insects, thus making them incapable of producing any offspring. In turn, the reproductive potential of the pest can be drastically reduced over several generations if enough sterile insects mate with normal insects. The landmark example of a successful sterile insect release was with the screwworm, *Cochliomyia hominivorax* (Coquerel) in the southwestern U.S. (Bushland, 1974;

Knipling, 1985). Since then, sterile insect releases have been successful for controlling many other pests, particularly fruit flies (Debach and Rosen, 1991). The pest's biology, natural history, and population dynamics limit the wide-scale use of sterile insect release.

7.6.4.2 Pheromones

Pheromones have proven invaluable for monitoring insect pest populations and disrupting the mating behavior of certain pests (Shorey, 1991). Many pheromone-based traps and mating disrupters are commercially available for managing pests. For example, the synthetic sex pheromone, gossyplure, has been an invaluable tool for disrupting the mating behavior of the pink bollworm, *Pectinophora gossypiella* (Saunders) in the southwestern U.S. (Gaston et al., 1967; Flint et al., 1974; Shorey et al., 1974; Gaston et al., 1977; Baker et al., 1990). The use of pest-specific pheromones is highly compatible with biological control (Shorey, 1991). Like many parasitoids and pathogens, pheromones are pest-specific and have no adverse effects on non-target organisms. Additionally, insects do not develop resistance to the pheromones. As with the other parabiologicals, pheromones are designed to be used as one of several components of an overall IPM program. As of 1995, sex pheromones had been formulated for almost two dozen lepidopteran pests (Cardé and Minks, 1995).

7.6.4.3 Insect Growth Regulators

Insect growth regulators (IGRs) have become popular for pest management over the past two decades; however, their potential has not yet been fully realized (Staal, 1975).

IGRs interfere with the endocrine system of the pest and affect their normal growth and development (Dhadialla et al., 1998). Most existing IGRs can be categorized into two major groups by their mode of action; juvenile hormone analogs or chitin synthesis inhibitors (Horowitz and Ishaaya, 1992; Plapp, 1991). For the most part, IGRs are thought to be compatible with biological control because they are pest-specific and they generally do not have any adverse effects on natural enemies. However, some recent studies have shown that certain IGRs are toxic to natural enemies (Croft, 1990; Biddinger and Hull, 1995; Delbeke et al., 1997). Additionally, as with the synthetic pesticides, insect pests can develop resistance to IGRs (Plapp and Vinson, 1973; Cerf and Georghiou, 1974; Brown and Brown, 1974; Wilson and Fabian, 1986; Horowitz and Ishaaya, 1994). To this end, IGRs should not be overused and they should be regarded as a single component used to complement an overall IPM program.

7.7   LIMITATIONS AND RISKS ASSOCIATED WITH THE VARIOUS BIOLOGICAL CONTROL APPROACHES

In many cases, biological control is a simple, effective, and environmentally sound pest management approach. However, biological control is not a panacea for all pest problems. Biological control requires patience. Even an effective natural enemy is almost always slower acting than an insecticide (U.S. Congress, 1995). Furthermore, a decision to commit to a biological control program might alter other pest management strategies, such as insecticides or cultural practices harmful to natural enemies. Research on the efficacy of many potential biological control agents is limited, as verification of the efficacy of a biological control agent requires considerable expertise.

Far too often, the efficacy of a biological control agent is compared with a chemical pesticide in terms of its direct capacity to kill a pest (Waage, 1996). Unfortunately, this standard of measurement is unfair because biological control is more difficult to assess. For example, most biological control agents (parasitoids and pathogens) do not immediately kill their hosts. However, they are not only compatible, but synergistic with most of the other IPM tactics. On the other hand, broad-spectrum pesticides rapidly kill pests, but are not compatible with most of the other IPM tactics. A key issue for biological control researchers is to document the long-term efficacy of biological control as a component to an overall area-wide IPM program (Knipling, 1979, 1980; U.S. Congress, 1995, Kogan, 1995, 1998; Wellings, 1996).

There are many reasons that biological control has not been used as frequently as pesticides. Pesticides are easy to apply and they produce rapid and dramatic results. In contrast, biological control is generally more difficult to apply, more expensive, slower acting, and more subtle than pesticides, but has several advantages over chemical pesticides (Debach and Rosen, 1991). For instance, unlike broad spectrum pesticides, biological control agents are usually pest-specific. Furthermore, many biological control agents need to be applied only once (or a few times) to become established and continue to work effectively. Unfortunately, while pest specificity, single or limited applications, and long-lasting pest control are positive qualities for biological control, these features inhibit their commercial development. Consequently, biological control agents are often unavailable to pest management personnel (Waage, 1996).

There are very few ecological risks associated with conservation or augmentation because they both use indigenous natural enemies (Wellings, 1996). However, the use of classical biological control has received some criticism. Proponents of classical biological control offer convincing arguments for its safety (if done properly), efficacy, and cost effectiveness. Opponents of classical biological control argue that it is not possible to predict if an agent will have any long-term, irreversible affects to non-target organisms (plants and animals). Some fear that a classical biological control agent might alter the composition of entire ecosystems. For example, it has been speculated that the introduction of parasitoids into Hawaii has had a negative impact on native butterfly and moth populations. However, biological control cannot be singled out as the sole cause of the decline in butterfly and moth populations because of habitat destruction, pesticide use, and other environmental problems found in Hawaii (U.S. Congress, 1995)

7.8 BIOTECHNOLOGY AND BIOLOGICAL CONTROL

Enormous progress has been made over the past decade toward advancing the role of biotechnology in biological control (Sheck, 1991). Biotechnology has and will continue to play a major role in biological control. Through genetic engineering, scientists have transferred genetic material from one organism to another. The classical example is with the insect pathogen, Bt. As mentioned above, Bt is a naturally occurring soil bacteria that has been formulated as a biological insecticide against a variety of pests. Scientists have genetically inserted toxic genes from Bt directly into plant tissue. The major advantage of having Bt directly inserted into the plant is that the plant receives continuous protection from certain pests (Kirschbaum, 1985). Additionally, the new growth of the plant is also protected, which is a problem when using spray formulations (U.S. Congress, 1995).

Despite the obvious advantages of using Bt crops, the possibility of their widespread use raises some potential problems. Already, there have been reports that certain pests have developed varying degrees of resistance to Bt crops (Gould, 1998). To this end, extensive resistance monitoring of genetically engineered crops used for pest control will be critical for further development of genetically altered organisms (Gould, 1994).

Insect pathogens are not the only natural enemies that have been genetically modified for biological control. Predators, parasitoids, and nematodes have also been modified to increase their potential for controlling pests (Hoy, 1986; Hokkanen, 1991; Gaugler et al., 1997). The major constraint on improving predators and parasitoids is accurately predicting which trait is helpful to improve (Sheck, 1991; Hopper et al., 1993). In the long history of biological control, it is difficult to pick out any single trait that a natural enemy has that makes it a successful natural enemy (Beddington et al., 1978; Hopper et al., 1993). Assuming that biotechnology advances to a point where we can easily produce genetically modified insects, then the "question" remains about which changes are needed to a natural enemy that will make it a more effective agent (e.g., dispersal capability, fecundity, diet breadth, etc.).

Some predators have been genetically modified so that they are resistant to certain insecticides. The most progress in this area has been breeding insecticide resistance into predatory mites (Hoy, 1985). The major drawback associated with having pesticide-resistant predators is that in order to maintain the resistance in the field, insecticides must be continuously applied to ensure that the selected strain does not breed with native (non-resistant) individuals. Ultimately, this practice could destroy the other natural enemies present in the field (Sheck, 1991). Additionally, pesticide resistant natural enemies are not resistant to all pesticides; therefore, pesticides must be chosen very carefully in order to avoid killing the resistant strain.

There are other possibilities for improving natural enemies using biotechnology (Hoy, 1989; Hoy, 1990). Heat or cold tolerance could be increased in certain natural enemies, allowing them to withstand greater climatic extremes and to inhabit a broader region for a greater length of time (Hoy, 1990; Gaugler et al., 1997). Another exciting possibility for genetic improvement on a natural enemy is to alter the venom of a parasitoid in such a way that it causes its host to stop feeding (due to paralysis). This would significantly reduce crop damage by reducing the pest feeding while it is parasitized (Beckage, 1990; Sheck, 1991).

The possibilities of using genetically engineered natural enemies are limitless. Certainly, more genetically altered natural enemies will be developed in the years to come. However, research in this area must proceed with caution. Projects that are initiated in the future toward genetically altering any natural enemy need to be scrutinized by researchers and the general public. Researchers need to be cautious when selecting desirable characteristics to be sure that these changes will not pose any long term consequences.

7.9 FUTURE OF BIOLOGICAL CONTROL

Biological control as a pest management strategy is gaining popularity. For the first time since the beginning of the pesticide revolution, biological control can play a major role in IPM because it is ecologically sound, environmentally benign, self-perpetuating, and inexpensive. These attributes, combined with changing consumer attitudes about food and fiber consumption, have influenced governments throughout the world to change their pest management policies. For instance, some European nations have set a goal to reduce their pesticide usage 50% by the year 2000 (Baerselman, 1995). In the U.S., a government initiative is underway to implement IPM on 75% of crops by the year 2000 (U.S. Congress, 1995). Furthermore, developing countries in Asia have declared IPM as their national crop protection strategy (Ooi et al., 1992). Clearly, biological control will be the backbone of these government initiatives. This global commitment to biological control provides biological control researchers with a unique opportunity to further develop safe, effective and user-friendly pest management.

Classical biological control will be used more than ever because of increased movement of alien pests due to technological advances in transportation and fewer restrictions on international trade. International trade agreements such as Europe's General Agreement on Trades and Tariffs (GATT) and North America's North American Free Trade Agreement (NAFTA) will increase the probability that alien pests will inhabit new regions of the world (Waage, 1996).

New information is needed to predict the effectiveness and safety of classical biological control agents. The ecological consequences of introducing an improper biological control agent, one that attacks insects other than the target pest, are irreversible. In the past, biological control researchers have had a difficult time estimating the long-term economic benefits derived from successful natural enemy introductions. Better economic assessments will increase the competitiveness of classical biological control with conventional pest control (i.e., pesticide applications) (Greathead, 1995; Hokkanen and Lynch, 1995).

Augmentative biological control will continue to gain popularity. There are an increasing number of small businesses dedicated to selling biological control agents (Hunter, 1994). Generally, these businesses sell relatively small quantities of macrobiological agents (e.g., predators and parasitoids) to home gardeners and greenhouse operators. Major breakthroughs are still needed toward developing artificial diets for macrobiologicals. Until these breakthroughs occur, predators and parasitoids will have limited use in large-scale augmentative biological control programs (Grenier et al., 1994).

The future's most promising augmentative biological control agents are pathogens, or biopesticides. In the past, many big businesses have developed useful biopesticides for controlling a variety of pests. This market will continue to grow in the near future. As of 1993, there were over 175 microbial products (mainly Bt products) on the market (Waage 1996). However, biopesticides still only constitute less that 5.0% of the pesticides sold worldwide (Ridgway and Inscoe, 1998).

Industry has been reluctant to invest in biopesticides for many reasons. Biopesticides often have a short shelf-life or are unstable in the field and have an uncertain market demand (Roberts et al., 1991). However, these shortcomings could be overcome with sufficient incentive and investment in research and development. Unfortunately, many of the reasons that biopesticides are not readily available for commercial use are of an economic nature. Many potentially marketable biopesticides have a narrow host range and are self-perpetuating, thus making them less profitable than chemical pesticides. Formulations of Bt and nematodes constitute over 90% of the biopesticide market. These two agents are known for their quick killing capacity, relatively wide host range, and non-persistence in the field. These products are on the forefront of commercial development because they have characteristics that are similar to chemical pesticides (Waage, 1996). Unfortunately, many other potentially useful biopesticides that are self-perpetuating, persistent in the field, and have a narrow host range (e.g., viruses, fungi, and protozoans) are not prime candidates for commercial development. In the future, we need to do a better job of developing biopesticides into useful products (Kirschbaum, 1985). We need to create incentives for industry to invest in the research and development of such products (Waage, 1996). Only then will we be able to develop biological control products that are truly compatible with IPM and not necessarily compatible with traditional pesticide development.

Conserving natural enemies as a pest management tactic needs to be practiced on a far greater scale than it is presently. Many more opportunities exist for conservation than for classical or augmentative biological control combined. Conservation is a pest management strategy that should be applied by everyone. For conservation to reach its full potential, several scientific, social, and economical barriers must be overcome.

Greater resources are needed to research various conservation tactics. Unfortunately, conservation is probably the least studied of the three biological control tactics because it does not lend itself easily to the research and extension activities of most academic institutions. We still lack a basic knowledge of the biology and ecology of most pests and their natural enemies. Moreover, we still do not fully understand the efficacy and economic impact that native natural enemies have on pest populations. Even farmers who use conservation tactics do not have clear-cut instructions on how to apply the tactics. Academic institutions need to design extension programs to educate farmers and the general public on conservation tactics. Extension personnel need to collaborate with entire communities and organize area-wide pest management programs. We need to empower individuals by training them to understand conservation tactics and to participate with scientists in selecting additional interventions when necessary to complement an areawide IPM approach to pest management (Knipling, 1979; Waage, 1996). A weak link in implementing conservation tactics (and the other biological control tactics) is convincing farmers to experiment with them. Area-wide extension programs are needed to provide farmers with unbiased technical support so that they can make informed decisions (Wearing, 1988). Presently, much of the technical support offered to farmers is from "private consultants" who are provided incentives from chemical companies to sell their products. In the future, we need to provide farmers with financial incentives to try IPM.

Producers of agricultural products will need to change their expectations concerning pest management if conservation is going to reach its full potential. Growers are accustomed to using fast-acting insecticides to "control" their pest problems. This monotactic approach to pest management must change if we expect to achieve environmentally friendly pest control. Conservation (and the other types of biological control) is not only slower acting than pesticides, but usually needs to be used with other environmentally friendly IPM tactics. Conservation of natural enemies is not usually a "silver bullet" approach to pest management, but it is compatible and often synergistic with other pest management tactics.

Consumers will also need to change their expectations concerning the produce they purchase if biological control is to succeed. Consumers' insistence that their produce must be "perfect" has severely deterred the use of biological control. Growers will require assistance from consumers to wean themselves off the "pesticide treadmill" (van den Bosch, 1978). To this end, consumers should be willing to accept some minor insect damage on their produce.

In summary, production agriculture throughout the world needs to maintain or increase its current level of production while providing the public with environmentally friendly and compatible pest management. The restoration of agricultural productivity by the reduction of pesticide use must be approached by using areawide IPM programs with biological control as their backbone. Ever-changing complexities to our landscape will increase the need for biological control. The enormous growth of urban areas is encroaching on agricultural land. This so-called "urban sprawl" has increased the concern of using insecticides near homes, schools, and commercial buildings. Urban sprawl will increase the attractiveness of biological control for pest management to farmers and suburban homeowners.

Enormous progress has been made toward improving the efficacy of biological control. Now it is critical that researchers engage in more extension activities that apply directly to consumer crops (Allen and Rajotte, 1990). For biological control to continue to prosper into the next millenium, entire communities need to become involved in the pest management decision-making process. This will require a dedicated effort from extension specialists, farmers, pest management regulators, and the general public.

ACKNOWLEDGMENTS

I thank Lindsey Flexner, Dawn Gouge, Debbie Hagler, Sujaya Udayagiri, and Robert Wiedenmann for providing helpful comments on earlier versions of this manuscript. I am grateful to Marla Lawrence and Debra Guerra for the help they provided in preparing the manuscript. Special thanks are extended to the Agriculture Research Service, USDA for providing the photographs used in this manuscript.

FOOTNOTE

Mention of a proprietary product does not constitute an endorsement or a recommendation for its use by the USDA-ARS.

REFERENCES

Akhurst, R. Safety to Nontarget Invertebrates of Nematodes of Economically Important Pests, in *Safety of Microbial Insecticides,* Laird, M., Lacey, L.A., and Davidson, E.W., Eds. CRC Press, Boca Raton, 1990.

Allen, W.A. and E.G. Rajotte. The Changing Role of Extension Entomology in the IPM Era. *Annual Review of Entomology.* 35, pp. 379–397, 1990.

Andreadis, T.G. Transmission, in *Epizootiology of Insect Diseases,* Fuxa, J.R. and Tanada, Y., Eds. John Wiley & Sons, New York, 1987.

Baerselman, F. The Dutch Attempt to Sustainable Plant Protection Practice, in *Policy Making, A Must for the Benefit of All, Policy Forum on Crop Protection Policy — 13th International Plant Protection Congress, The Hague, Netherlands, 2-7 July,* Maan, G.C. and Zadoks, J.E., Eds. 1995.

Baker, T.C., R.T. Staten, and H.M. Flint. Use of Pink Bollworm Pheromone in the Southwestern United States, in *Behavior-Modifying Chemicals for Insect Management-Applications of Phermones and Other Attractants,* Ridgway, R.L., Silverstein, R.M. and Inscoe, M.N., Eds. Marcel Dekker, New York, 1990.

Bateman, R.P., M. Carey, D. Moore, and C. Prior. The Enhanced Infectivity of *Metarhizium flavoviride* Oil Formulations to Desert Locusts at Low Humidities. *Annals of Applied Biology.* 122, pp. 145–152, 1993.

Beckage, N.E. Parasitic Effects on Host Development, in *New Directions in Biological Control, UCLA Symposium on Molecular and Cellular Biology,* Baker, R. and Dunn P., Eds. Alan R. Liss, NY, 1990.

Beddington, J.R., C.A. Free, and J.H. Lawton. Characteristics of Successful Natural Enemies in Models of Biological Control of Insect Pests. *Nature.* 273, p. 513–519, 1978.

Biddinger, D.J. and L.A. Hull. Effects of Several Types of Insecticides on the Mite Predator, *Stethorus punctum* (Coleoptera: Coccinellidae), Including Insect Growth Regulators and Abamectin. *Journal of Economic Entomology.* 88, pp. 358–366, 1995.

Bonning, B.C. and B.D. Hammock. Development of Recombinant Baculoviruses for Insect Control. *Annual Review of Entomology.* 41, pp. 191–210, 1996.

Brooks, W.M. Entomogenous Protozoa, in *CRC Handbook of Natural Pesticides,* Vol 5; *Microbial Pesticides,* Ignoffo, C.M., Ed. CRC Press, Boca Raton, 1988.

Brown, T.M. and A.W.A. Brown. Experimental Induction of Resistance to a Juvenile Hormone Mimic. *Journal of Economic Entomology.* 67, pp. 799–801, 1974.

Bugg, R.L., F.L. Wackers, K.E. Brunson, J.D. Dutcher, and S. C. Phatak. Cool-Season Cover Crops Relay Intercropped with Cantaloupe: Influence on a Generalist Predator, *Geocoris punctipes* (Hemiptera: Lygaeidae). *Journal of Economic Entomology.* 84, pp.408–416, 1991.

Bushland, R.C. Screwworm Eradication Program. *Science.* 184, pp. 1010–1011, 1974.

Caltagirone, L.E. Landmark Examples in Classical Biological Control. *Annual Review of Entomology.* 26, pp. 213–232, 1981.

Caltagirone, L.E. and R.L. Doutt. The History of the Vedalia Beetle Importation to California and its Impact on the Development of Biological Control. *Annual Review of Entomology.* 34, pp. 1–16, 1989.

Canning, E.U. An Evaluation of Protozoan Characteristics in Relation to Biological Control of Insects. *Parasitology,* 84, pp. 119–149, 1982.

Cardé, R.T. and A.K. Minks. Control of Moth Pests by Mating Disruption: Successes and Constraints. *Annual Review of Entomology.* 40, pp. 559–585, 1995.

Carson, R. *Silent Spring.* Houghton Mifflin Company, Boston, MA, 1962.

Casida, J.E. and G.B. Quistad. Golden Age of Insecticide Research: Past, Present or Future? *Annual Review of Entomology.* 43, pp. 1–16, 1998.

Cerf, D. and G.P. Georghiou. Cross-Resistance to an Inhibitor of Chitin Synthesis in Insecticide Resistant Strains of the Housefly. *Journal of Agriculture and Food Chemistry.* 22, pp. 1145–1146, 1974.

Chapple, A.C., R.A. Downer, T.M. Wolf, R.A.J. Taylor, and F.R. Hall. The Application of Biological Pesticides: Limitations and a Practical Solution. *Entomophaga.* 41, 465–474, 1996.

Cloarec, A. Handling Time and Multi-Prey Capture by a Water Bug. *Animal Behavior.* 42, pp. 607–613, 1991.

Cohen, A.C. Solid-to-Liquid Feeding: The Inside(s) Story of Extra-oral Digestion in Predaceous Arthropoda. *American Entomologist.* 3, pp. 103–117, 1998.

Cook, R.J., W.L. Bruckhart, J.R. Coulson, M.S. Goettel, R. A. Humber, R.D. Lumsden, J. V. Maddox, M.L. McManus, L. Moore, S.F. Meyer, P.C. Quimby, Jr., J.P. Stack, and J.L. Vaughn. Safety of Microorganisms Intended for Pest and Plant Disease Control: A Framework for Scientific Evaluation. *Biological Control.* 7, pp. 333–351, 1996.

Coulson, J.R., W. Klaasen, R.J. Cook, E.G. King, H.C. Chiang, K.S. Hagen, and W.G. Tendol. Notes on Biological control of Pests in China, 1979, in *Biological Control of Pests in China.* United States Department of Agriculture, Washington, D.C., 1982.

Croft, B.A. *Arthropod Biological Control Agents and Pesticides.* Wiley, New York, 1990.

Debach, P. and D. Rosen. *Biological Control by Natural Enemies,* 2nd Edition. Cambridge University Press, Cambridge, NY, 1991.

Delbeke, F., P. Vercruysse, L. Tirry, P. De Clercq, and D. Degheele. Toxicity of Diflubenzuron, Pyriproxyfen, Imidacloprid, and Diafenthiuron to the Predatory bug *Orius laevigatus* (Heteroptera: Anthocoridae). *Entomophaga.* 42, pp. 349–358, 1997.

Dhadialla, T.S., G.R. Carlson, and D.P. Le. New Insecticides with Ecdysteroidal and Juvenile Hormone Activity. *Annual Review of Entomology.* 43, pp. 545–569, 1998.

Elkinton, J.S. and A.M. Liebhold. Population Dynamics of Gypsy Moth in North America. *Annual Review of Entomology.* 35, pp. 571–596, 1990.

Feener, D.H., Jr. and B.V. Brown. Diptera as Parasitoids. *Annual Review of Entomology.* 42, pp. 73–97, 1997.

Ferron, P. Biological Control of Insect Pests by Entomogenous Fungi. *Annual Review of Entomology.* 23, pp. 409–442, 1978.

Flint, H.M., S. Kunn, B. Horn, and H.A. Saalam. Early Season Trapping of Pink Bollworm with Gossyplure. *Journal of Economic Entomology.* 67, pp. 738–740, 1974.

Flint, M.L. and S.H. Dreistadt. *Natural Enemies Handbook.* University of California Statewide Integrated Pest Management Project, University of California Press, Berkeley, CA., 1998.

Fuxa, J.R. Ecological Considerations for the Use of Entomopathogens in IPM. *Annual Review of Entomology.* 32, pp. 225–251, 1987.

Gaston, L.K., H.H. Shorey, and C.A. Saario. Insect Population Control by use of Sex Pheromone to Inhibit Orientation Between the Sexes. *Nature.* 213, pp. 1155, 1967.

Gaston, L.K., R.S. Kaae, H.H. Shorey, and D. Sellers. Controlling the Pink Bollworm by Disrupting Sex Pheromone Communication Between Adult Moths. *Science.* 196, pp. 904–905, 1977.

Gaugler, R. Ecological Considerations in the Biological Control of Soil-Inhabiting Insects with Entomopathogenic Nematodes. *Agriculture, Ecosystems and Environment.* 24, pp. 351–360, 1988.

Gaugler, R., A. Bednarek, and J.F. Campbell. Ultraviolet Inactivation of Heterorhabditid and Steinernematid Nematodes. *Journal of Invertebrate Pathology.* 59, pp. 155–160, 1992.

Gaugler, R., M. Wilson, and P. Shearer. Field Release and Environmental Fate of a Transgenic Entomopathogenic Nematode. *Biological Control.* 9, pp. 75–80, 1997.

Georgis, R. The Role of Biotechnology Companies in Commercialization of Entomopathogenic Nematodes, in *Proceedings of the International Congress of Nematologists.* 2, pp. 294–306, 1990.

Georgis, R. Present and Future Prospects for Entomopathogenic Nematode Products. *Biocontrol Science and Technology.* 2, pp. 83–99, 1992.

Georgis, R. and N.G.M. Hague. Nematodes as Biological Insecticides. *Pesticide Outlook.* 2, pp. 29–32, 1991.

Georgis, R., H.K. Kaya, and R. Gaugler. Effect of Steinernematid and Heterorhabditid Nematodes (Rhabditida: Steinernematidae and Heterorhabditidae) on Nontarget Arthropods. *Environmental Entomology.* 20, pp. 815–822, 1991.

Gill, S.S., E.A. Cowles, and P.V. Pietrantonio. The Mode of Action of *Bacillus Thuringiensis* Endotoxins. *Annual Review of Entomology.* 37, pp. 615–636, 1992.

Godfray, H.C.J. *Parasitoids: Behavioral & Evolutionary Ecology,* University Press, Princeton, NJ, 1994.

Gould, F. Potential and Problems with High-Dose Strategies for Pesticidal Engineered Crops. *Biocontrol Science and Technology.* 4, pp. 451–461, 1994.

Gould, F. Sustainability of Transgenic Insecticidal Cultivars: Integrating Pest Genetics and Ecology. *Annual Review of Entomology.* 43, pp. 701–726, 1998.

Greathead, D.J. Benefits and Risks of Classical Biological Control, in *Biological Control: Benefits and Risks,* Hokkanen, H.T.M. and Lynch, J.M., Eds. Cambridge University Press, Cambridge, U.K., 1995.

Greathead, D.J. and A.H. Greathead. Biological Control of Insect Pests by Parasitoids and Predators: The BIOCAT Database. *Biocontrol News and Information.* 13, pp. 61–68, 1989.

Grenier, S., P.D. Greany, and A.C. Cohen. Potential for Mass Release of Insect Parasitoids and Predators Through Development of Artificial Culture Techniques, in *Pest Management in the Subtropics, Biological Control — A Florida Perspective,* Rosen, D., Bennett, F.D. and Capinera, J.L., Eds. Intercept Ltd., Andover, U.K., 1994.

Hagler, J.R. and A.C. Cohen. Prey Selection by *In Vitro-* and Field-Reared *Geocoris punctipes. Entomologia Experimentalis et Applicata.* 59, pp. 201-205, 1991.

Hagler, J.R. and S.E. Naranjo. Determining the Frequency of Heteropteran Predation on Sweetpotato Whitefly and Pink Bollworm Using Multiple ELISAs. *Entomologia Experimentalis et Applicata.* 72, pp. 59–66, 1994.

Hagler, J.R. and S.E. Naranjo. Using Gut Content Immunoassays to Evaluate Predaceous Biological Control Agents: A Case Study, in *The Ecology of Agricultural Pests,* Symondson, W.O.C. and Liddell, J.E., Eds. Chapman & Hall, London, 1996.

Hagler, J.R., A.C. Cohen, F.J. Enriquez, and D. Bradley-Dunlop. An Egg-Specific Monoclonal Antibody to *Lygus hesperus. Biological Control.* 1, 75-80, 1991.

Hajek, A.E. and R.J. St. Leger. Interactions Between Fungal Pathogens and Insect Hosts. *Annual Review of Entomology.* 39, pp. 293–322, 1994.

Harris, P. Environmental Impact of Introduced Biological Control Agents, in *Critical Issues in Biological Control,* Mackauer, M., Ehler, L.E., and Roland, J., Eds. Intercept: Andover, Hants, 1990.

Hassell, M.P. and H.C.J. Godfray. The Population Biology of Insect Parasitoids, in *Natural Enemies: The Population Biology of Predators, Parasites, and Diseases,* Crawley, M.J., Ed. Blackwell Scientific Publications, Oxford, 1992.

Hazard, E.I., E.A. Ellis, and D.J. Joslyn. Identification of Microsporidia, in *Microbial Control of Pests and Plant Diseases 1970-1980,* Burges, H.D., Ed. Academic Press, London, 1981.

Heimpel, G.E. and T.R. Collier. The Evolution of Host-Feeding Behavior in Insect Parasitoids. *Biological Reviews.* 71, pp. 373–400, 1996.

Henry, J.E. Natural and Applied Control of Insects by Protozoa. *Annual Review of Entomology.* 26, pp. 49–73, 1981.

Hentz, M.G., P.C. Ellsworth, S.E. Naranjo, and T.F. Watson. Development, Longevity, and Fecundity of *Chelonus* sp. nr. *curvimaculatus* (Hymenoptera: Brachonidae), an Egg-Larval Parasitoid of Pink Bollworm (Lepidoptera: Gelechiidae). *Environmental Entomology.* 27, pp. 443–449, 1998.

Hoffmann, M.P., R.L. Ridgway, E.D. Show, and J. Matteoni. Practical Application of Mass-Reared Natural Enemies: Selected Case Histories, in Mass-Reared Natural Enemies: Application, Regulation, and Needs, Ridgway, R.L, M.P. Hoffmann, M.N. Inscoe, and C.S. Glenister, Eds. Thomas Say Publications in Entomology, Entomological Society of America, Lanham, Maryland, 1998.

Hokkanen, H.M. New Approaches in Biological Control, in *CRC Handbook of Pest Management in Agriculture,* 2nd Edition, Volume II, Pimentel, D., Ed. CRC Press, Boca Raton, 1991.

Hokkanen, H.M. and J.M. Lynch, Eds. *Biological Control — Benefits and Risks,* Cambridge University Press, Cambridge, U.K., 1995.

Hopper, K.R., R.T. Roush, and W. Powell. Management of Genetics of Biological Control Introductions. *Annual Review of Entomology.* 38, pp. 27–51, 1993.

Horowitz, A.R. and I. Ishaaya. Susceptibility of the Sweetpotato Whitefly (Homoptera: Aleyrodidae) to Buprofezin During the Cotton Season. *Journal of Economic Entomology.* 85, pp. 318–324, 1992.

Horowitz, A.R. and I. Ishaaya. Managing Resistance to Insect Growth Regulators in the Sweetpotato Whitefly (Homoptera: Aleyrodidae). *Journal of Economic Entomology.* 87, pp. 866–871, 1994.

Hoy, M.A. Recent Advances in Genetics and Genetic Improvement of the Phytoseiidae. *Annual Review of Entomology.* 30, pp. 345–370, 1985.

Hoy, M.A. Use of Genetic Improvement in Biological Control. *Agriculture, Ecosystems and Environment.* 15, pp. 109, 1986.

Hoy, M.A. Biological Control of Arthropod Pests: Traditional and Emerging Technologies. *American Journal of Alternative Agriculture.* 3, pp. 63–68, 1988.

Hoy, M.A. Genetic Improvement of Arthropod Natural Enemies: Becoming a Conventional Tactic? In *New Directions in Biological Control,* UCLA Symp. Molec. Cell. Biol., Baker, R. and Dunn, P., Eds. New Series, Vol 112, Alan R. Liss, New York, 1990.

Hoy, M.A., R.M. Nowierski, M.W. Johnson, and J.L. Flexner. Issues and Ethics in Commercial Releases of Arthropod Natural Enemies. *American Entomologist.* 37, pp. 74-75, 1991.

Hunter, C.D. Suppliers of Beneficial Organisms in North America. California Environmental Protection Agency, Department of Pesticide Regulation, Environmental Monitoring and Pest Management Branch, 1994.

Jaronski, S.T. New Paradigms in Formulating Mycoinsecticides, *Pesticide Formulations and Application Systems, 17th Volume,* ASTM STP 1328, Goss, G.R., Hopkinson, M.J. and Collins, H.M., Eds. American Society for Testing and Materials, pp. 99–112, 1997.

Jervis, M.A. and N.A.C. Kidd. Host-Feeding Strategies in Hymenopteran Parasitoids. *Biological Reviews*. 61, pp. 395–434, 1986.

Jones, W.A., M.A. Ciomperlik, and D.A. Wolfenbarger. Lethal and sublethal effects of insecticides on two parasitoids attacking *Bemisia argentifolii* (Homoptera: Aleyrodidae). Biological Control. 11, pp. 70–76, 1998.

Kaya, H.K. and R. Gaugler. Entomopathogenic Nematodes. *Annual Review of Entomology*. 38, pp. 181–206, 1993.

Kirschbaum, J.B. Potential Implication of Genetic Engineering and Other Biotechnologies to Insect Control. *Annual Review of Entomology*. 30, pp. 51–70, 1985.

Knipling, E.F. *The Basic Principles of Insect Population Suppression and Management*. USDA Agriculture Handbook, Washington, D.C. 512, 1979

Knipling, E.F. Areawide Pest Suppression and Other Innovative Concepts to Cope with our more Important Insect Pest Problems, in *Minutes 54th Annual Meeting of the National Plant Board*, National Plant Board, pp. 68-97, Sacramento, CA, 1980.

Knipling, E.F. Sterile Insect Technique as a Screwworm Control Measure: The Concept and its Development, in *Symposium on Eradication of the Screwworm from the United States and Mexico*, Graham, O.H., Ed. Miscellaneous Publications of the Entomological Society of America, College Park, MD, 1985.

Knutson, A. and J. Ruberson. *Field Guide to Predators, Parasites and Pathogens Attacking Insect and Mite Pests in Cotton*, in Texas Agricultural Extension Service, The Texas A&M University System, College Station, Texas, 1996.

Kogan, M. Areawide Management of Major Pests: Is the Concept Applicable to the *Bemisia* Complex?, in *Bemisia: 1995 Taxonomy, Biology, Damage, Control and Management*, Gerling, D., Mayer, R.T., Eds. Intercept, Andover, U.K., 1995.

Kogan, M. Integrated Pest Management: Historical Perspectives and Contemporary Developments. *Annual Review of Entomology*. 43, pp. 243–270, 1998.

Leppla, N.C. and E.G. King. The Role of Parasitoid and Predator Production in Technology Transfer of Field Crop Biological Control. *Entomophaga*. 41, pp. 343-360, 1996.

Lofgren, C.S., W.A. Banks, and B.M. Glancey. Biology and Control of Imported Fire Ants. *Annual Review of Entomology*. 20, pp. 1–30, 1975.

Lucas, E., D. Coderre, and J. Brodeur. Intraguild predation among aphid predators: characterization and influence of extraguild prey density. *Ecology*. 79, pp. 1084–1092, 1998.

Maddox, J.V. Possibilities for Manipulating Epizootics Caused by Protozoa: A Representative Case-History of *Nosema pyrausta*, in *Proc. 4th International Colloquium of Invertebrate Pathology*, Veldhoven, The Netherlands, 1986.

Maddox, J.V. Protozoan Diseases, in *Epizootiology of Insect Diseases*, Fuxa, J.R. and Tanada, Y., Eds. John Wiley & Sons, New York, 1987.

McGaughey, W.H. Problems of Insect Resistance to *Bacillus thuringiensis*. *Agriculture, Ecosystems and Environment*. 49, pp. 95-102, 1994.

McGaughey, W.H. and R.W. Beeman. Resistance to *Bacillus thuringiensis* in Colonies of Indian Meal Moth and Almond Moth (Lepidoptera: Pyralidae). *Journal of Economic Entomology*. 81, pp. 28–33, 1988.

McGaughey, W.H. and M.E. Whalon. Managing Insect Resistance to *Bacillus thuringiensis* Toxins. *Science*. 258, pp. 1451-1455, 1992.

Meeusen, R.L. and G. Warren. Insect Control with Genetically Engineered Crops. *Annual Review of Entomology*. 34, pp. 373–381, 1989.

Milner, R. J. Prospects for Biopesticides for Aphid Control. *Entomophaga*. 42, pp. 227–239, 1997.

Naranjo, S.E. and J.R. Hagler. Characterizing and Estimating the Effect of Heteroptera Predation, in *Predatory Heteroptera: Their Ecology and Use in Biological Control*, Coll, M. and Ruberson, J.R., Eds. Proceedings of the Thomas Say Publications in Entomology, Entomological Society of America, Lanham, Maryland, 1998.

National Academy of Sciences. *Principles of Plant and Animal Control,* Vol. 3. Washington, D.C., 1969.

National Research Council. Report of the Research Briefing on Biological Control on Managed Ecosystems. National Academy Press, Washington, D.C., 1987.

Obrycki, J.J. and T.J. Kring. Predaceous Coccinellidae in Biological Control. *Annual Review of Entomology.* 43, pp. 295–321, 1998.

Obrycki, J.J., L.C. Lewis, and D.B. Orr. Augmentative Releases of Entomophagous Species in Annual Cropping Systems. *Biological Control.* 10, pp. 30–36, 1997.

Ooi, P.A, G.S. Lim, T.H. Ho, P.L. Manalo, and J.K. Waage, Eds. Integrated Pest Management in the Asia-Pacific Region, in *Proceedings of the Conference on Integrated Pest Management in the Asia-Pacific Region, Kuala Lumpur, Malaysia, 23-27 September, 1991.* CAB International/Asian Development Bank, 1992.

O'Neil, R.J., K.L. Giles, J.J. Obrycki, D.L. Mahr, J.C. Legaspi, and K. Katovich. Evaluation of the Quality of Four Commercially Available Natural Enemies. *Biological Control.* 11, pp. 1–8, 1998.

Paré, P.W. and J.H. Tumlinson. *De Novo* Biosynthesis of Volatiles Induced by Insect Herbivory in Cotton Plants. *Plant Physiology.* 114, pp. 1101–1106, 1997.

Payne, C.C. Insect Pest Management Concepts: The Role of Biological Control, in *Proceedings Conference on Biotechnology, Biological Pesticides & Novel Plant-Resistance for Insect Pest Management.* Roberts, D.W. and Granados, R.R., Eds. Boyce Thompson Institute, Ithaca, NY, 1989.

Pinnock, D.E., J.E. Milstead, M.E. Kirby, and B.J. Nelson. Stability of Entomopathogenic Bacteria, in *Environmental Stability of Microbial Insecticides,* Hostetter, D.L. and Ignoffo, C. M. Eds. *Miscellaneous Publications of the Entomological Society of America.* 10, pp. 77–97, 1977.

Plapp, F.W., Jr. and S.B. Vinson. Juvenile Hormone Analogs: Toxicity and Cross-Resistance in the Housefly. *Pesticide Biochemistry and Physiology.* 3, pp. 131–136, 1973.

Plapp, F.W., Jr. The Nature, Modes of Action, and Toxicity of Insecticides. *CRC Handbook of Pest Management in Agriculture,* 2nd Edition, Volume II, Pimentel, D., Ed. CRC Press, Boca Raton, 1991.

Poinar, G.O., Jr. Taxonomy and Biology of Steinernematidiae and Heterorhabditidae, in *Entomopathogenic Nematodes in Biological Control,* Gaugler, R. and Kaya, H.K., Eds. CRC Press, Boca Raton, 1990.

Polis, G.A. Food Webs, Trophic Cascades & Community Structures. *Australian Journal of Ecology.* 19, pp. 121-36, 1994.

Poppy, G.M. Tritrophic Interactions: Improving Ecological Understanding and Biological Control? *Endeavour.* 21, pp. 61–65, 1997.

Rice, M.E. and C.D. Pilcher. Potential Benefits and Limitations of Transgenic Bt Corn for Management of the European Corn Borer (Lepidoptera: Crambidae). *American Entomologist.* 3, pp. 75–78, 1998.

Ridgway, R.L, M.P. Hoffmann, M.N. Inscoe, and C.S. Glenister, Eds. Mass-Reared Natural Enemies: Application, Regulation, and Needs. Thomas Say Publications in Entomology, Entomological Society of America, Lanham, Maryland, 1998.

Ridgway, R.L. and M.N. Inscoe. Mass-Reared Natural Enemies for Pest Control: Trends and Challenges, in Mass-Reared Natural Enemies: Application, Regulation, and Needs, Ridgway, R.L, M.P. Hoffmann, M.N. Inscoe, and C.S. Glenister, Eds. Thomas Say Publications in Entomology, Entomological Society of America, Lanham, Maryland, 1998.

Roberts, D.W., J.R. Fuxa, R. Guagler, M. Goettel, R. Jaques, and J. Maddox. Use of Pathogens in Insect Control, in *CRC Handbook of Pest Management in Agriculture,* 2nd Edition, Volume II, Pimentel, D., Ed. CRC Press, Boca Raton, 1991.

Rosenheim, J.A. Higher-Order Predators and the Regulation of Insect Herbivore Populations. *Annual Review of Entomology*. 43, pp. 421–447, 1998.

Sabelis, M.W. Predatory Arthropods, in *Natural Enemies: The Population Biology of Predators, Parasites, and Diseases*, Crawley, M.J., Ed. Blackwell Scientific Publications, Oxford, 1992.

Sailer, R.I. Extent of Biological and Cultural Control of Insect Pests of Crops, in *CRC Handbook of Pest Management in Agriculture*, 2nd Edition, Volume II, Pimentel, D. Ed. CRC Press, Boca Raton, 1991.

Sheck, A.L. Biotechnology and Plant Protection, in *CRC Handbook of Pest Management in Agriculture*, 2nd Edition, Volume II, Pimentel, D., Ed. CRC Press, Boca Raton, 1991.

Shorey, H.H. The Use of Chemical Attractants in Insect Control, in *CRC Handbook of Pest Management in Agriculture*, 2nd Edition, Volume II, Pimentel, D. Ed. CRC Press, Inc., Boca Raton, FL, 1991.

Shorey, H.H., R.S. Kaae, and L.K. Gaston. Sex Pheromones of Lepidoptera, Development of a Method for Pheromonal Control of *Pectinophora gossypiella* in Cotton. *Journal of Economic Entomology*. 67, pp. 347–350, 1974.

Siegel, J.P., J.V. Maddox, and W.G. Ruesink. Impact of *Nosema pyrausta* On a Brachonid, *Macrocentrus grandii* in Central Illinois. *J. Invertebrate Pathology*. 47, pp. 271–276, 1986.

Smith, S.M. Biological Control with *Trichogramma*: Advances, Successes, and Potential of Their Use. *Annual Review of Entomology*. 41, pp. 375–406, 1996.

Staal, G.B. Insect Growth Regulators with Juvenile Hormone Activity. *Annual Review of Entomology*. 20, pp. 417–460, 1975.

Stern, V.M., R.F. Smith, R. van den Bosch, and K.S. Hagen. The Integrated Control Concept. *Hilgardia*. 29, pp. 81–101, 1959.

Sullivan, D.J. Insect Hyperparasitism. *Annual Review of Entomology*. 32, pp. 49–70, 1987.

Sunderland, K.D. Progress in Quantifying Predation Using Antibody Techniques, in *The Ecology of Agricultural Pests*, Symondson, W.O.C. and Liddell, J.E., Eds. Chapman & Hall, London, 1996.

Sunderland, K.D., J.A. Axelsen, K. Dromph, B. Freier, J.L. Hemptinne, N.H. Holst, P.J.M. Mols, M.K. Petersen, W. Powell, P. Ruggle, H. Triltsch, and L. Winder. Pest Control by a Community of Natural Enemies, in *Arthropod Natural Enemies in Arable Land-III, The Individual, the Population and the Community*, Powell, W., Ed. *Acta Jutlandica*. 72, pp. 271–326, Aarhus University Press, Denmark, 1997.

Tabashnik, B.E. Evolution of Resistance to *Bacillus Thuringiensis*. *Annual Review of Entomology*. 39, pp. 47–79, 1994.

Theiling, K. M. and B.A. Croft. Pesticide Side-Effects on Arthropod Natural Enemies: A Database Summary. *Agricultural Ecosystems and the Environment*. 21, pp. 191–218, 1988.

Tisdell, C. Economic Impact of Biological Control of Weeds and Insects, in *Critical Issues in Biological Control*, Mackauer, M., Ehler, L.E. and Roland, J., Eds. Intercept, Andover, U.K., 1990.

Tumlinson, J.H., W.J. Lewis, and E.M. Vet. How Parasitic Wasps Find Their Hosts. *Scientific American*. 268, pp. 100-106, 1993.

Udayagiri, S. and R.L. Jones. Variation in flight response of the specialist parasitoid *Macrocentrus grandii* Goidanich to odours from food plants of its European corn borer host. Entomologia Experimentalis et Applicata. 69, pp. 183–193, 1993.

U.S. Congress, Office of Technology Assessment. From Research to Implementation, *Biologically Based Technologies for Pest Control*, OTA-ENV-636, Washington, D.C. U.S. Government Printing Office, September, 1995.

van den Bosch, R. *The Pesticide Conspiracy.* University of California Press, Berkeley & Los Angeles, CA, 1978.

Van Driesche, R.G. and T.S. Bellows, Jr., Eds. Steps in Classical Arthropod Biological Control. Thomas Say Publications in Entomology, Entomological Society of America, Lanham, Maryland, 1993.

Van Driesche, R.G. and T.S. Bellows, Jr. *Biological Control.* Chapman & Hall, New York, 1996.

van Lenteren, J.C. and J. Woets. Biological and Integrated Pest Control in Greenhouses. *Annual Review of Entomology.* 33, pp. 239–269, 1988.

van Lenteren, J.C., M.M. Roskam, and R. Timmer. Commercial Mass Production and Pricing of Organisms for Biological Control of Pests in Europe. *Biological Control.* 10, pp. 143–149, 1997.

Vet, L.E.M. and M. Dicke. Ecology of Infochemical Use by Natural Enemies in a Tritrophic Context. *Annual Review of Entomology.* 37, pp. 141–172, 1992.

Waage, J.K. and D.J. Greathead. Biological Control: Challenges and Opportunities, in *Biological Control of Pests, Pathogens, and Weeds,* Wood, R.K.S. and Way, M.J., Eds. Cambridge University Press, Cambridge, 1989.

Waage, J.K. and N.J. Mills. Biological Control, in *Natural Enemies: The Population Biology of Predators, Parasites, and Diseases,* Crawley, M.J., Ed. Blackwell Scientific Publications, Oxford, 1992.

Waage, J. "Yes, But Does It Work in the Field?" The Challenge of Technology Transfer in Biological Control. *Entomophaga.* 41, pp. 315-332, 1996.

Way, M.J. and K.C. Khoo. Role of Ants in Pest Management. *Annual Review of Entomology.* 37, pp. 479–503, 1992.

Wearing, C.H. Evaluating the IPM Implementation Process. *Annual Review of Entomology.* 33, pp. 17–38, 1988.

Webster, J.M. Nematology: From Curiosity to Space Science in Fifty Years. *Annals of Applied Biology.* 132, pp. 3-11, 1998.

Wellings, P.W. The Role of Public Policy in Biological Control: Some Global Trends. *Entomophaga.* 41, pp. 435-441, 1996.

Whitcomb, W.H. and K.E. Godfrey. The Use of Predators in Insect Control, in *CRC Handbook of Pest Management in Agriculture,* 2nd Edition, Volume II, Pimentel, D. Ed. CRC Press, Boca Raton, 1991.

Wiedenmann, R.N. and R.J. O'Neil. Searching Behavior and Time Budgets of the Predator *Podisus maculiventris. Entomologia Experimentalis et Applicata.* 60, pp. 83–93, 1991.

Wilson, E.O. First Word. *Omni.* 12, pp. 6, 1990.

Wilson, T.G. and J. Fabian. A *Drosophila melanogaster* Mutant Resistant to a Chemical Analog of Juvenile Hormone. *Dev. Biol.* 118, pp. 190–201, 1986.

CHAPTER **8**

Biological Control by
Bacillus thuringiensis
subsp. *israelensis*

Yoel Margalith and Eitan Ben-Dov

CONTENTS

1-56670-478-2/00/$0.00+$.50
© 2000 by CRC Press LLC

8.1   INTRODUCTION

It is estimated that after nearly half a century of synthetic pesticide application, mosquito-borne epidemic diseases such as malaria, filariasis, yellow fever, dengue and encephalitis are still affecting over two billion people. Malaria remains one of the leading causes of morbidity and mortality in the tropics. An estimated 300 to 500 million cases of malaria each year result in about one million deaths, mainly children under five, in Africa alone (WHO, 1997).

The introduction of synthetic pesticides and prophylactics initially resulted in a drop in malaria cases. However, resistance of mosquitoes to synthetic insecticides, coupled with resistance developed by the malaria-causing pathogen, *Plasmodium* spp., to various anti-malaria drugs, resulted in a dramatic increase of malaria in the tropical world (Olliaro amd Trigg, 1995; WHO, 1997). The very properties that made chemical pesticides useful — long residual action and toxicity to a wide spectrum of organisms — have brought about serious environmental problems (Van Frankenhuyzen, 1993). The emergence and spread of insecticide resistance in many species of vectors, safety risks for humans and domestic animals, the concern with environmental pollution, and the high cost of developing new chemical insecticides, made it apparent that vector control can no longer depend upon the use of chemicals

(Lacey and Lacey, 1990; Margalith, 1989; Mouchès et al., 1987; Wirth et al., 1990). An urgent need has thus emerged for environmentally friendly pesticides, to reduce contamination and the likelihood of insect resistance (Margalith et al., 1995; Van Frankenhuyzen, 1993).

Thus, increasing attention has been directed toward biological control agents, natural enemies such as predators, parasites, and pathogens. Unfortunately, none of the predators or parasites can be mass-produced and stored for long periods of time. They all must be reared *in vivo*. The ideal properties of a biological agent are: high specific toxicity to target organisms; safety to non-target organisms; ability to be mass produced on an industrial scale; long shelf life; and application using conventional equipment and transportability (Federici, 1995; Lacey and Lacey, 1990; Margalith, 1989; McClintock et al., 1995; Van Frankenhuyzen, 1993).

8.1.1 *Bacillus thuringiensis* (Bt) as an Environmentally Safe Biopesticide

Bacillus thuringiensis (Bt) fulfills the requisites of an "ideal" biological control agent better than all other biocontrol agents found to date, thus leading to its widespread commercial development. Bt is a gram-positive, aerobic, endospore-forming saprophyte bacterium, naturally occurring in various soil and aquatic habitats (Aronson, 1994; Kumar et al., 1996; Lacey and Goettel, 1995; Van Frankenhuyzen, 1993). Bt subspecies are recognized by their ability to produce large quantities of insect larvicidal proteins (known as δ-endotoxins) aggregated in parasporal bodies (Bulla et al., 1980; Kumar et al., 1996). These insecticidal proteins, synthesized during sporulation, are tightly packed by hydrophobic bonds and disulfide bridges (Bietlot et al., 1990). The transition to an insoluble state presumably makes the δ-endotoxins protease-resistant and allows them to accumulate inside the cell. The high potencies and specificities of Bt's insecticidal crystal proteins (ICPs) have spurred their use as natural pest control agents in agriculture, forestry and human health (Kumar et al., 1996; Van Frankenhuyzen, 1993). The gene codings for the ICPs, that are normally associated with large plasmids, direct the synthesis of a family of related proteins that have been classified as *cryI–VI* and *cytA* classes (the old nomenclature), depending on the host specificity (lepidoptera, diptera, coleoptera, and nematodes) and the degree of amino acid homology (see Table 8.1 and Feitelson et al., 1992; Höfte and Whiteley, 1989; Tailor et al., 1992). The current classification (*cry1–28* and *cyt1–2* group genes) is uniquely defined by the latter criterion(Crickmore et al.,1998; http://www.biols.susx.ac.uk/home/Neil_ Crickmore /Bt/index.html).

8.1.2 *Bacillus thuringiensis* subsp. *israelensis* (Bti)

Biological control of diptera in general and mosquitoes in particular has been the subject of investigation for many years. Biocontrol agents found to date which are active against diptera larvae include several species of larvivorous fish, mermithid nematode, fungi, protozoa, viruses, the bacteria, Bt, *B. sphaericus*, and *Clostridium bifermentis* (Delecluse et al., 1995a; Federici, 1995; Lacey and Goettel, 1995; Lacey and Lacey, 1990). *Bacillus thuringiensis* subsp. *israelensis* (Bti) was the first subspecies of Bt,

Table 8.1   Current and Original Nomenclature of *cry* Genes and Host Specicfity

Original (based on host specificity and degree of amino acid homology)	Current (based solely on amino acid identity)	Host Specificity
cryI	*cry1, cry2, cry9, cry15*	Lepidoptera
cryII	*cry1, cry2*	Lepidoptera, Diptera
cryIII	*cry3, cry7, cry8, cry14, cry18, cry 23*	Coleoptera
cryIV	*cry4, cry10, cry11, cry16, cry17, cry19, cry20*	Diptera
cryV	*cry1*	Lepidoptera, Coleoptera
cryVI	*cry5, cry6, cry12, cry13, cry21*	Nematode
	cry5, cry22	Hymenoptera
cytA	*cyt1, cyt2*	Diptera; cytolitic *in vitro*

which was found to be toxic to diptera larvae. In the summer of 1976, as part of an ongoing survey for mosquito pathogens, we came across a small pond in a dried-out river bed in the north central Negev Desert near Kibbutz Zeelim (Goldberg and Margalith, 1977; Margalith, 1990). A dense population of *Culex pipiens* complex larvae were found dying, on the surface, in an epizootic situation. The etiological agent was later identified and designated by Dr. de Barjac of the Pasteur Institute of Paris (Barjac, 1978) as a new (H-14) serotype.

Bti was found to be much more effective against many species of mosquito and black fly larvae than any previously known biocontrol agent. Bti in addition to being biologically effective, possesses all of the desirable properties of an "ideal" biocontrol agent as mentioned above (Becker and Margalith, 1993; Federici et al., 1995). Bti has been shown to be completely safe to the user and the environment. Extensive mammalian toxicity studies clearly demonstrate that the tested isolates are not toxic or pathogenic (McClintock et al., 1995; Murthy, 1997; Siegel and Shadduck, 1990). The extensive laboratory studies, coupled with no reported cases of human or animal disease after more than 15 years of widespread use, clearly argue for the safety of this active microbial biocontrol agent (McClintock et al., 1995; Siegel and Shadduck, 1990). Due to its high specificity, Bti is remarkably safe to the environment; it is non-toxic to non-target organisms (except for a few other nematocerous Diptera and only when exposed to much higher than recommended rates of application) (Margalith et al., 1985; Mulla, 1990; Mulla et al., 1982; Painter et al., 1996; Ravoahangimalala et al., 1994). No resistance has been detected to date toward Bti in field populations of mosquitoes despite 15 years of extensive field usage (Becker and Ludwig, 1994; Georghiou et al., 1990; Becker and Margalith, 1993; Margalith et al., 1995). Bti has been proven over the years to be a highly successful control agent against mosquito and black fly larvae and has been integrated into vector control programs at the national and international levels.

8.1.3 Mosquitocidal Bt and Other Microbial Strains

Recent extensive screening programs (Ben-Dov et al., 1997; Ben-Dov et al., 1998; Prieto-Samsonov et al., 1997) have expanded the number of novel microbial strains active against diptera. The current status of microbial mosquitocidal strains which harbor diptera-specific Cry toxins fall into three groups of Bt and one other group of *Clostridium*, based on the classification suggested by Delécluse et al., 1995a.

1. Bt strains which demonstrate larvicidal activity as potent as Bti and contain all four major Bti toxins Cry4A, Cry4B, Cry11A and Cyt1Aa, but belong to different serotypes (Delecluse et al., 1995a; Lopez-Meza et al., 1995; Ragni et al., 1996); Bt *kenyae* (serotype H4a, 4c), Bt *entomocidus* (serotype H6), Bt *morrisoni* (serotype H8a, 8b), Bt *canadensis* (serotype H5a, 5c), Bt *thompsoni* (serotype H12), Bt *malaysiensis* (serotype H36), Bt AAT K6 and Bt AAT B51 (two last autoagglutinated strains that cannot be serotyped). These results demonstrate that the 125 kb transmissible plasmid (Gonzalez and Carlton, 1984) bearing these insecticidal genes occurs in ecologically diverse habitats as well as in different subspecies of Bt. Moreover, the latter finding in conjunction with previous studies shows further that the serotype/subspecies designation used to classify isolates of this bacterium is not a definitive indicator of the insecticidal spectrum of activity.
2. Bt strains producing different toxins nearly as active as Bti (Delecluse et al., 1995b; Kawalek et al., 1995; Orduz et al., 1996, 1998; Rosso and Delecluse, 1997a; Thiery et al., 1997); Bt *jeguthesan* (H28a, 28c) and Bt *medellin* (H30).
3. Bt strains synthesizing different toxins but displaying weak activity (Drobniewski and Ellar, 1989; Held et al., 1990; Ishii and Ohba, 1997; Lee and Gill, 1997; Ohba et al., 1995; Smith et al., 1996; Yomamoto and McLaughlin, 1981; Yu et al., 1991); Bt *kurstaki* (H3a 3b), Bt *fukuokaensis* (H3a, 3d, 3e), Bt *canadensis* (serotype H5a, 5c), Bt *aizawai* (H7), Bt *darmstadiensis* (H10a, 10b), Bt *kyushuensis* (H11a, 11c), and Bt *higo* (H44).
4. Anaerobic bacterium which produce mosquitocidal toxins; *Clostridium bifermentas* subsp. *malaysia* (CH18), *C. bifermentas* subsp. *paraiba*, *C. septicum* strain 464 and *C. sordelli* strain A1 (Barloy et al., 1996; Barloy et al., 1998; Delecluse et al., 1995a; Seleena et al., 1997). Existence of *cry* genes associated with transposable elements may indicate that transfer of these genes occurs from one bacterial species to another and suggests that cry-like genes are widely distributed between bacterial species (Barloy et al., 1998).

A second *Bacillus* species, *B. sphaericus*, has potential as a mosquito larvicide. Bs contains binary toxin and Mtx toxins, but its host range is considerably narrower, being toxic mostly against *Culex* species (Porter et al., 1993). Resistance has recently been demonstrated to *B. sphaericus* in a laboratory colony of *Culex quinquefasciatus* (Rodcharoen and Mulla, 1996) and under natural conditions (Silva-Filha et al., 1995). Production costs are higher for *B. sphaericus* than for Bti since carbohydrates cannot be utilized as a carbon source, and production relies upon more expensive amino acids. Recently, a third crystal forming *Bacillus* species, *Bacillus laterosporus*, has been found to be effective against *Aedes aegypti*, *Anopheles stephensi* and *Culex pipiens* (Orlova et al., 1998).

Among the above mosquitocidal isolates, Bti remains the most potent against the majority of the mosquito species. Microbial agents in groups 2, 3, 4 (see above) and Bs are not as toxic as Bti, but produce toxins related to those found in Bti, and therefore these toxic genes may prove useful for recombinant strain improvement for overcoming potential problems associated with resistance (Lee and Gill, 1997).

It has recently been reported that Bt strains, such as Bti (HD567), Bt *kurstaki* (HD1) and Bt *tenebrionis* (NB-125), which were isolated from various food items and are used commercially for insect pest management (Damgard et al., 1996) demonstrated enterotoxin activity very similar to that of *B. cereus* FM1 (Asano et al., 1997). However, these Bt strains have been used for decades as insecticides, and have been applied on a large scale to food crops and unlike *B. cereus* (which contains enterotoxin-causing diarrhea in higher animals); there is no report that substantiates the human health problem caused by Bt (McClintock et al., 1995).

8.1.4   Expanded Host Range of Bti

Horak et al. (1996) recently demonstrated that the water-soluble metabolite of Bti (M-exotoxin, which belongs to same class as β-exotoxin, but has shown no activity in animal tests) was toxic to aquatic snails, including *Biomphalaria glabrata* and on cercariae of seven trematode species including a human parasitic species, *Schistosoma mansoni* and an avian parasite, *Trichobilharzia szidati*.

An expanded host range of Bti was recently found by several investigators: larvicidal activity was demonstrated against *Tabanus triceps* (Thunberg) (Diptera: Tabanidae) (Saraswathi and Ranganathan, 1996), Mexican fruit fly, *Anastrepha ludens* (Loew) (Diptera: Tephritidae) (Robacker et al., 1996), fungus gnats, *Bradysia coprophila* (Diptera: Sciaridae) (Harris et al., 1995), *Rivellia angulata* (Diptera: Platystomatidae) (Nambiar et al., 1990) and root-knot nematode, *Meloidogyne incognita* on barley (Sharma, 1994). Recently, Bti has been used for the control of nuisance chironomid midges (Ali, 1996; Kondo et al., 1995a; Kondo et al., 1995b).

8.1.5   Limited Application of Bti

Application of Bti for mosquito control is limited by short residual activity of current preparations, under field conditions (Becker et al., 1992; Eskils and Lovgren, 1997; Margalith et al., 1983; Mulla, 1990; Mulligan et al., 1980). The major reasons for this short residual activity are: (a) sinking to the bottom of the water body (Rashed and Mulla, 1989); (b) adsorption onto silt particles and organic matter (Margalith and Bobroglo, 1984; Ohana et al., 1987); (c) consumption by other organisms to which it is nontoxic (Blaustein and Margalith, 1991; Vaishnav and Anderson, 1995); and (d) inactivation by sunlight (Cucchi and Sanchez de Rivas, 1998; Hoti and Balaraman, 1993; Liu et al., 1993). In order to overcome these disadvantages, efforts are being made to improve effectiveness of Bti by prolonging its activity as well as targeting delivery of the active ingredient in the feeding zone of the larvae. These improvements are being facilitated by development of new formulations utilizing conventional and advanced tools in molecular biology and genetic engineering.

Originally isolated from a temporary pond with *Cx. pipiens* larvae (Goldberg and Margalith, 1977), Bti seems able to reproduce and survive under natural conditions, but the actual reproduction cycle is still a mystery. Recycling of ingested spores in the carcasses of mosquito larvae (Aly et al., 1985; Barak et al., 1987; Khawaled et al., 1988; Zaritsky and Khawaled, 1986) and pupae (Khawaled et al., 1990) was demonstrated for Bti in the laboratory. Manasherob et al. (1998b) recently described a new possible mode of Bti recycling in nature by demonstrating that, at least under laboratory conditions, the bacteria can recycle in climate protozoan *Tetrahymena pyriformis* food vacuoles. Recycling is thus not restricted to carcasses of its target organisms: *B. thuringiensis* subsp. *israelensis* can multiply in non-target organisms as well.

8.2   STRUCTURE OF TOXIN PROTEINS AND GENES

The family of related ICPs, encoded by genes that are normally associated with large plasmids (Lereclus et al., 1993), have been classified as *cryI–VI* and *cytA* classes on the basis of their host specificity (lepidoptera, diptera, coleoptera and nematodes; the old nomenclature) (Feitelson et al., 1992; Höfte and Whiteley, 1989) and depending on the degree of amino acid homology as *cryI–22* and *cytI–2* classes (the current classification) (Crickmore et al., 1998; http://www.biols.susx.ac.uk/home/ Neil_Crickmore/Bt/index.html). The ICPs of Bt strains contains two classes of toxins Cry: insecticidal and the Cyt, cytolytic δ-endotoxins. Cyt δ-endotoxins are found only in Dipteran-specific Bt strains. Although these toxins are not related structurally, they are functionally related in their membrane-permeating activities.

8.2.1   The Polypeptides and Their Genes

The larvicidal activity of Bti is localized in a parasporal, proteinaceous crystalline body (δ-endotoxin) synthesized during sporulation (Porter et al., 1993) and is composed of at least four major polypeptides (δ-endotoxins), with molecular weights of about 27, 72, 128 and 135 kDa (as calculated from the derived amino acid sequences of the genes), encoded by the following respective genes: *cytIAa*, *cryIIA*, *cry4B* and *cry4A* (see Table 8.2 and Federici et al., 1990; Höfte and Whiteley, 1989). The specific mosquitocidal properties are attributed to complex, synergistic interactions between the four proteins, Cry4A, Cry4B, Cry11A and CytIAa, but still the whole crystal is much more toxic than combination of these four proteins (Crickmore et al., 1995; Federici et al., 1990; Poncet et al.,1995; Tabashnik, 1992). In addition, the Bti parasporal body contains at least three minor polypeptides: Cry10A, Cyt2Ba, and 38 kDa protein (Table 8.2) which might contribute to the overall toxicity of Bti (Guerchicoff et al., 1997; Lee et al., 1985; Thorne et al., 1986). Expression in recombinant bacteria and sequence determinations yielded the following information:

1. Cry4A protoxin is encoded by a sequence of 3543 bp (1180 amino acids) and determined by SDS-PAGE as 125 kDa (Sen et al., 1988; Ward and Ellar,1987) Cry4A toxin (48 to 49 kDa) is toxic to the larvae of all three mosquito species: *Ae. aegypti*, *An. stephensi* and *Cx. pipiens* (Angsuthanasombat et al., 1992; Poncet

Table 8.2   δ-endotoxin Proteins of *B. thuringiensis* subsp. *israelensis* Parasporal Inclusion Body

Major Toxins and % in a Crystal[a]	Predicted Mol Mass (kDa)	Predicted No. of Amino Acids	Raning by SDS-PAGE (kDa)	Activated Toxin (kDa)	Transcriptional σ-Factors	Toxicity (function)[b]
Cry4A (12–15%)	134.4	1180	125	48–49	σ^H, σ^E, σ^K	Cx > Ae > An Synergistic
Cry4B (12–15%)	127.8	1136	135	46–48	σ^H, σ^E	An > Ae > Cx[c] Synergistic
Cry11A (20–25%)	72.4	643	65–72	30–40	σ^H, σ^E, σ^K	Ae > Cx > An Synergistic
Cyt1Aa (45–50%)	27.4	248	25–28	22–25	σ^E, σ^K	Ae > Cx > An (in high con.) Highly synergistic; Suppress resistance; Haemo and cytolytic *in vitro*
Minor Toxins						
Cry10A	77.8	675	58	?	ND[d]	Ae > Cx[c] Synergistic
Cyt2Ba	29.0	263	25	22.5	σ^E	Haemolytic; Potentially synergistic
38 kDa			38	ND	ND	Non-toxic to Ae larvae

[a] Six genes encoding these polypeptides are located on a plasmid 125 kb (75 MDa; see Figure 8.1). Gene encoding the 38 kDa protein is located on a 66 MDa plasmid (Purcell and Ellar, 1997).
[b] Toxicity of δ-endotoxin proteins against Cx, *Culex pipiens*; Ae, *Aedes aegypti* and An, *Anopheles stephensi*.
[c] Both polypeptides Cry4B and Cry10A are needed for the toxicity against *Cx. pipiens*.
[d] Not determined.

et al.,1995). The gene, *cry4A*, is carried on a 14 kb-*Sac*I fragment, which contains two insertion sequences (ISs) — namely IS240A and B — lying in opposite orientations and forming a composite transposon-like structure (Bourgouin et al., 1988). The ISs of 865 bp, each differing in six bases only, contain 16 bp of identical terminal inverted repeats and an open-reading-frame (Orf), encoding 235 amino acids of putative transposase (Delecluse et al., 1989). Six copies of ISs were found on the 125 kb plasmid, the Orfs of which differ in five amino acids only (Bourgouin et al., 1988; Rosso and Delecluse, 1997b).

2. Cry4B is encoded by a sequence of 3408 bp (1136 amino acids) and determined by SDS-PAGE as 135 kDa (Chungiatupornchai et al., 1988; Sen et al., 1988). Its gene, *cry4B*, is found on a 9.9 kb-*Sac*I (Bourgouin et al., 1988) or on 9.6 kb-*Eco*RI fragment. Two Orfs: Cry10A (58 to 65 kDa, Orf1) and Orf2 (56 kDa) (Thorne et al., 1986; Delecluse et al., 1988) are found 3 kb downstream from *cry4B*. Cry4B is a protoxin, which is cleaved by proteolysis in the gut of the mosquito larva to polypeptides (46 to 48 kDa) having high larvicidal activity against *Ae. aegypti* and *An. stephensi*, and very low activity against *Cx. pipiens* (Delecluse et al., 1988; Angsuthanasombat et al., 1992). Both Cry4B and Cry10A are needed for the toxicity against *Cx. pipiens* (Delecluse et al., 1988). There is a high level of homology (40%) between the carboxylic ends of Cry4A and Cry4B, while the amino acid identity is only 25% in their amino end (Sen et al., 1988).

3. Cry10A is encoded by a sequence of 2025 bp (675 amino acids) and determined by SDS-PAGE as 58 to 65 kDa (Thorne et al., 1986). The sequence of Cry10A differs markedly from that of Cry4A and Cry4B. Cry10A shows a 65% homology to Cry4A only in the first 58 amino acids on the amino end (Delecluse et al., 1988). Cry10A contains two potential trypsin cleavage sites. The first site is homolgous to that of Cry4A, whereas it is identical in only two amino acids in Cry4B. The second site is homologous in all three proteins. The *orf2* is located 66 bp downstream from *cry10A* (Thorne et al., 1986) and is highly homologous (over 65%) to sequences at the carboxylic end of Cry4A and Cry4B (Delecluse et al., 1988; Sen et al., 1988). There is a theory that *cry10A* (*orf1*) and *orf2* are modifications of the *cry4* genes (Delecluse et al., 1988). When Cry10A is produced in a recombinant *B. subtilis*, *Escherichia coli* or in a Bti mutant without the 125 kb plasmid, it is converted to a 58 kDa toxin, (probably as a result of proteolysis) and demonstrate low mosquitocidal activity (Thorne et al., 1986). The 53 to 58 kDa polypeptide is also found in minor amounts in Bti crystals (Garguno et al., 1988; Lee et al., 1985)

4. Cry11A is encoded by a sequence of 1929 bp (643 amino acids) and determined by SDS-PAGE as 65 to 72 kDa (Donovan et al., 1988). It is found on a 9.7 kb-*Hind*III fragment. Cry11A is cleaved by proteolysis into two small fragments of about 30 kDa, both of which are needed for full toxicity (Dai and Gill, 1993). This polypeptide is not highly homologous to the other toxic Bti polypeptides; it rather shows some homolgy to the Cry2-type polypeptides (Höfte and Whiteley, 1989; Porter et al., 1993). The 72 kDa protein isolated from the crystal has the highest larvicidal activity against *Ae. aegypti, Cx. pipiens* and less against *An. stephensi* (Poncet et al., 1995).

5. Cyt1Aa is encoded by a sequence of 744 bp (248 amino acids), localized on a 9.7 kb-*Hind*III fragment (Waalwijck et al., 1985). It is toxic to some vertebrate and invertebrate cells and causes lysis of mammalian erythrocytes (Thomas and Ellar, 1983a). The cytotoxicity seems to derive from an interaction between its hydrophobic segment and phospholipids in the membrane, which is thus perforated. Recombinant *E. coli* cells expressing *cyt1Aa* lose viability, probably as a result of an immediate inhibition of DNA synthesis (Douek et al., 1992). Cyt1Aa has low

Table 8.3   Sequence Alignment of the Cyt1Aa1 from Bti to Cytolitic Toxins from Different Bt Strains[a]

Cyt-type Toxin	Seq. Similarity to Cyt1Aa (%)	Seq. Identity to Cyt1Aa (%)	Bt Strains and Their Serotypes
Cyt1Aa3	99.6	99.6	Bt *morrisoni* (H14)
Cyt1Ab1	90.7	86.3	Bt *medellin* (H30)
Cyt1Ba1	74.5	65.0	Bt *neoleoensis* (H24)
Cyt2Aa1	53.9	46.1	Bt *kyushuensis* (H11a, 11c), *darmstadiensis* (H10a, 10b)
Cyt2Ba1	50.8	43.5	Bt *israelensis* (H14)
Cyt2Bb1	51.1	42.1	Bt *jegathesan* (H28a, 28c)
CytC not sequenced			Bt *fukuokaensis* (H3a, 3d, 3e)

[a] Alignment and comparisons of amino acid sequences of cytolitic toxins were performed with the Genetic Computer Group package (BestFit program; creates an optimal alignment of the best segment of similarity between two sequenses). GenBank accession number of Cyt sequences were as follows: X03182 for Cyt1Aa1; Y00135 for Cyt1Aa3; X98793 for Cyt1Ab1; U37196 for Cyt1Ba; Z14147 for Cyt2Aa; U52043 for Cyt2Ba; and U82519 for Cyt2Bb.

larvicidal activity, but in combination with Cry4A, Cry4B and/or Cry11A toxins, a synergistic effect is achieved. This synergistic effect is greater than that obtained by a combination of three Cry polypeptides only (Crickmore et al., 1995; Wirth et al., 1997). The sequence of Cyt1Aa does not show any homology to genes encoding other δ-endotoxin polypeptides (Porter et al., 1993) but play a critical role in delaying the development of resistance to Bti's Cry proteins (Georghiou and Wirth, 1997; Wirth and Georghiou, 1997; Wirth et al., 1997). To date, seven cytolitic, mosquitocidal specific toxins from different Bt strains are known (see Table 8.3 and Cheong and Gill, 1997; Drobniewski and Ellar, 1989; Earp and Ellar, 1987; Guerchicoff et al., 1997; Koni and Ellar, 1993; Thiery et al., 1997; Yu et al., 1997). These toxins demonstrate cytolitic activity *in vitro* and highly specific mosquitocidal activity *in vivo* which imply a specific mode of action. Moreover, these Cyt toxins contain several conserved regions observed in loop regions as well as in α-helices and β-strands (Cheong and Gill, 1997; Thiery et al., 1997).

6. A new gene, *cyt2Ba* encoding for the 29 kDa (263 amino acids) cytolytic toxin and run by SDS-PAGE as 25 kDa, has recently been detected in Bti and other mosquitocidal subspecies (Guerchicoff et al., 1997). It is found on a 10.5 kb-*Sac*I about 1 kb upstream from *cry4B*. The toxin, Cyt2Ba, was found at very low concentrations in their crystals. Cyt2Ba is highly homologous (67.6%) to the Cyt2Aa toxin from Bt subsp. *kyushuensis*. In addition, a stabilizing sequence at the 5′ mRNA of *cyt2Ba,* which resembled that described for *cry3* genes, was found (Guerchicoff et al., 1997). Truncated 22.5 kDa Cyt2Ba (by *Ae. aegypti* gut extract) was shown to be hemolytic against human erythrocytes. A synergistic effect was demonstrated when Cyt2Ba was combined with Cry4A, Cry4B, and Cry11A, respectively; therefore, Cyt2Ba may also contribute to the overall toxicity of Bti (Purcell and Ellar, 1997).

7. A gene encoding a 38 kDa protein is located on a 66 MDa plasmid (and not on 75 MDa which contains all other δ-endotoxin genes). This protein is found in the Bti inclusion body (Lee et al., 1985; Purcell and Ellar, 1997) and its function is still unknown (38 kDa protein alone was not toxic to *Ae. aegypti* larvae) (Lee et al., 1985).

8.2.2   Accessory Proteins (P19 and P20)

Large ICPs (130 to 140 kDa) have conserved C-terminal halves participating in spontaneous crystal formation via inter and intra-molecular disulphide bonds (Bietlot et al., 1990; Couche et al., 1987). The smaller ICPs, which do not possess the conserved C-terminal domain, may require assistance in crystal formation. Cry2A, Cry11A and Cyt1Aa indeed require the presence of accessory proteins for assembly of an inclusion body (Adams et al., 1989; Crickmore and Ellar, 1992; McLean and Whiteley, 1987; Visick and Whiteley, 1991; Wu and Federici, 1995) and the genes of two former proteins are organized in operons; they are co-transcribed with genes not involved in toxicity *orf1/orf2* and *p19/p20*, respectively (Agaisse and Lereclus, 1995; Baum and Malvar, 1995; Widner and Whiteley, 1989).

At least two accessory proteins (P19 and P20) seem to be involved in Bti's δ-endotoxin production, as follows:

1. The 20 kDa product of *p20* stabilizes both Cyt1Aa and Cry11A in recombinant *E. coli* and Bt by a post-transcriptional mechanism (Adams et al., 1989; McLean and Whiteley, 1987; Visick and Whiteley, 1991; Wu and Federici, 1993; Wu and Federici, 1995). Substantially more Cry11A was produced in recombinant *E. coli* carrying the 20 kDa protein gene than in those without it (Visick and Whiteley, 1991). Induction of *cry11A* alone in *E. coli* resulted in no larvicidal activity, but when expressed together with 20 kDa protein gene, some toxicity was obtained (Ben-Dov et al., 1995). Cry11A is thus apparently degraded in *E. coli*, and partially stabilized by the 20 kDa regulatory protein. The combination of Cry11A and 20 kDa protein was larvicidal in *B. megaterium* but not in *E. coli* (Donovan et al., 1988; Chang et al., 1992). Cry11A alone was produced and formed parasporal inclusions in an acrystalliferous Bt species, but higher levels were observed in the presence of the 20 kDa protein (Chang et al., 1992; Chang et al., 1993; Wu and Federici, 1995).

    Expression of *p20* (in *cis* or in *trans*) significantly increases the amount of Cyt1Aa in *E. coli*, but not of its mRNA, implying that the effect of P20 is exerted after transcription (Adams et al., 1989; Visick and Whiteley, 1991). Expression of *cyt1Aa* alone in acrystalliferous strains of Bt was poor and no obvious inclusions were observed, but in the presence of the 20 kDa protein relatively large (larger than those of wild-type Bt) ovoidal, lemon-shaped inclusions of Cyt1Aa were produced (Crickmore et al., 1995; Wu and Federici, 1993). In the absence of P20, recombinant cells of *E. coli* and of an acrystalliferous Bt *kurstaki* lost its colony-forming ability (Douek et al., 1992; Wu and Federici, 1993). Expression of *cyt1Aa* in the presence of P20, however, preserved cell viability (Manasherob et al., 1996a; Wu and Federici, 1993). Proteolysis of Cyt1Aa in *E. coli* occurs during its synthesis or before completing its tertiary stable structure. The protein-protein interaction between P20 and Cyt1Aa occurs while Cyt1Aa is synthesized. P20 therefore protects unfolded and nascent peptide from proteolysis (Adams et al., 1989; Visick and Whiteley, 1991). These results suggest that the 20 kDa protein promotes crystal formation, perhaps by chaperoning Cyt1Aa molecules during synthesis and crystallization, concomitantly preventing them from a lethal interaction with the host.

    A chimera of *cry4A* with Δ*lacZ* (on a high copy number pUC-type plasmid) in *E. coli* when expressed with the 20 kDa protein gene *in trans* (on another compatible low copy number pACYC-type plasmid) resulted in an increased production of the fused Cry4A (Yoshisue et al., 1992). However, other researchers who cloned *cry4A* and *p20 in cis* on the high copy number plasmid (so that *cry4A* was expressed under a strong promoter and *p20* with its own promoter) in *E. coli*, did not obtain increased toxicity (Ben-Dov et al., 1995). Likewise, inclusion formation of Cry4A was not induced in acrystalliferous Bti in the presence of *p20* (Crickmore et al., 1995). Low levels of expression of the *p20* were more effective than high levels in assisting the production of Cyt1Aa (Adams et al., 1989). The balance of intracellular concentrations of the Cyt1Aa and P20 proteins could thus be important. It is conceivable that P20 increases production of the major crystal components such as Cyt1Aa and Cry11A to a greater extent than that of the minor components such as Cry4A (Yoshisue et al., 1992).

It has recently been shown that expression of *p20* could increase the rate of production of heterogenous truncated Cry1C proteins in acrystalliferous Bt *kurstaki*, and that this is apparently due to protection from endogenous proteases (Rang et al., 1996). A new finding has been reported of a P21 protein from Bt subsp. *medellin* (located upstream of *cyt1Ab* and transcribed in the same direction) which has 84% similarity to the P20 and may potentially have same chaperone-like activity (Thiery et al., 1997).

2. P19 may play a role in protein-protein interactions (as another chaperone; 11.7% of its amino acids are cysteine residues) necessary for assembly of the crystal (Dervyn et al., 1995) and stabilization by disulfide bonds (Gill et al., 1992). If P19 is involved in the crystallization process of Cyt1Aa, it is predicted to protect host cells from the lethal action of Cyt1Aa, as does P20 (Manasherob et al., 1996a; Wu and Federici, 1993). When *p19* was cloned in a pairwise combination with *cyt1Aa* using inducible expression vectors in *E. coli*, P19 did not prevent lethal action as predicted (Manasherob et al., 1996a).

P19 and Orf1 from Bt *kurstaki* are homologous (33%), but their roles in crystallization are not known yet. The electrophoretic mobility of the expression product of cloned *p19* in *E. coli* and acrystalliferous Bti corresponds to a molecular mass of about 30 kDa rather than 19 kDa (Manasherob et al., 1997a), as predicted from the coding sequence. The same slow migration anomaly was also demonstrated with Orf2 (29 kDa) from Bt *kurstaki* which has an electrophoretic mobility corresponding to a molecular mass of 50 kDa (Widner and Whiteley, 1989). This phenomenon is known to occur in small spore-coat proteins of *B. subtilis* (Zhang et al., 1993) and may shed light on the nature of P19 and its function.

8.2.3   Extra-Chromosomal Inheritance

Bti harbors eight circular plasmids, ranging in size from 5 to 210 kb (3.3 to 135 MDa) and a linear replicon of approximately 16 kb. One of the largest plasmids (125 kb) contains all genetic information for mosquitocidal activity (Gonzalez and Carlton, 1984; Sekar, 1990). The genes encoding toxic proteins have been cloned and expressed, their sequences deciphered and toxicities examined, yielding much information (see below Section 8.5, and Sekar, 1990). Toxic proteins are produced during sporulation, but the plasmid is not required for the sporulation process.

A partial restriction map was constructed and all currently known genes located (Figure 8.1) (Ben-Dov et al., 1996). The two linkage groups (with sizes of about 56 and 76 kb) have recently been aligned and full circularity proved

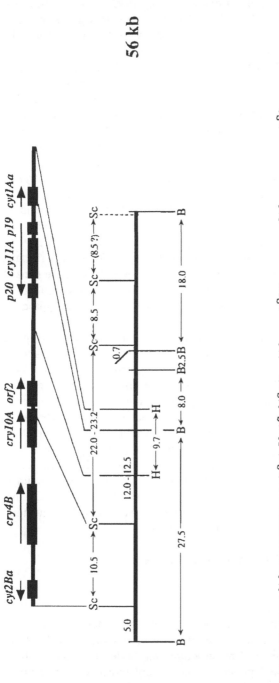

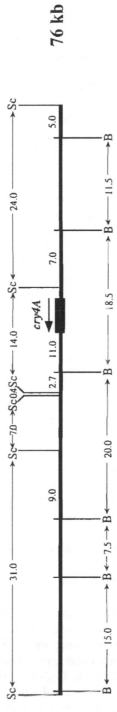

Figure 8.1   Partial restriction map of the *B. thuringiensis* subsp. *israelensis* 125 kb plasmid. Numbers indicate sizes of the relevant fragments, some of which (*Bam*HI [B], *Sac*I [Sc], and one *Hind*III [H]) are enclosed by double-headed, thin arrows and fragments of BamHI-SacI are on black thick line. Genes are indicated by black boxes and their transcription direction by thick arrows. The 26 kb (*Sac*I-HindIII) region with most of the known genes is enlarged about 2.5-fold. Based on Ben-Dov et al., 1996.

(http://www.bgu.ac.il/life/zaritsky.html; Ben-Dov et al., 1999). Five δ-endotoxin genes (*cry4B*, *cry10A*, *cry11A*, *cyt1Aa* and *cyt2Ba*), two regulatory genes (*p19* and *p20*) and another gene with an unknown function (*orf2*) were localized on a 23 kb stretch of the plasmid; however, without *cyt1Aa*, they are placed on a single 27 kb *Bam*HI fragment (Figure 8.1). This convergence enables sub-cloning of δ-endotoxin genes (excluding *cry4A*, localized on the other linkage group) as an intact natural fragment (Ben-Dov et al., 1996). The two accessory protein genes (*p19* and *p20*) are linked to *cry11A* on an operon (organized as a single transcriptional unit; Dervyn et al., 1995). *p19* is the first, *cry11A* is the second and the last, *p20*, is located 281 bp downstream from *cry11A*. *p20* is located 4 kb upstream from *cyt1Aa* and is transcribed in opposite orientation (Adams et al., 1989). All four genes occupy 5.2 kb on a single 9.7 kb *Hind*III fragment. Four additional genes (*cyt2Ba*, *cry4B*, *cry10A* and *orf2*) occupy about 11.5 kb (Ben-Dov et al., 1996; Guerchicoff et al., 1997). Several insertion sequences (IS*231*F, V, W and IS*240*A and B) have been found on the plasmid, which seem to allow transposition, duplication, rearrangement, and modification of the genes for the crystal polypeptides (Ben-Dov et al., 1999; Mahillon et al., 1994). The coding information on this plasmid, known to date, accounts for less than 20% its length. The role of the remaining 80% of the genetic information on this plasmid is still unknown and its elucidation will contribute to the understanding of the genetic interactions important for developing mosquitocidal crystal proteins.

The 125 kb plasmid can be mobilized naturally to acrystalliferous recipient strains (Cry⁻) (Gonzalez and Carlton, 1984) converting them to Cry⁺ strains. Andrup et al. (1993), distinguished between two phenotypes of aggregation, Agr⁺ and Agr⁻, which depend on the presence of a conjugative plasmid in Bti and is expressed after mixing cells of both phenotypes in exponential phase in liquid medium. Transfer of small plasmids from the Agr⁺ to the Agr⁻ cells of Bti is accompanied by formation of aggregates between donor and recipient cells (Andrup et al., 1993; Andrup et al., 1995). The genetic basis of this aggregation system and Agr⁺ phenotype is associated with the presence of the large 135 MDa self-transmissible plasmid (Andrup et al., 1998; Jensen et al., 1995; Jensen et al., 1996). Furthermore, the large plasmid is efficient in mobilizing the small "nonmobilizable" plasmids. It was suggested that this is a new mobilization mechanism of the aggregation-mediated conjugation system of Bti (Andrup et al., 1996; Andrup et al., 1998).

8.2.4   Three-Dimensional Structure of Bt Toxins

8.2.4.1   *Cry* δ-*endotoxins*

Basic studies of genetic structure and mode of action of δ-endotoxins and him receptors are very important for future development of biopesticides and for combating insect resistance mechanisms. The structure and mode of action has been studied in some depth only for the lepidoptera- and coleoptera-active toxins belonging to the Cry1 and Cry3 classes and, to a lesser extent, for the lepidoptera- and diptera-specific Cry2 and Cry4 classes. However, because the mosquitocidal proteins, particularly Cry4A, Cry4B and Cry10A, show significant amino acid sequence and secondary-structure homology with Cry1 and Cry3, and they all contain five conserved sequence blocks, it is likely that their mechanisms of action and tertiary conformations are similar (Porter et al., 1993). A major advance toward the understanding of the three-dimensional structure of Bt crystal proteins (Cry3A) was achieved by Li et al. (1991) and recently those results were complemented by Grochulski et al. (1995), who determined the tertiary structure of the Cry1Aa. The structure of Cry toxins consists of three distinct domains (I to III) which are from N- to C-terminal:

a). Domain I consists of seven-α helix bundle (for Cry1Aa, eight-α helices) (hydrophobic and amphipatic helices) arranged in an $\alpha5$-helix in the center and clearly adapted for pore formation in the insect membrane (Dean et al., 1996).

b) Domain II consists of three anti-parallel β-sheets arranged in common "Greek key" motifs (eleven β-strands) packed around a hydrophobic core (α-helix) and three surface-exposed loops at the apex of the domain. Domain II is responsible for receptor binding and host specificity determination (Dean et al., 1996).

c) Domain III consists of two anti-parallel sheets packed in a β-sandwich (twelve β-strands) and two loops which provide the interface for interactions with Domain I. It may be essential for maintaining the structural integrity of the toxins (Li et al., 1991; Nishimoto et al., 1994), and may play a role in regulation of pore-forming activity by conductance effect (Wolfersberger et al., 1996). Domain III also may contribute to the initial, specific reversible binding to the receptors (Aronson et al., 1995; Dean et al., 1996; Flores et al., 1997).

Three domains are closely packed due to van der Waals, hydrogen bond (salt bridges), and electrostatic interactions, where the largest number of interactions occur between Domains I and II (Li et al., 1991; Grochulski et al., 1995).

The crystal structure of a representative Cry toxin consists three domains, including a helix bundle able to function in pore formation and a β-sheet prism whose apical loops are probably responsible for receptor binding (Li et al., 1991). The structure of a Cyt δ-endotoxin, however, is entirely distinct from this three-domain mode (Li et al., 1996).

8.2.4.2 Cyt δ-endotoxins

The structure and function of Cyt δ-endotoxin has recently been investigated by a number of researchers. The crystal structure of Cyt2Aa (CytB) toxin was determined by isomorphous replacement using heavy-atom derivatives (Li et al., 1996). The three dimensional structure of Cyt2Aa has a single pore-forming domain, composed of two outer layers of α-helix hairpins, wrapped around mixed β-sheets (Li et al., 1996). Due to the high similarity (70% in their amino acid sequences) between Cyt1Aa and Cyt2Aa (the existence and positioning of α-helices and β-sheets in Cyt1Aa was predicted from the alignment sequences of these two genes), it was supposed that Cyt1Aa would show a similar folding pattern (Li et al., 1996; Gazit et al., 1997).

8.3   MODE OF ACTION

Early studies investigating the mode of action of Bti toxicity revealed that the primary target is the midgut epithelium, where the enzymatic systems transforms the protoxin into an active toxin under alkaline conditions. After liberation of crystal proteins by dissolution, proteolytic enzymes cleave the four major protoxins Cyt1Aa, Cry11A, Cry4A, and Cry4B to yield the active δ-endotoxin polypeptides of 22 to 25 kDa, 30 to 40 kDa, 48 to 49 kDa and 46 to 48 kDa, respectively (Al-yahyaee and Ellar, 1995; Dai and Gill, 1993; Anguthanasombat et al., 1992). These toxins act coordinately and synergistically to disrupt the epithelial cells of the larval gut (midgut cells vacuolize and lyse) (Lahkim-Tsror et al., 1983). The symptoms caused by the Cry and Cyt toxins of Bti are similar to those caused by toxins in other Bt strains, i.e., larvae become paralyzed and die within a short time.

In fact, Cry polypeptides of Bti and Cyt1Aa are not structurally related, and inevitably form pores with different structures; however, they are functionally related in their membrane-permeating ability. They also differ in their requirement of essential membranal components; the Cry toxins of Bti bind to membranal proteins (receptors) while Cyt1Aa binds to the unsaturated phospholipids acting as "binding sites" (Federici et al., 1990; Feldmann et al., 1995; Gill et al., 1992; Gazit et al., 1997; Porter et al., 1993).

8.3.1   Cry δ-endotoxins

Basically, a two-step model was proposed for the mode of action of Bt toxins by Knowles and Ellar (1987). This model consists of the δ-endotoxin binding to a cell receptor and subsequent pore formation. The δ-endotoxin is released as protoxin, which is solubilized in the midgut of insects and activated by gut proteases. It is assumed that the trigger for the insertion of the pore-forming domain (Domain I) into the epithelial cell membrane is a conformational change in the toxin. This change occurs when Domain II of the toxin binds to a receptor present on the brush-border membranes (Dean et al., 1996; Flores et al., 1997). Binding involves two steps: reversible and irreversible binding to a receptor. The irreversible binding occurs when Domain I is inserted into the plasma membrane of the cell, leading to pore formation, and is more critical than reversible binding for determining ICP specificity (Chen et al., 1995; Flores et al., 1997; Ihara et al., 1993; Rajamohan et al., 1995). Gazit and Shai (1998) recently demonstrated that only helices α4 and α5 (Domain I) of Cry3A insert into the membrane as a helical hairpin in an antiparallel manner, while the other helices lie on the membrane surface like ribs of an umbrella (the "umbrella model" Li et al., 1991), and α7 serves as a binding sensor to initiate the structural rearrangement of the pore-forming domain (Gazit et al., 1994; Gazit and Shai 1995; Gazit and Shai 1998). It was recently demonstrated that unfolding of the Cry1Aa protein around a hinge region linking Domain I and II is a necessary step for pore formation, and that membrane insertion of α4 and α5 helices (Domain I) plays a critical role in the formation of a functional pore (Schwartz et al., 1997). The suggested role for the α5 helix is consistent with the recent finding that the cleavage site of Cry4B protoxin (cut by exposure to gut enzymes *in vitro*) was found in an inter-helical loop between α5 and α6 and is extremely important for its larvicidal activity (Angsuthanasombat et al., 1993). The α4 helix (Domain I) of the Cry4B δ-endotoxin was recently demonstrated to play a crucial role in membrane insertion and pore formation. The substitution of glutamine 149 by proline in the center of helix 4 resulted in a nearly complete loss of toxicity against *Ae. aegypti* mosquito larvae (Uawithya et al., 1998).

The production of truncated proteins was achieved by sequential deletions of *cry4A* and *cry4B* genes, which resulted in minimum 75 kDa and 72 kDa active proteins, respectively (Yoshida et al., 1989a; Pao-intara et al., 1988). However, Cry4A and Cry4B protoxins digested by mosquito gut extracts were truncated to active toxins sized 48 to 49 kDa and 46 to 48 kDa, respectively (Anguthanasombat et al., 1992). Specific toxicity *in vitro* was dependent on the type of gut extract used to activate the protoxin. For example, Cry4B toxin was very toxic to *Ae. aegypti* cells when activated by gut extract from *Ae. aegypti* and was non-toxic to the same cells when treated with *Culex* gut proteases (Angsuthanasombat et al., 1992).

Mechanism of action of the Cry11A is significantly different than Cry4A and Cry4B. Cry11A has a specific pattern of proteolytic cleavage into two small fragments of about 30 kDa, which occurs even prior to solubilization, whereas proteolytic products of the solubilized protein were 40 and 32.5 kDa. The 40 kDa N-terminal fragment then further degraded to 30 kDa (Dai and Gill, 1993). It was demonstrated that cleaved Cry11A toxin has a somewhat higher toxicity than uncleaved solubilized toxin; however, the N- and C-terminal moieties of the cleaved toxin have none or very marginal larvicidal activity when applied individually. It was proposed that the N- and C-terminal fragments of cleaved Cry11A toxin probably held together as aggregate in conformation, resulting in slightly greater toxicity than the intact Cry11A polypeptide (Dai and Gill, 1993). Ligand-blotting experiments on dipteran brush border membrane vesicles (BBMVs) showed binding of Cry11A to 148 kDa and 78 kDa protein in *An. stephensi* and *Tipula oleracea*, respectively (Feldmann et al., 1995). The specific receptors for Cry4A and Cry4B still remain to be determined.

8.3.2   Cyt1Aa δ-endotoxin

Histopathological and biochemical studies investigating the mode of action of activated toxin on cultured insect cells have provided evidence that the cellular targets of the 27 kDa cytolytic toxin are the plasma-membrane liposomes containing phospholipids (Thomas and Ellar, 1983b). Toxin binding leads to a detergent-like rearrangement of the bound lipids, resulting in hypertrophy, disruption of membrane integrity, and eventually cytolysis. The binding affinity of the crystalline polypeptides to lipids containing unsaturated fatty acids is higher than that to lipids with saturated fatty acids. Incubation of the Cyt1Aa with lipids extracted from *Ae. albopictus* larvae neutralized its activity, while incubation with *B. megaterium* membranes, which do not contain suitable unsaturated phospholipids, did not neutralize toxin activity (Thomas and Ellar, 1983b). The mechanism of Cyt1Aa toxicity begins with primary binding of Cyt1Aa, as a monomer, followed after a time lag by aggregation of several molecules of Cyt1Aa which are produced in the membrane of the epithelium cells; pores are formed and, finally, cytolysis occurs (Gill et al., 1992).The pores that Cyt1Aa forms (1 to 2 nm in diameter) are selective channels to cations as K^+ and Na^+ in the phosphatidyl-ethanolamine planar bilayer with fast cooperative opening and closing. Equilibrium of these ions across the insect cell membrane results in an influx of water which leads to a colloid osmotic lysis (Knowles et al., 1989). Alkali soluble Cyt1Aa (27 kDa) is active *in vitro* against mosquito cell lines and erythrocytes, but proteolytic cleavage by trypsin and proteinase K, as well as endogenous proteases from both the N and C-termini to polypeptides of 22 to 25 kDa, enhances toxicity (Al-yahyaee and Ellar, 1995; Gill et al., 1987). Recent studies demonstrate that both Cyt1Aa and its proteolytically active form (24 kDa) are very effective in membrane permeabilization of unilamellar lipid vesicles. The 24 kDa form was about three times more effective than the protoxin (Butko et al., 1996). At least 311 and 140 aggregate-forming molecules of protoxin and Cyt1Aa activated toxin, respectively, must bind to unilamellar lipid vesicles which subsequently lose their contents via the "all-or-none mechanism." This suggests that the effect of Cyt1Aa is a general, detergent-like, perturbation of membrane rather than creation of ion-specific proteinaceous channels (Butko et al., 1996; Butko et al., 1997). Recently contradictory results were reported by Gazit et al. (1997) who demonstrated that membrane permeability of unilamellar vesicles induced by the Cyt1Aa is via formation of distinct trans-membrane pores rather than by a detergent-like effect. It is still possible that cation-selective channels and detergent-like effect in permeabilization of the membrane occur at different steps in the mode of action.

Recent studies of membrane permeation experiments suggest that Cyt1Aa toxin (with four major helices A to D and seven β1 to β7 strands) exerts its activity by aggregation of several toxin monomers (Gazit et al., 1997). Furthermore they suggest that Cyt1Aa toxin self-assembles within phospholipid membranes, and helices A and C are major structural elements involved in the membrane interaction (strong membrane permeating agents). Helices A and C, but not the β-strands and helix D, caused a large increase in the fluorescence of membrane-bound fluorescein-labeled Cyt1Aa, whereas helix B had only a slight effect. These results demonstrate that helices A and C interact specifically with Cyt1Aa and suggest that they both serve as structural elements in the oligomerization process. Intermolecular aggregation of several toxin monomers may have a direct role in the formation of pores by Cyt1Aa toxin (Gazit and Shai, 1993; Gazit et al., 1997).

In vitro binding of Bti toxins to midgut cells of *An. gambiae* larvae by immunodetection demonstrate that Cry4A, Cry4B, Cry11A, and Cyt1Aa were detected on the apical brush border of midgut cells (rich in specific receptors), in the gastric caecae and posterior stomach. Cyt1Aa was also detected in anterior stomach cells which could be related to the ability of the toxin to induce pores without requiring the participation of any specific receptor (Ravoahangimalala and Charles, 1995). A relatively higher proportion of unsaturated phospholipids in dipteran insects (as compared to other insects) can be expected to lead to a greater affinity of the Cyt δ-endotoxins to their cell membranes and activity *in vivo*. This implies a specific mode of action; however, an insect-specific protein receptor may still be essential for this toxin specificity (Koni and Ellar, 1993; Li et al., 1996). Furthermore, specificity of Cyt1Aa to certain cells may be enhanced, for example, by linking Cyt1Aa to insulin. This insulin-Cyt1Aa conjugate was toxic to cells bearing an insulin receptors (Al-yahyaee and Ellar, 1996).

8.3.3 Synergism

The insecticidal activity of Bti derives from a parasporal proteinaceous inclusion body (δ-endotoxin) which is synthesized during sporulation. The δ-endotoxin proteins differ qualitatively and quantitatively in their toxicity levels and against different species of mosquitoes (Table 8.2) (Federici et al., 1990; Poncet et al., 1995). The crystal is much more toxic than each of the polypeptides alone. The toxic activity of Cry11A, Cry4B and Cry4A is much greater than that of Cyt1Aa (Crickmore et al., 1995; Delecluse et al., 1991), but this alone does not explain the high larvicidal activity of the crystal. Different combinations of these four proteins display synergistic effects. For example, Cry4A and Cry4B display a synergistic effect against *Culex*, *Aedes*, and *Anopheles* mosquito larvae (Anguthanasombat et al., 1992; Delecluse et al., 1993; Poncet et al., 1995). Mixtures of purified Cry4A and Cry11A display significant synergy against three mosquito species (Poncet et al., 1995). Furthermore, the combination of *cry4A* and *cry11A* cloned into *E. coli* demonstrate a synergistic activity, seven-fold higher than that of *cry4A* alone, against *Ae. aegypti*, probably due to cross-stabilization of the polypeptides (Ben-Dov et al., 1995). Contradictory results regarding the synergistic activity between Cry4B and Cry11A have been reported. Crickmore et al. (1995) reported synergism between these proteins against *Ae. aegypti*, while no synergism against *Ae. aegypti* and simple additive effect against *Cx. pipiens* were reported by Poncet et al. (1995). These differences may be explained by recombinant strains used, methods of purification of inclusions, different proportions of combined toxins, and mosquito-larvicidal bioassays. Mixtures of three Cry4A, Cry4B, and Cry11A protoxins display expanded synergism against mosquito species (Crickmore et al.,1995; Poncet et al., 1995).

Cyt1Aa is the least toxic of the four δ-endotoxin proteins, but is the most active synergist with any of the other three polypeptides and their combinations (Crickmore et al.,1995; Tabashnik, 1992; Wirth et al., 1997; Wu and Chang, 1985; Wu et al., 1994). This may be related to the possible differences in the mechanism of action of Cyt1Aa and the Cry toxins. Moreover, different binding behavior of Cyt1Aa was demonstrated when it was used alone or in combination with Cry toxins of Bti, apparently due to different conformations of Cyt1Aa in the presence of Cry toxins of Bti (Ravoahangimalala and Charles, 1995; Ravoahangimalala et al., 1993). Cyt1Aa preferentially bind in the same region as the Cry toxins and this might explain the mechanism of synergism between Cry and Cyt1Aa toxins. It has already been suggested that Cyt1Aa may synergize the activity of Cry11A by facilitating the interactions between Cry11A and the target cell or the translocation of the corresponding toxic fragment (Chang et al., 1993). The mechanism responsible for synergism has not yet been clarified; however, it may be due to hydrophobic interactions between different toxins, cooperative receptor binding, and/or formation of hybrid pores, allowing a more efficient membrane permeability breakdown (Poncet et al., 1995). Because whole crystals demonstrate insecticidal activity greater even than the combinations of the four major polypeptides, several additional factors associated with the native crystal for example, minor components like Cry10A, Cry2Ba, and 38 kDa protein may induce the overall toxicity, and affect ingestion and solubilization of the whole crystal.

Recently it was demonstrated that *cryIIA* and *cytIAa* cloned into Bt field strain with dual activity against lepidoptera and diptera, are stably expressed. Diptera toxicity was enhanced by a synergistic effect between introduced and resident crystal proteins (Park et al., 1995).

8.3.4   The Properties of Inclusions and Their Interactions

Crystallization of δ-endotoxin into the inclusion body and its solubility is a main characteristic of Bt and is an important factor in susceptibility. Amount of toxin within the cell and the particular combinations of toxins depend on the following factors: availability of accessory proteins (see accessory proteins above, 8.2.2); intra- and inter-toxin bonding as disulfide bonds and salt bridges; host strain effects; sporulation dependent proteases; and growth and storage conditions of the product. These all affect the production of the crystal, proteolytic stability and its resulting solubility profile (Angsuthanasombat et al., 1992; Aronson et al., 1991; Ben-Dov et al., 1995; Bietlot et al.,1990; Chilcott et al., 1983; Couche et al., 1987; Delecluse et al., 1993; Donovan et al., 1997; Kim and Ahn, 1996; Kraemer-Schafhalter and Moser, 1996; Li et al., 1991).

For example, inclusion bodies were not formed when *cry4A* was weakly expressed, but formed when expressed at a high level or with Cry4B which could promote crystallization of Cry4A (Delecluse et al., 1993). Cry4B produced inclusions when *cry4B* was cloned on a low-copy-number plasmid in a crystal-negative strain (4Q7) of Bti (Delecluse et al., 1993); however, not as a native crystal protein body, but as a large loosely amorphous inclusion (Panjaisee et al., 1997).

The δ-endotoxins of Bt closely packed by several types of bonding like van der Waals, hydrogen bond, electrostatic interactions, and covalent disulfide bonds can affect the solubility of an inclusion body (Grochulski et al., 1995; Couche et al., 1987). Solubilization of Cry4A and Cry4B proteins (3.24 disulfide bonds per 100 kDa) occurs at pH 11.25 or higher required disulfide cleavage, where the disulfide bonds are responsible for the biphasic solubility properties of the crystal (Couche et al., 1987). CytIAa protein contains two cysteine residues and interchain disulfide bonds responsible for 52 kDa CytIAa dimer even after solubilization at pH 12 (Couche et al., 1987). Alkali-solubilized Bti δ-endotoxins contained both intra- and interchain disulfide bonds which have structural significance; it is unlikely that disulfide bonds participate in larvicidal activity (Couche et al., 1987).

Cry4A and Cry4B inclusions had different solubility when synthesized in Bti acrystalliferous strain. Solubility of Cry4A was also dependent on acrystalliferous host strain; in Bt *kurstaki* it was two-fold higher than in Bti (Angsuthanasombat et al., 1992). The combination of toxins present can affect the solubility profile of an inclusion body; for example, the absence of a Cry1Ab toxin in Bt *aizawai* has been shown to dramatically affect the solubility of inclusions, but the solubility and toxicity properties of the inclusions were restored upon reintroduction of *cry1Ab* gene (Aronson et al., 1991; Aronson, 1995). During sporulation, Bti synthesizes proteolytic enzymes (Chilcott et al., 1983; Hotha and Banik, 1997; Reddy et al., 1998) and a certain percentage of crystal proteins are susceptible to degradation by neutral protease A. In neutral protease A-deficient strains, this susceptible protein improved stability and is detected as increased full-length crystal protein (Donovan et al., 1997).

Differential solubility of distinct toxins may be used for partial separation of the toxins. Cry4A, Cry4B, and Cyt1Aa are soluble at pH 9.5 to 11, while the Cry11A toxin requires a pH of 12 (Gill et al., 1992). Furthermore, the processing of Cry11A is affected both by the physical configuration and the pH. At pH 10, (the pH of mosquito midguts), solubilization of the Cry11A parasporal crystal proceeds slowly, but proteolytic cleavage occurs simultaneously in the midgut of mosquito larvae (Dai and Gill, 1993; Feldmann et al., 1995). Different mechanisms in toxin processing in the gut are affected by pH and protease activity and may therefore explain the differences in specificity and level of toxicity against mosquito species. These differential toxin processing mechanisms may also imply a synergistic mode of action for the whole crystal.

8.4   REGULATION OF SYNTHESIS AND TARGETING

Crystal formation involves accumulation of toxin proteins. Accumulation of toxin proteins is achieved in Bt by gene expression with a strong promoter in non-dividing cells, thus avoiding protein dilution by cell division. A *Bacillus* endospore develops in a sporangium consisting of two cellular compartments, mother cell and forespore. In *B. subtilis,* the process is temporally and spatially regulated at the transcriptional level by successive activation of 5 σ factors, σ^H, σ^F, σ^E, σ^G and σ^K, respectively (σ^A is active in vegetative cells only); σ^H functions primarily during stationary phase, prior to septation (Baum and Malvar, 1995; Errington, 1993; Haldenwang, 1995). Transcription of genes within the forespore compartment required for early and late prespore development depends upon σ^F and σ^G, respectively, while early (mid-sporulation) and late (late-sporulation) transcription in the mother cell are controlled by σ^E and σ^K, respectively (Agaisse and Lereclus, 1995; Baum and Malvar, 1995). This timing and compartmentalization of σ activities in *B. subtilis* ensures precise control over gene expression during spore development (Errington, 1993; Haldenwang, 1995).

Many of the proteins that regulate sporulation in *B. subtilis* are present and appear to function similarly in Bt, including σ^E (homologous to σ^{35} of Bt) and σ^K (homologous to σ^{28} of Bt) (Adams et al., 1991; Agaisse and Lereclus, 1995; Baum and Malvar, 1995; Bravo et al., 1996). The production of ICPs in Bt normally coincides with sporulation, resulting in the appearance of parasporal crystalline inclusions within the mother cell. The dependence of δ-endotoxin gene transcription on σ^E and σ^K links its expression to sporulation to the mother-cell compartment and ensures its production throughout much of the sporulation process which contributes to the large amounts of ICP produced by Bt (Agaisse and Lereclus, 1995). The promoters of most gene codings for ICPs are dual, including one (proximal) strong

σ^E-dependent promoter, and another (distal) weak σ^K-dependent promoter (Yoshisue et al., 1997). Some of the genes are preceded only by σ^E-dependent promoters (Baum and Malvar, 1995).

To date, sporulation-specific ICP genes of Bti appear to be transcribed generally by either or both of the σ^E and σ^K forms of RNA polymerase (Table 8.2). The genes *p19, cry11A, p20* (three-gene operon) and *cytIAa* are under σ^E and σ^K transcriptional control (Dervyn et al., 1995; Baum and Malvar, 1995). Analysis of the promoter region for *cry4A* found that the gene transcribed by RNA polymerases contains σ^H, σ^E (with overlapping consensus sequences), and σ^K (Yoshisue et al., 1993a; Yoshisue et al., 1995; Yoshisue et al., 1997). While *cry4B* and *cyt2Ba* have only σ^E-dependent transcription (Guerchicoff et al., 1997; Yoshisue et al., 1993b; Yoshisue et al., 1995). Recently, it was demonstrated that *cry4B* and *cry11A* are also expressed during the transition phase by RNA polymerases associated with the σ^H, but were weaker than the *cry4A* gene (Poncet et al., 1997a). The σ^H-specific promoters for *cry4A, cry4B,* and *cry11A* overlap with σ^E-specific promoters. The 38 kDa protein begins to be synthesized during the first hour after onset of sporulation (sigma factors used still unknown) and the polypeptide accumulates as small "dot" inclusions (Lee et al., 1985). Both CytIAa and Cry11A which form rounded large and bar-shaped inclusions, respectively, are synthesized during middle and late stages of sporulation, whereas Cry4A and Cry4B, which form hemispherical to spherical body, are synthesized during midsporulation (Lee et al., 1985; Federici et al., 1990). These differences apparently indicate that the quantitative accumulation of Cry protoxins in the parasporal body of Bti, which are synthesized and assimilated in a stepwise manner, depend more on promoter strength and less on the number of promoters existing.

8.5 EXPRESSION OF BTI δ-ENDOTOXINS IN RECOMBINANT MICROORGANISMS

Expression of Bt δ-endotoxins in recombinant microorganisms is used to evaluate the toxicities of the individual proteins and to study their structure-function relationships. In addition this tool can be used to improve toxin stability in the environment, enhance expression levels, increase reproduction levels under field conditions, improve toxicity, and expand host spectrum.

The Bti toxin genes have already been expressed, in previous studies, individually or in combinations in *E. coli* (Adams et al., 1989; Angsuthanasombat et al., 1987; Ben-Dov et al., 1995; Bourgouin et al., 1986; Bourgouin et al., 1988; Chungiatupornchai et al., 1988; Delecluse et al., 1988; Donovan et al., 1988; Douek et al., 1992; McLean and Whiteley, 1987; Thorne et al., 1986; Visick and Whiteley, 1991; Ward and Ellar, 1988; Yoshisue et al., 1992), *B. subtilis* (Thorne et al., 1986; Ward et al., 1986; Ward et al., 1988; Ward and Ellar, 1988; Yoshida et al., 1989b) *B. megaterium* (Donovan et al., 1988; Sekar and Carlton 1985), *B. sphaericus* (Bar et al., 1991; Poncet et al., 1994; Poncet et al., 1997b; Servant et al., 1999; Trissicook et al., 1990), *B. thuringiensis* (Angsuthanasombat et al., 1992; Angsuthanasombat et al., 1993; Chang et al., 1992; Chang et al., 1993; Crickmore et al., 1995; Delecluse et al., 1993; Panjaisee et al., 1997; Park et al., 1995; Roh et al., 1997; Wu and Federici, 1993;

Wu and Federici, 1995), different Cyanobacteria (Angsuthanasombat and Panyim, 1989; Chungjatupornchai, 1990; Murphy and Stevens, 1992; Soltes-Rak et al., 1993; Soltes-Rak et al., 1995; Stevens et al., 1994; Wu et al., 1997), *Caulobacter crescentus* (Thanabalu et al., 1992; Yap et al., 1994a), *Ancylobacter aquaticus* (Yap et al., 1994b), Baculoviruses (Pang et al., 1992), *Bradyrhizobium* (Nambiar et al., 1990), and *Rhizobium* spp (Guerchicoff et al., 1996). Moreover, an attempt was made to obtain a broader spectrum of activity against mosquito larvae, using Bti as a heterologous host for *B. sphaericus* binary toxin genes, but without success (Bourgouin et al., 1990). Crystal negative strains of Bti can also be used as a host for expressing mosquitocidal toxin genes from other sources; for example, *cryIIBb* gene encoding the 94 kDa toxin from Bt. *medellin* was cloned and expressed in such a strain (Orduz et al., 1998; Restrepo et al., 1997). Recently, it was demonstrated that efficient synthesis of mosquitocidal toxins (binary toxin of *B. sphaericus*) in *Asticcacaulis excentricus* gram-negative bacteria has potential for mosquito control. Genetically engineered *A. excentricus* has potential advantages as a larvicidal agent especially with regard to persistence in the larval feeding zone, resistance to UV light, lack of toxin-degrading proteases, and low production costs (Liu et al., 1996).

The amount of active heterologous protein expressed depends on various factors including: regulation of replication (plasmid copy number); transcription (promoter strength, tandem promoters and σ factors); translation (efficiency of ribosomal binding site, U-rich sequence and codon usage); and mRNA stability (stem-loop structure at the 3′ end, and 5′ mRNA stabilizer) (Agaisse and Lereclus, 1995; Baum and Malvar, 1995; Chandler and Pritchard, 1975; Dong et al., 1995; Ikemura, 1981; Nordström, 1985; Soltes-Rak et al., 1995; Studier and Moffatt, 1986; Vellanoweth and Rabinowitz, 1992; Yap et al., 1994a).

8.5.1 Expression of Bti δ-endotoxins in *Escherichia coli*

Toxicity of the recombinant *E. coli* in contrast with the recombinant *Bacillus* spp, is usually poor due to weak expression of Bti δ-endotoxin genes (Bti's promoters for *cry* genes are weakly expressed in *E. coli*), low stability and proteolytic cleavage of polypeptides, and nonformation or malformation of crystals. Furthermore, the expression of *cytIAa* into *E. coli* and acrystalliferous Bt *kurstaki* kills the cells by a lethal interaction of CytIAa molecules with the host (Douek et al., 1992; Wu and Federici, 1993) and/or spore formation in latter bacteria was aberrant (Chang et al., 1993). The *cry4B* gene, however, was efficiently expressed in *E. coli* and form phase-bright insoluble inclusions which were highly toxic to *Ae. aegypti* larvae (Angsuthanasombat et al., 1987; Chungiatupornchai et al., 1988; Delecluse et al., 1988; Ward and Ellar, 1988). The best expression and highest toxicity in recombinant *E. coli* was achieved when the combination *cry4A + cryIIA*, with or without the 20 kDa protein gene was cloned under a stronger resident promoter (Ben-Dov et al., 1995). Values of LC_{50} against third instar *Ae. aegypti* larvae for these clones were about $3 \cdot 10^5$ cells ml^{-1} after 4 h induction.

8.5.2   Expression of Bti δ-endotoxins in Cyanobacteria

To overcome the low efficacy and short residual activity in nature of current formulations of Bti, and to create more stable and compatible agents for toxin delivery, toxin genes should be cloned into alternative hosts that are eaten by mosquito larvae and multiply in their habitats. Photosynthetic cyanobacterial species are attractive candidates for this purpose (Boussiba and Wu, 1995;, Boussiba and Zaritsky, 1992; Porter et al., 1993; Zaritsky, 1995): they are ubiquitous, float in the upper water layer and resist adverse environmental conditions. They are used as natural food sources for mosquito larvae (Avissar et al., 1994; Merritt et al., 1992; Stevens et al., 1994), can be cultured on a large scale (Boussiba, 1993), and are genetically manipulatable (Elhai, 1993; Elhai and Wolk, 1988; Shestakov and Khyen, 1970; Wolk et al., 1984; Wu et al., 1997). Several attempts have been made during the last decade to produce transgenic mosquito larvicidal cyanobacteria (Angsuthanasombat and Panyim, 1989; Chungjatupornchai, 1990; Murphy and Stevens, 1992; Soltes-Rak et al., 1993; Soltes-Rak et al., 1995; Stevens et al., 1994; Tandeau de Marsac et al., 1987; Wu et al., 1997; Xudong et al., 1993). Some success has been achieved in expressing single *cry* genes in unicellular species, but larvicidal activity was limited. For example, recombinant cyanobacterium *Agmenellum quadruplicatum* PR-6, bearing *cryIIA*,with its own strong phycocyanin promoter (P_{cpcB}) had very limited mosquitocidal activity against *Cx. pipiens* larvae (Murphy and Stevens, 1992). Transgenic *A. quadruplicatum* PR-6 expressing *cry4B* under the same (P_{cpcB}) promoter produced a maximum of 45% mortality against second instar *Ae. aegypti* larvae after 48 h exposure (Angsuthanasombat and Panyim, 1989). When *cry4B* was expressed in *Synechocystis* PCC 6803 from P_{psbA}, levels of the toxic polypeptide were very low and whole cells were not mosquitocidal at $4 \cdot 10^8$ cell ml^{-1} (Chungjatupornchai, 1990). Using tandem promoters for expression of *cry4B* (its own and P_{lac}) in *Synechococcus* PCC 7942 slightly improved mosquitocidal activity against first instar larvae of *Cx. restuans*, but was still insufficient (Soltes-Rak et al., 1993).

Very high mosquito larvicidal activities were achieved in the cyanobacterium *Anabaena* PCC 7120 when *cry4A* + *cryIIA*, with and without *p20*, were expressed by the dual constitutive and very efficient promoters P_{psbA} and P_{AI} (Wu et al., 1997). An additional reason that high activities were obtained is because codon usage of *Anabaena* resembles that of the four *cry* genes of Bti. The LC$_{50}$ of these clones against third instar *Ae. aegypti* larvae is ca. $9 \cdot 10^4$ cells ml^{-1}, which is the lowest reported value for engineered cyanobacterial cells with Bti toxin genes (Wu et al., 1997). In addition, the recombinant plasmids are stable inside the transgenic *Anabaena* PCC 7120; the constitutive expression of Bti Cry toxins is apparently not harmful to the host cells. Preliminary results indicate that toxicities of these clones were retained following irradiation by high doses of UV-B (at wavelengths of 280 to 330 nm), which is an important asset for Bti formulations (Manasherob et al., 1998a). These transgenic strains are thus of high potential value and have recently been patented (Boussiba et al., 1997; Boussiba et al., 1998).

8.5.3   Expression of Bti δ-endotoxins in Photoresistant *Deinoccocus radiodurans*

Commercial Bti preparations undergo rapid deactivation by sunlight in the field (Hoti and Balaraman, 1993; Liu et al., 1993); therefore, it was recently proposed to clone Bti δ-endotoxin genes into the extremely photoresistant bacterium *Deinoccocus radiodurans* R1 (Manasherob et al., 1997b). The species *Deinoccocus radiodurans* is extremely resistant to ionizing and UV radiation (Battista, 1997; Moseley, 1983) and desiccation (Mattimore and Battista, 1995). It is a gram-positive, non-sporulating and nonpathogenic diplococcus containing a red pigment. Its resistance is acquired by an exceedingly efficient DNA repair mechanism, which extends to resident plasmids with a similar efficacy (Daly et al., 1994). The characteristic pigmentation of *D. radiodurans* may play a role in resistance on the protein level by the free radical scavenging potential of its carotenoids (Carbonneau et al., 1989), which might exert protection on heterologous proteins. Additional factors which may contribute to the extreme radiation protection of proteins are: its unusually complex cell wall (Battista, 1997), UV screening by high concentrations of sulfhydryl groups, and unique lipids (Reeve et al., 1990). Indeed, cells of *D. radiodurans* R1 were found to be much more photoresistant to UV-B (280 to 330 nm) than spores of Bti (Manasherob et al., 1997b). Cloning Bti's mosquito larvicidal genes for expression in *D. radioduran* R1 is thus expected to protect them as well as their products from the harmful affects of sunlight.

8.5.4   Molecular Methods for Enhancing Toxicity of Bti

Despite the fact that no resistance has been detected to date toward Bti in field populations, laboratory-reared *Cx. quinquefasciatus* develop different levels of resistance to individual Bti toxins under heavy selection pressure (Georghiou and Wirth, 1997). Various approaches that utilize the tools of molecular biology and genetic engineering will be developed to lessen the chance of resistance development in the future. Engineered toxins with improved efficacy by differing modes of action or receptor-binding properties may be used for recombinant cloning. For example, Cry4B mutant toxin inclusion (site-directed mutagenesis of *cry4B* for replacement of arginine-203 by alanine) was twice more toxic to *Ae. aegypti* larvae than the wild-type toxin inclusion (Angsuthanasombat et al., 1993), and toxicity of Cyt1Aa mutant (lysine124 replaced by alanine) increased cytolytic activity *in vitro* by threefold (Ward et al., 1988). Recently, hyper-toxic mutant strains of Bti were isolated by mutagenising the parent strain which produce more toxin (6- to 25-fold) than the parent (Bhattacharya, 1995).

On the other hand, co-expression of natural toxins from different origins by unique combinations of their genes, chimeric toxins, or replacement of one gene on another more potent gene in the same bacterial strain may enhance larvicidal activity by a synergistic effect between them. In addition, it can delay or prevent the development of resistance and expand the host spectrum. Mosquitocidal strains Bt *medellin* and Bt *jegathesan* are less potent than Bti, but they harbor the CryIIBb and Cry11Ba toxins, respectively, which are more toxic than any of the individual Bti toxins (Delecluse et al., 1995b; Orduz et al., 1996, 1998) making good candidates for use in genetic improvement efforts.

8.6   RESISTANCE OF MOSQUITOES TO BTI δ-ENDOTOXINS

Resistance to microbial insecticides was detected in several species in the laboratory as well as in field strains of diamondback moth *Plutella xylostella* and mosquito species *Cx. quinquefasciatus* and *Cx. pipiens*. A knowledge of resistance mechanisms in insect pests is therefore very important for developing resistance management programs (Georghiou, 1994).

Resistance mechanisms include: toxin receptors, binding characteristics, competition aspects of different toxins, and physiological changes. Altered and/or slower protoxin activation by midgut proteinases, lack of major gut proteinase that activate Bt protoxins, or decreased solubility (by change in pH) in resistant subspecies (Dai and Gill, 1993; McGaughey and Whalon, 1992; Oppert et al., 1994; Oppert et al., 1997) are potential mechanisms for insect resistance to Bt toxins. The characterization of some lepidopteran-active Bt toxin receptors previously showed that the mechanism of resistance was based on reduction in binding affinity to the membrane receptor and/or decrease in receptor concentration (Ferre et al., 1991; Tabashnik et al., 1994; Van Rie et al., 1990). However, in some cases no difference in affinity of toxin to receptor was observed in susceptible and resistant larvae. The hypothesis is that the insect could attain resistance by altering toxin binding without eliminating toxin binding proteins (Gould et al., 1992; Luo et al., 1997). Other aspects of the possible mechanisms of resistance remain to be explored, including post-binding events such as membrane insertion and pore formation (Marrone and MacIntosh, 1993). In addition, it was recently demonstrated that spores of Bt increased the toxicity of Bt δ-endotoxins to both resistant and susceptible larvae of *P. xylostella*. The role of spores, therefore, may also be helpful for understanding and managing pest resistance to Bt (Liu et al., 1998).

The phenomenom of cross-resistance to Bt has been recorded in several studies (Gould et al., 1992; Tabashnik, 1994; Tabashnik et al., 1993; Tang et al., 1997). Cross-resistance may develop against Bt toxins that are similar or differ in structure and activity. Cross-resistance in the former case may be due to several Bt toxins with the same binding site, and in the latter case receptors which bind multiple toxins. However, in some cases resistance to Cry1C is inherited independently and differently than resistance to Cry1Ab and cross-resistance is not conferred (Luo et al., 1997). Multiple toxin genes with different modes of action or receptor-binding properties may reduce the chances of insect developing resistance (Tabashnik, 1994). To date, there have been no reported cases of cross-resistance between Bt toxins and synthetic insecticidal pesticides (Georghiou, 1994; Marrone and MacIntosh, 1993).

Partially or fully recessive inheritance of resistance by an autosomal trait (primarily controlled by one or few genes) to Bt was reported (Marrone and MacIntosh, 1993; McGaughey, 1985; Tabashnik, 1994; Tang et al., 1997). Recently the partially dominant inheritance of resistance (Gould et al., 1992; Liu and Tabashnik, 1997) was observed. Selection for resistance by partial dominance is affected by either high or low concentrations of toxin, depending on the Cry protein. In addition, other environmental factors may affect resistance inheritance and complicate resistance management programs (Gould et al., 1992; Liu and Tabashnik, 1997).

Significant levels of resistance and cross-resistance to different strains of *B. sphaericus* which harbor binary toxin, have already been demonstrated for *Cx. quinquefasciatus* and *Cx. pipiens* larval populations, both in the laboratory and in the field (Nielsen-LeRoux et al., 1995; Nielsen-LeRoux et al., 1997; Rodcharoen and Mulla, 1996; Silva-Filha et al., 1995). Some cases of resistance are related to a loss of the crystal toxin's binding ability (Nielsen-LeRoux et al., 1995; Rodcharoen and Mulla, 1996). One possible explanation for this is that the functionality of the receptor has been altered, and another is that a reduction in active receptor sites occurred. In other cases of resistance, the binding step remains unchanged (Nielsen-LeRoux et al., 1997; Silva-Filha et al., 1995; Silva-Filha et al., 1997). In both cases, the inheritance of the resistance was recessive, and due to a single gene (Nielsen-LeRoux et al., 1995; Nielsen-LeRoux et al., 1997). No competition between the *B. sphaericus* binary toxin and the δ-endotoxin of Bti on binding-site was observed in resistant *Cx. quinquefasciatus* populations indicating the involvement of different specific receptors (Nielsen-LeRoux et al., 1995; Rodcharoen and Mulla, 1996). The fact that the *B. sphaericus* crystal toxin binds to a single type of receptor (Nielsen-LeRoux and Charles, 1992; Nielsen-LeRoux et al., 1995), means that it is possible to obtain quick development of mosquito resistance, while this is not the case for Bti (Cheong et al., 1997; Georghiou and Wirth, 1997; Wirth et al., 1997; Wirth and Georghiou, 1997).

No resistance has been detected to date toward Bti in field populations of mosquitoes despite 15 years of extensive field usage (Becker and Ludwig, 1994; Georghiou, 1990; Becker and Margalith, 1993; Margalith et al., 1995). Selection attempts in the laboratory with natural Bti toxins have produced no resistance in *Ae. aegypti* (Goldman et al., 1986) and negligible levels of resistance in *Cx. quinquefasciatus* under heavy selection pressure (Georghiou, 1990). Resistance of *Cx. quinquefasciatus* was obtained, however, by selection to the polypeptides Cry4A, Cry4B, and Cry11A alone or in combination (Georghiou and Wirth, 1997; Wirth et al., 1997). These strains retained their original wild-type sensitivity levels to the above polypeptide combinations in the presence of moderate concentrations of Cyt1Aa, thus resistance was completely suppressed by Cyt1Aa (Wirth et al., 1997). Moreover, cross-resistance was observed between resistant strains; for example, a strain resistant to Cry11A demonstrated significant cross-resistance to Cry4A + Cry4B, and vice versa (Wirth and Georghiou, 1997). Extremely low resistance was obtained to the toxin mixture (Cry4A, Cry4B, Cry11A plus Cyt1Aa) but moderate cross-resistance levels were detected toward individual Cry toxins or their combinations (Georghiou and Wirth, 1997; Wirth and Georghiou, 1997). Despite the presence of resistance and cross-resistance to Cry proteins, all of the selected strains remained sensitive to the three Cry toxin mixture plus Cyt1Aa. In addition, resistant laboratory strains of *Cx. quinquefasciatus* to single or multiple toxins of Bti, showed cross-resistance to Cry11Ba from Bt *jegathesan* (Wirth et al., 1998). In the same study, it was found that Cyt1Aa combined with Cry11Ba can suppress most of the cross-resistance to Cry11Ba in the resistant strains. Cyt1Aa has been shown to be toxic to the Cottonwood Leaf beetle, *Chrysomela scripta*, and it also suppressed high levels of resistance to Cry3Aa found in Bt *tenebrionis* (Federici and Bauer, 1998). All the above findings suggest that the Cyt1Aa toxin may play a critical role in suppressing resistance to the Cry toxins and may be useful in managing resistance to bacterial insecticides (Cheong et al., 1997; Federici and Bauer, 1998; Wirth and Georghiou, 1997; Wirth et al., 1998).

Recently, the gene encoding cytolytic Cyt1Ab protein from Bt *medellin* (Thiery et al., 1998), Cry11A from Bti and Cry11Ba from Bt *jegatheson* (Servant et al., 1999) were introduced into *B. sphaericus* toxic strains. Production of these proteins in *B. sphaericus* partially restored susceptibility of resistant mosquito populations to the binary toxin (Servant et al., 1999; Thiery et al., 1998). Furthermore, to date several Cyt toxins have been identified (all of them host specific against mosquito larvae, Table 8.2).

Bti contains a unique natural Cry protein complex of δ-endotoxin which confers an effective defense against the development of resistance in the target organisms, and which ensures successful biological control over many years to come.

8.7   USE OF BTI AGAINST VECTORS OF DISEASES

Bti is a highly selective biological larvicide used to control mosquitoes and black fly larvae. Because of its selective activity, Bti does not harm mosquito and black fly predators such as fish, frogs, insects and crustaceans that contribute significantly to larval control. Bti is also non-hazardous to human, livestock, pets, and other forms of beneficial organisms (McClintock et al., 1995; Murthy, 1997; Siegel and Shadduck, 1990). One other significant edge for this biological larvicide is the absence of field resistance to Bti products, even when used repeatedly for over 15 years (Becker and Ludwig, 1994; Becker and Margalith, 1993; Margalith et al., 1995).

Many environmental factors affect control performance of Bti, such as water quality, solar radiation, high organic content, suspended material, water current, and weather conditions, as well as larval age and mosquito species (surface or bottom feeder). Bti has a short residual activity, often only 1 to 2 days due to adsorption to particles and sinking to the bottom, out of the feeding zone of larval mosquitoes and black flies (Blaustein and Margalith, 1991; Liu et al., 1993; Vaishnar and Anderson, 1995). Denaturation of the crystal by sunlight and engulfment of Bti by filter feeding fauna are also factors in reducing the control efficacy of Bti. Since water filters out much of the UV radiation, sunlight is much less important in aqueous habitats, where adsorption to particles and settling are the main factors in efficacy reduction (Ignoffo et al., 1981).

Presently, Bti is used in all continents. Over 300,000 liters of Bti are applied annually in Western Africa against black flies — vectors of onchocerciasis (Margalith et al., 1995). In Europe, along the Upper Rhine Valley, over 100 communities with a population of over 2.5 million people are protected through mosquito abatement programs utilizing Bti (Becker and Margalith, 1993). The number of abatement districts using Bti is rising steadily in the U.S. and Canada. With the development of appropriate formulations, effective and economic control of mosquitoes and black flies is now generally possible. Worldwide usage of Bti is increasing year by year. It is estimated that over 1000 metric tons (mt) of Bti preparations are used annually (Keller et al., 1994). So far, there have been no negative effects. In suitable formulations, this microbial agent is a useful supplement to, or replacement for, broad-spectrum chemicals in larval mosquito control programs. Further improvements, particularly to extend their residual activity and to enhance *Anopheles* control, will increase Bti consumption still further. In temperate regions, Bti offers an ecologically defensible compromise between the desire of man to protect himself from troublesome mosquitoes or black flies and the requirements of current environmental policies to protect sensitive ecosystems by the use of selective methods.

8.7.1 Formulations

Many different formulations of Bti have been developed since its establishment as a commercially viable and promising alternative to conventional pesticides several years after its discovery in 1976 (Goldberg and Margalith, 1977). Differently formulated products are required for mosquito and black fly larvae of different feeding types and habitats. Preparations, which persist at the bottom of a water container, are suited for the control of so-called bottom-feeding larvae like *Aedes* spp. The larvae of anophelines, on the other hand, are controlled most effectively by granules, which float on the water surface and release the toxin slowly. Black fly larvae live in fast-moving water courses and are controlled by pouring bacterial suspensions into the water at consecutive points, from which they are carried downstream. Development of effective formulations suited to the biology and habitats of the target organisms is the basic requirement for the successful use of Bti (Ali et al., 1994; Becker et al., 1991; Ravoahangimalala et al., 1994).

The major limiting factor in the further development of Bti is its short residual activity (Becker et al., 1992; Blaustein and Margalith, 1991; Eskils and Lovgren, 1997; Margalith and Bobroglo, 1984; Margalith et al., 1983; Mulla, 1990; Mulligan et al., 1980; Ohana et al., 1987; Rashed and Mulla, 1989). To increase environmental stability of the Bti, several studies were carried out by encapsulation with latex beads, emulsions, polyethylene, and starch-based matrices (Cheung and Hammock, 1985; Margalith et al., 1984; Schnell et al., 1984).

Efforts are being made not only to improve the efficacy of Bti by prolonging its residual effect, but also targeting delivery of the active ingredient in the feeding zone of the larvae. These improvements are primarily based on development of a variety of formulations (Ali et al., 1994). To date several commercial formulations are available: liquid concentrates, wettable powders, granules, pellets, dunks and briquets, tablets, polymer matrix, and ice granules.

8.7.1.1 Production Process

Liquid concentrate represents the most widely used and largest volume of formulation. Bti, the naturally occurring bacterium, is grown commercially in fermenting vats. During the production process, variables such as nutrients, temperature and supply of oxygen can affect bacterial growth, sporulation, and yield of crystal toxins

(Kim and Ahn, 1996; Kraemer-Schafhalter and Moser, 1996). Once the desired insecticidal activity is achieved, the bacterial cells are allowed to lyse. Spores and insecticidal crystalline proteins are harvested after approximately 24 h of fermentation time by precipitation, centrifugation, or ultrafiltration. At the end of this process, preservatives and dispersing agents are added to obtain the final liquid concentrate formulation product.

Primary powder ("technical powder") of Bti is produced by spray drying of the concentrated culture medium. *Wettable powder* formulations are produced by adding dispersing and stabilizing inert ingredients, such as bentonites and diatomes to the primary powder. Primary powder is the active ingredient used in production of granules, pellets, briquets, tablets, polymer matrix (Culigel®), and ice formulations.

Granules consist of 1 to 3 mm particles of ground corn cob impregnated with Bti.

Sand granules consist of primary powder mixed with fire-dried quartz and vegetable oil as a binding agent.

Pellets consist of ground corn cob mixed with Bti and compressed into 5 to 10 mm pellets.

Ice granule formulations have recently been developed and produced by the German Mosquito Control Association (KABS) (Becker and Mercatoris, 1997). An aqueous suspension of primary powder is sprayed into special ice machines, which transforms the Bti water suspension into ice granules using liquid nitrogen. When applied in breeding sites, melting ice granules gradually release Bti into the feeding zone of mosquito larvae, allowing a more cost-effective control operation.

Culigel®, a granulated controlled release formulation, has been developed by Lee County Mosquito Abatement District, U.S. It consists of crosslinked polyacrylamide superabsorbant matrix, capable of absorbing over $\times 100$ to *ca* $\times 5000$ their weight of water-containing primary powder. This formulation is reported to be active for well over one month under field conditions (Burges and Jones, 1998; Levy, 1989). These air-dried granules of 4 to 5 mm diameter contain up to 50% technical bacterial powder and about 50% Culigel® polymer.

Tablets of 1 cm diameter, containing Bti primary powder, have been developed in Germany by KABS in order to provide a suitable formulation for controlling *Ae. aegypti* and *Cx. pipiens* mosquitoes which breed in containers around households. A long-term effect of about 30 days was achieved in trials in Jakarta, Indonesia (Becker, 1996).

Briquets and dunks were designed by Summit Chemical Co., Baltimore, MD, U.S., as floating, sustained-release larvicides for long-term control of mosquito larvae. They consist of ground cork particles mixed with primary powder and are compressed into donut-shaped dunks measuring 5 cm in diameter with a hole in the center for anchoring. These formulations are designed to release effective levels of Bti for a period of 30 days or more (Kase and Branton, 1986).

8.7.1.2 Application Methods

For the use of Bti against larval black flies in turbulent and flowing waters, highly concentrated liquid formulations had to be produced that float as long as possible in the river current.

Microbial control formulations have been developed to improve the efficacy of Bti by prolonging its activity as well as targeting delivery of the active ingredient in the feeding zone of the larvae.

Liquid concentrate (LC) and aqueous suspensions of wettable powder (WP) are sprayed in open breeding habitats, in situations where dense weed or tree canopy does not obstruct delivery of the active ingredient. In these situations, the active biocontrol agent can easily reach the feeding zone of the target larval mosquitoes. These formulations can be applied effectively with ground application equipment such as conventional back pack; truck or tractor mounted sprayers; or by aerial application with helicopter and fixed winged aircraft. Bti formulations are applied at the rate of 2.5 liter of LC per hectare or 1 kg of WP per hectare, suspended in 400 to 500 liters water.

The corn cob granules, pellets, granular sand formulations, as well as the recently developed ice granules ("IcyCube") are dropped through tree canopy and dense vegetation, penetrating to the obstructed breeding habitats, using specially devised applicators either mounted on trucks and tractors, or helicopters and fixed winged aircraft.

Briquets, dunks, tablets and Culigel® are designed to provide prolonged residual activity, usually 30 days or more. Briquets and dunks can be used in ponds, cisterns, ditches, and other breeding places of limited dimensions. Tablets are suitable for small containers such as drinking jars, barrels, and other breeding places in and around the household. Culigel® has been designed to be used in most of the above described habitats.

Ultralow volume (ULV) technology has been utilized with considerable success in a recent effort to develop a more effective technique in dispersing Bti against larval mosquitoes of several *Aedes* species — potential vectors of diseases — that breed in confined habitats such as crevices along the tidal zones and in containers, drinking jars and tires, in and around the household, mostly in tropical and sub-tropical countries (Lee et al., 1996; Tidwell et al., 1994). Successful mosquito control operations under canopies using ULV aerial sprays have been reported as well (Cyanamid, 1992). Lim and Poorani (1998) have successfully used microdroplet ULV application techniques with Bti, using cold fog ULV generators and mist blowers — substantially enhancing the effectiveness and coverage of Bti, especially in the control of dengue vectors. They also applied an aqueous suspension mixture of Bti with organophosphate insecticides in order to control both larval and adult stages in a single application.

8.7.1.3   Encapsulation

Encapsulated formulations have been developed to provide protection against destructive UV radiation and to provide sustained release of the active ingredient for prolonged control efficacy. Several types of encapsulated products have been developed: matrices, microencapsulation, lipid droplets, and bioencapsulation. Margalith et al., 1984 made two types of capsules: 1) by stirring a mixture of Bti and dried yeast in a solution of low density polyethylene in cyclohexane; and 2) by stirring the bacteria into a slowly cooling fine emulsion of a fatty acid (decanoic, palmitic, or stearic). Both types increased the flotation coefficient and improved the insecticidal activity against *Culex* and *Aedes* larvae in glass containers with mud on the bottom.

Bti was encapsulated in alginate microcapsules in an attempt to develop durable formulations resistant to detrimental environmental conditions (Elçin, 1995). A mixture of Bti spores and crystals was also encapsulated in an insolubilized carboxymethylcellulose (CMC)-aluminum matrix to obtain a controlled-release system. This system could improve stability of the δ-endotoxin to environmental factors during and following application (Cokmus and Elçin, 1995).

Cheung and Hammock (1985) formed lipid capsules of different buoyancy. These microcapsules of 3 to 12 μm in diameter remained in the upper water column at the depth of 0.2 to 12 inches up to 24 h. Efficacy improved two to three times against *Culex* and *Aedes* larvae and twenty times against *Anopheles*.

An innovative approach has recently been devised, to concentrate and deliver Bti and its δ-endotoxin to the larvae by a ciliate protozoan, *T. pyriformis* (Manasherob et al., 1994; Manasherob et al., 1996b; Zaritsky et al., 1991; Zaritsky et al., 1992). Spores and δ-endotoxin of Bti are not destroyed during the digestion process in the food vacuoles of the protozoan *T. pyriformis*; in fact, spores germinate in excreted food vacuoles, develop to vegetative bacteria and complete a full sporulation cycle in them (Ben-Dov et al., 1994; Manasherob et al., 1998b). This approach was termed "bioencapsulation" and proved efficient under laboratory conditions (Zaritsky et al., 1991), but its usefulness in the field is still to be explored.

8.7.1.4 Standardization

Standardized methods have been developed to determine the LC_{50} values using standard formulations for comparative purposes (e.g., IPS-82 produced by Institut Pasteur).The active ingredient may be stated in several ways: 1) in International Toxic Units (ITUs); and 2) in AA units, with AA standing for *Aedes aegypti*. The original measure of Bti activity was the ITU, which was established several years ago by Pasteur Institut in France and the World Health Organization (deBarjac, 1983). A product's ITU is established by conducting mosquito bioassays using *A. aegypti* as the test species. Each AA unit (based on tests using larval instar II), however, represents less activity than an ITU (based on tests using larval instars III and IV), differing by a factor of 2.5. Dividing the AA unit value by 2.5 gives the ITU equivalent (Wassmer, 1995). Potency is usually expressed in ITU and is calculated according to the following formula:

$$\text{Potency sample } (\text{ITU/mg}) = \frac{LC_{50} \text{ standard}}{LC_{50} \text{ sample}} \times \text{potency standard } (\text{ITU/mg})$$

8.7.2 Worldwide Use of Bti Against Mosquitoes and Black Flies

Numerous strains of Bt have been isolated and tested against mosquito larvae; however, the discovery of Bti availed an agent that was highly effective against mosquito species under field conditions (Mulla, 1990). Bti is used for mosquito and black fly control on a large scale worldwide; at present, more than 1000 mt are used annually. The dramatic increase in utilization of this biopesticide is due to its properties, e.g., efficacious, safe to the environment and users, simple to produce and apply, availability of formulations, and no build-up of resistance in mosquito populations (Becker and Ludwig, 1994; Becker and Margalith, 1993; Margalith, 1989).

8.7.2.1 U.S.

Larviciding with Bti is an important component of most U.S. mosquito control programs.

Based on a U.S. EPA pesticide usage survey for 1987 though 1996, an average of 1.5 million acres of surface water had been treated annually with Bti throughout the U.S., compared to approximately 2.1 million acres of agricultural land treated with Bt. Production quantities of Bti in the U.S. were generally not available since these data are confidential business information (CBI).

The number of abatement districts using Bti is rising steadily in the U.S. and Canada. Two examples illustrate Bti usage: 1) in Massachusetts, approximately 70 mt of Bti corn cob granules were applied by helicopter in one week during August 1993 to 20,000 acres of surface water, mostly against *Ae. vexans,* one of the suspected zoonotic vectors of Eastern Equine Encephalitis (EEE) (Margalith, unpublished data); 2) in Florida, ca. 800 kg of active ingredient (primary powder) was used in 1995.

8.7.2.2 Germany

Due to public concern and growing awareness of environmental protection in Europe, microbial agents are being used increasingly in integrated control programs. Among the European countries, Germany leads the way in controlling mosquitoes with Bti.

In Germany, 89 cities and municipalities along a 300 km stretch of the Upper Rhine River have joined forces to form an agency whose purpose is to control mosquitoes, mainly *Ae. vexans* over a breeding area of over 500 square km. The responsible agency, German Mosquito Control Association (GMCA), has set up a program to address all aspects of mosquito control with emphasis on environmental protection. The control of *Aedes* mosquitoes is based solely on the use of Bti products. The program involves mapping of all mosquito breeding places, monitoring water flow in the Rhine and in the flood plains, sampling mosquito populations to determine success of control measures, documenting the environmental impact, and monitoring resistance to Bti (Becker and Ludwig, 1983; Becker and Margalith, 1993).

In the past years, approximately 50 mt of Bti powder and 25 mt of Bti liquid concentrates have been used for treating over 700 square km of breeding area. Since the widespread application of Bti, initiated in 1981, more than 90% of the *Ae. vexans* population has been successfully controlled each year. Despite extreme flooding in the past few years, mass occurrences of mosquitoes have successfully been averted. Environmental impact is minimal since all other invertebrates are unaffected and *Aedes* mosquitoes form a negligible part of the food chain. No resistance to Bti has been detected over the past 10 years (Becker and Ludwig, 1994).

All studies carried out to date have shown that since the introduction of Bti in the mosquito control program in Germany, mosquitoes are successfully controlled with minimal impact on the environment with no foreseeable resistance development (Margalith et al., 1995).

8.7.2.3   People's Republic of China

The Hubei Province, with a population of more than 20 million people on both sides of the Yangtze River is threatened by malaria, caused by *Plasmodium vivax*. Control of the main vector, *An. sinensis*, with insecticides has become increasingly difficult, due to developing resistance and ecological and toxicological risks. In routine treatments, liquid formulations of microbial control agents are applied using high pressure sprayers attached to a 600-liter tank pulled by a mini-tractor. In the years 1986–1990, about 25 tons of microbial agents (Bti and *B. sphaericus*, local strains) have been produced annually (using locally available culture media) in the Hubei Province, which was enough to treat approximately 12,000 hectares of mosquito breeding sites. The impact of the treatments was recorded by measuring the density of adult mosquitoes and the incidence of malaria before and after the campaign. Both the mosquito population density and the malaria incidence were reduced by more than 90%, from 8.2 cases per 10,000 people in 1986 to 0.8 per 10,000 in 1989 (Becker, 1996). The great success of this campaign is based on an adequate infrastructure and defined breeding sites inside or near the human settlements.

8.7.2.4   Peru, Ecuador, Indonesia, and Malaysia

In these countries, the potential of Bti to control malaria vectors such as *An. albimanus, An. rangeli, An. nigerrimus* and *An. sundaicus* was evaluated in field studies. Adult mosquito density (measured by bites per person per hour on human baits) was reduced by more than 70% using Bti at 1 to 2 kg/hectare at weekly intervals (Becker, 1996).

In Indonesia, pellets were introduced by Becker (1995, personal communication) in household water containers such as barrels, concrete cisterns, etc. where *Aedes* spp dengue vectors breed. Nearly a 100% larval mortality was obtained. In Malaysia, Lim and Poorani (1998) have successfully used microdroplet ULV application techniques with Bti in an urban situation, using cold fog ULV generators and mist blowers — substantially enhancing the effectiveness and coverage of Bti, especially in the control of dengue vectors.

8.7.2.5 Israel

Mosquito control in Israel is directed against five major mosquito species. Although malaria had been eradicated from Israel 50 years ago, two species are vectors of the West Nile Fever disease and others are a public nuisance. The growing awareness of environmental pollution, as well as development of resistance to synthetic pesticides, have been the motivating force, as in many other countries, to develop integrated pest control methods which rely more on selective biological pesticides. Bti has successfully been used in Israel to control populations of nuisance mosquitoes, allowing the continued regulation by natural enemies. A typical use pattern of Bti against mosquitoes in Israel includes a series of 45 consecutive applications at 10-day intervals. The first spray usually reduces mosquito populations by ca. 90%, the remaining treatments prevent population build-up, and from that point natural enemies suffice in keeping populations below threshold levels. Bti is used predominantly in nature reserves and other ecologically sensitive breeding sites (Margalith et al., 1995).

8.7.2.6 West Africa

The onchocerciasis control program (OCP) — Human onchocerciasis, also called river blindness, is a parasitic disease caused by a filarial worm, *Onchocerca volvulus*. In West Africa, this disease is transmitted by adult black fly females belonging to the *Simulium damnosum* complex (Diptera: Simuliidae). The larvae of these species breed in the rapids of small to large rivers. In the 1960s, human onchocerciasis was identified as a concern for public health and as a major obstacle to social and economic development in West Africa, since it prevented the settlement of fertile river valleys (Samba, 1994). International commitment led to the creation of the Onchocerciasis Control Programme in West Africa (OCP) in 1974, with the objective of eliminating this threat to health and development (Kurtak, 1986). The funds contributed by donor countries and agencies are administered by the World Bank, and WHO is the executing agency (Samba, 1994). The initial area of OCP (654,000 km^2) has been gradually expanded and now covers 11 countries of West Africa, for a total area of 1,235,000 km^2.

Strategy — Over most of this area, the main strategy of the OCP has been to interrupt transmission of the disease by controlling vector populations for a period longer than the lifespan of the adult worm in man, now known to be approximately 12 years. Vector control operations consisted of weekly aerial treatments with insecticides of river rapids where the black fly larvae breed (Agoua et al., 1991). In addition to vector control, ivermectin (a microfilaricide) is now distributed yearly to nearly two million people within the OCP area; this very safe drug provides relief from onchocerciasis symptoms, prevents the development of blindness, and consolidates the results obtained by vector control.

Between 1974 and 1979, OCP exclusively used temephos, an organophosphorous (OP) compound, for black fly control. Resistance appeared in rivers of southern Côte d'Ivoire, and in 1980 a second OP insecticide, chlorphoxim, was introduced, to which resistance appeared as well the following year (Guillet et al., 1990; Kurtak, 1986). In the absence of suitable or ecologically acceptable alternative chemicals, Bti was identified as a potential candidate for OCP operations (Guillet et al., 1990). Based on successful experimental treatments in 1980 (Lacey et al., 1982), Bti was first used on a large scale by OCP in 1981 (Akpoboua et al., 1989), when 8000 L of a commercial formulation produced by Sandoz (Teknar WDC) were applied in the Bandama River basin (southern Côte d'Ivoire). This water-dispersible formulation had to be mixed with a minimum of 20% of water, and it was applied at the rate of 1.2 liters per m³/s of river discharge (1.2 L/m³s). This dosage was eight times higher than that of temephos, which put a heavy burden on aerial operations, but OP-resistant black fly populations were effectively controlled by Bti. Based on this first success and due to the spread of resistance to temephos and chlorphoxim in southern Côte d'Ivoire, the use of Teknar WDC increased up to 233,000 L in 1982, and to 310,000 L in 1983. In 1984, it decreased to 257,000 L, due to an increase in chlorphoxim use and because of river discharges too high for Bti treatments.

Since 1985, the annual use of Bti by OCP has varied between 210,000 L and 406,500 L. This wide range has been due to several factors including the introduction of more potent Bti formulations: shifts in the susceptibility of S. damnosum s.l. to OPs; changes in the area under vector control; and introduction of new insecticides. In 1985, Sandoz produced a more potent formulation, Teknar HPD, which could be used without dilution at the dosage rate of 0.72 L/m3/s. Concerning chemical insecticides, OCP introduced in 1985 two additional compounds (permethrin and carbosulfan) for the treatment of rivers at high discharges. This marked the onset of the rotation strategy still used by OCP. In 1987, OCP started using Vectobac 12AS, a formulation of Bti equivalent to Teknar HPD, which could be used at the same dosage rate; Vectobac 12AS has been used exclusively by OCP since 1990. The number of insecticides used by OCP has now increased from five to seven, with the introduction of pyraclofos (an OP compound) in 1990, the replacement of chlorphoxim by phoxim in 1991, and the introduction of etofenprox (a pseudo-pyrethroid) in 1994.

Success of program — OCP is now considered a great success of international cooperation. Onchocerciasis has been eliminated from the initial area, enabling the settlement of 13 million hectares of arable land, and with the more recent extensions an additional 10 million hectares have been opened to settlement. Due to OCP, 1.5 million people have recovered from early onchocerciasis symptoms and sight impairment, and it is estimated that 265,000 cases of blindness have been avoided since 1974. Thirty million people are now protected from onchocerciasis, and the 10 million children born since the start of the program have been spared the risk of contracting the disease. These major achievements are largely the result of an efficient vector control, of which Bti has been a major tool. Bti has been used mostly along the upstream rivers and rivulets, the tributaries in the Volta River basin, preserving all the non-target hydrofuana which was able to replenish the high discharge downstream rivers where the aquatic fauna had been exterminated by OP chemical pesticides.

It is now scheduled that OCP will draw to an end around 2002, when blinding onchocerciasis will have been eliminated as a public health concern and as an obstacle to social and economic development in West Africa.

8.7.2.7 Temperate Climate Zones

In those regions of the world with temperate climate, where black flies do not transmit pathogens that affect human health, these insects cause considerable nuisance (Molloy, 1990), to man and cattle. Temperate regions subject to these problems generally are ecologically sensitive as, for example, mountain resorts. The quantity of Bti to be applied in mountain streams varies with discharge rate of the water, profile of the stream, and degree of vertical mixing, turbidity and presence of pollutants, water temperature and pH, settling due to presence of pools, and characteristics of the substrate. Although the ideal particle size seems to be about 35 μm, the rate of application varies several-fold depending on these characteristics (Molloy et al., 1984). Larger particles settle faster than smaller particles. This provides liquid formulations with a distinct advantage over powders; the small particles in a liquid formulation carry better, a crucial factor in the treatment of fast-moving streams. Some 5 to 30 ppm of Bti per minute generally provide satisfactory control of larval black flies over a span of 50 to 250 m in moderate sized streams (Knutti and Beck, 1987).

The non-target effects of Bti in these sites are minimal. Of the wide variety of organisms present, only filter-feeding chironomids were found to be sensitive to Bti, and at very high rates of application. Bti is thus the sole larvicide that can be used in such sensitive ecosystems.

8.8   CONTROL OF OTHER DIPTERA

Due to the increasing eutrophication of many rivers and lakes, aquatic midges such as *Chironomus* species often find optimal developmental conditions in aquatic sediments. Although the large-scale occurrence of flying insects can be very unpleasant (Ali, 1996; Mulla et al., 1990), chironomids occupy an important position in the food chain because of their large biomass. Furthermore, they play a critical role in self-cleaning of waterways, leading to reduction in the degree of eutrophication. Chironomids are neither vectors of diseases nor biting pests. Their ecological value far outweighs the disadvantages of their occurrence and, therefore, one should generally refrain from controlling them. Recently, Bti has been used for the control of filter feeding nuisance chironomid midges deriving from eutrophication of inland, urban, and suburban natural and man-made aquatic ecosystems (Ali, 1996; Kondo et al., 1995a; Kondo et al., 1995b).

In Germany, studies have shown that larval *C. plumosus* and *C. annularius*, which occur in aquatic sediments, are only affected by Bactimos at doses of above 4 ppm, approximately 10 times the dosage used in mosquito control. Mulla et al. (1990) achieved similar results in California, where they found that *C. decorus* could be controlled successfully with a dosage of 5 kg Vectobac TP or 10 liters Vectobac 12 AS per hectare.

Because of their ecological value, chironomids and similar insects are protected in Germany. Due to their low sensitivity towards Bti and occurrence mostly in permanent waters with silty sediments, in which *Aedes* mosquitoes generally do not occur, such protection is easily achieved. Additionally, other species of midges, such as *Smittia, Procladius* and *Tanypus*, are unaffected by Bti; as a result of this selective activity, a rich and diverse aquatic fauna remain as prey for predators, birds, and other higher trophic-level organisms after application.

Psychodidae, particularly *Psychoda alternata,* can occasionally become trouble-some in sewage treatment plants. In laboratory experiments, 1 ppm of Bactimos was sufficient to kill larval *P. alternata.* In the field, however, a dose of 100 ppm was required to achieve nearly 100% mortality. Economic control of *Psychoda* therefore seems unfeasible. Even if Bti was active against the related phlebotomine vectors of Leishmaniasis, similar problems in their control could be anticipated (de Barjac and Larget, 1981).

Tipulid larvae (e.g., *Tipula paludosa* and *T. oleracea)* are the main insect pests in grasslands in Europe. Experiments in the laboratory, greenhouses, and in the field showed that first instar tipulid larvae could be successfully controlled by spraying Bti (Feldmann et al., 1995; Smits and Vlug, 1990). To avoid the high doses of Bti which are usually necessary for tipulid control, food baits containing high concentrations of Bti can be used as an alternative method.

8.9   FUTURE PROSPECTS

Numerable challenges lie ahead in the further development and commercialization of Bti as an environmentally safe and economically feasible alternative to chemical pesticides for the control of mosquitoes and black flies. During the past two decades, since the discovery of Bti, increasing environmental awareness has gained tremendous momentum, providing impetus to the exponential growth in scientific and technological advancements in the fields of fermentation, formulation, microbiology, and genetic engineering.

Various approaches that utilize the tools of molecular biology and genetic engineering will be further developed to lessen the chance of resistance development in the future. As genetically engineered and formulated products become more acceptable, biotechnology will continue to provide new and improved products. Engineered toxins with improved efficacy, achieved by differing modes of action or receptor-binding properties, may be used for recombinant cloning. Mutagenic studies will also be used to produce more toxic Bti strains. Co-expression of natural toxins from different origins by unique combinations of their genes, chimeric toxins, or replacement of one gene with another more potent gene in the same bacterial strain may enhance larvicidal activity by a synergistic effect, delay or prevent the development of resistance, and expand the host spectrum.

Advances in fermentation technology and downstream processes in producing primary powder and the entire range of formulations will reduce production costs, as well as improve the quality of Bti-related products. Formulations of aqueous suspensions avoid the cost of drying; however, they have a shorter shelf life than the powder and granular formulations. Shelf life of dry products is comparable with chemical pesticides, giving them a competitive edge over liquid formulations.

Bti has a unique mode of action which prevents the development of resistance. This property derives from four major and three minor proteins which are different in homology and mode of action, especially with regard to the Cyt1Aa protein which exhibits a very high degree of synergism with the other toxin proteins, which has also been found to almost completely suppress resistance development. This finding has significant implications for future resistance management programs.

Since the gram-positive, non-sporulating, and non-pathogenic diplococcus *Deinoccocus radiodurans* is extremely resistant to ionizing and UV radiation and desiccation, it would be advisable to attempt to clone Bti δ-endotoxin genes into this extremely photoresistant bacterium to protect them as well as their products from the harmful effects of sunlight.

New formulations and new isolates which can overcome the short residual activity will allow its future utilization in a variety of habitats and conditions. Development of effective formulations suited to the biology and habitats of the target organisms is the basic requirement for the successful use of Bti. Encapsulation technology, including microencapsulation and bioencapsulation which is continuously undergoing improvement, will be utilized in the development of durable slow release formulations, resistant to detrimental environmental conditions with improved residual activity.

Success in meeting these challenges provides opportunities for controlling insect vectors of human and animal diseases and nuisance pests of public health importance.

ACKNOWLEDGMENTS

We wish to thank Dr. Bert Schneider for continuous assistance in preparation of the manuscript and Prof. Yechiel Shai of the Weizmann Institute for the thorough scrutiny of the article and critical — helpful comments. We thank Dr. Bob Rose of the United States Environmental Protection Agency for providing data from a survey of Bti usage in the U.S.

This review would not be possible without the support of the U.S. AID CDR grant TA-MOU-CA13-067, and by a post-doctoral fellowship (to E.D.-B.) from the Israel Ministry of Science

REFERENCES

Adams, L. F., J. E. Visick, and H. R. Whiteley. A 20-kilodalton protein is required for efficient production of the *Bacillus thuringiensis* subsp. *israelensis* 27-kilodalton crystal protein in *Escherichia coli*. *Journal of Bacteriology*. 171, pp. 521-530, 1989.
Adams, L. F., K. L.Brown, and H. R. Whiteley. Molecular cloning and characterization of two genes encoding sigma factors that direct transcription from a *Bacillus thuringiensis* crystal protein gene promoter. *Journal of Bacteriology*. 173, pp. 3846-3854, 1991.

Agaisse, H., and D. Lereclus. How does *Bacillus thuringiensis* produce so much insecticidal crystal protein? *Journal of Bacteriology.* 177, pp. 6027-6032, 1995.

Agoua, H., D. Quillévéré, C. Back, P. Poudiougo, P. Guillet, D. G. Zerbo, J. E. E. Henderickx, A. Sékétéli, and S. Sowah. Evaluation des moyens de lutte contre les simulies dans le cadre du Programme OCP (Onchocerciasis Control Programme). *Annals Society Belgique Médicine Tropical.* 71, (Suppl. 1), pp. 49-63, 1991.

Akpoboua, L. K. B., P. Guillet, D. C. Kurtak, and P. Pangalet. Le rôle du *Bacillus thuringiensis* H14 dans la lutte contré *Simulium damnosum* Theobald (Diptera: Simuliidae), vecteur de l'onchocercose en Afrique occidentale. *Naturaliste Canadien.* 116, pp. 167-174, 1989.

Al-yahyaee, S. A. S., and D. J. Ellar. Maximal toxicity of cloned CytA δ-endotoxin from *Bacillus thuringiensis* subsp. *israelensis* requires proteolytic processing from both the N- and C-termini. *Microbiology.* 141, pp. 3141-3148, 1995.

Al-yahyaee, S. A. S. and D. J. Ellar. Cell targeting of a pore-forming toxin, CytA δ-endotoxin from *Bacillus thuringiensis* subspecies *israelensis*, by conjugating CytA with anti-Thy 1 monoclonal antibodies and insulin. *Bioconjugate Chemistry.* 7, pp. 451-460, 1996.

Ali, A. A concise review of chironomid midges (Diptera: Chironomidae) as pests and their management. *Journal of Vector Ecology.* 21, pp. 105-121, 1996.

Ali, A., R.-D. Xue, R. Lobinske, and N. Carandang. Evaluation of granular corncob formulations of *Bacillus thuringiensis* serovar *israelensis* against mosquito larvae using a semi-field bioassay method. *Journal of the American Mosquito Control Association.* 10, pp. 492-495, 1994.

Aly, C., M. S. Mulla, and B. A. Federici. Sporulation and toxin production by *Bacillus thuringiensis* var. *israelensis* in cadavers of mosquito larvae (Diptera: Culicidae). *Journal of Invertebrate Pathology.* 46, pp. 251-258, 1985.

Andrup, L., J. Damgaard, and K. Wassermann. Mobilization of small plasmids in *Bacillus thuringiensis* subsp. *israelensis* is accompanied by specific aggregation. *Journal of Bacteriology.* 175, pp. 6530-6536, 1993.

Andrup, L., H. H. Bendixen, and G. B. Jensen. Mobilization of *Bacillus thuringiensis* plasmid pTX14-3. *Plasmid.* 33, pp. 159-167, 1995.

Andrup, L., O. Jorgensen, A. Wilcks, L. Smidt, and G. B. Jensen. Mobilization of "nonmobillizable" plasmids by the aggregation-mediated conjugation system of *Bacillus thuringiensis*. *Plasmid.* 36, pp. 75-85, 1996.

Andrup, L., L. Smidt, K. Andersen, and L. Boe. Kinetics of conjugative transfer: A study of the plasmid pXO16 from *Bacillus thuringiensis* subsp. *israelensis*. *Plasmid.* 40, pp. 30-43, 1998.

Angsuthanasombat, C. and S. Panyim. Biosysnthesis of 130-kilo-dalton mosquito larvicide in the cyanobacterium *Agmenellum quadruplicatum* PR-6. *Applied and Environmental Microbiology.* 55, pp. 2428-2430, 1989.

Angsuthanasombat, C., W. Chungjatupornchai, S. Kertbundit, P. Luxanabil, C. Settasatian, P. Wilairat, and S. Panyim. Cloning and expression of 130-kd mosquito-larvicidal δ-endotoxin gene of *Bacillus thuringiensis* var. *israelensis* in *Escherichia coli*. *Molecular and General Genetics.* 208, pp. 384-389, 1987.

Angsuthanasombat, C., N. Crickmore, and D. J. Ellar. Comparison of *Bacillus thuringiensis* subsp. *israelensis* CryIVA and CryIVB cloned toxins reveals synergism in vivo. *FEMS Microbiology Letters.* 94, pp. 63-68, 1992.

Angsuthanasombat, C., N. Crickmore, and D. J. Ellar. Effects on toxicity of eliminating a cleavage site in a predicted interhelical loop in *Bacillus thuringiensis* CryIVB δ-endotoxin. *FEMS Microbiology Letters.* 111, pp. 255-262, 1993.

Aronson, A. I. *Bacillus thuringiensis* and its use as a biological insecticide. *Plant Breeding Reviews.* 12, pp. 19-45, 1994.

Aronson, A. I. The protoxin composition of *Bacillus thuringiensis* insecticidal inclusions affects solubility and toxicty. *Applied and Environmental Microbiology.* 61, pp. 4057-4060, 1995.

Aronson, A. I., E.-S. Han, W. McGaughey, and D. Johnson. The solubility of inclusion proteins from *Bacillus thuringiensis* is dependent upon protoxin composition and is a factor in toxicity to insects. *Applied and Environmental Microbiology.* 57, pp. 981-986, 1991.

Aronson A. I., D. Wu, and C. Zhang. Mutagenesis of specificity and toxicity regions of *Bacillus turingiensis* protoxin gene. *Journal of Bacteriology.* 177, pp. 4059-4065, 1995.

Asano, S.-I., Y. Nukumizu, H. Bando, T. Iizuka, and T. Yamamoyo. Cloning of novel enterotoxin genes from *Bacillus cereus* and *Bacillus thuringiensis. Applied and Environmental Microbiology.* 63, pp. 1054-1057, 1997.

Avissar, V.J., Y. Margalith, and A. Spielman. Incorporation of body components of diverse microorganisms by larval mosquitoes. *Journal of the American Mosquito Control Association.* 10, pp. 45-50, 1994.

Bar, E., J. Lieman-Hurwitz, E. Rahamim, A. Keynan, and N. Sandler. Cloning and expression of *Bacillus thuringiensis israelensis* δ-endotoxin DNA in *Bacillus sphaericus. Journal of Invertebrate Pathology.* 57, pp. 149-158, 1991.

Barak, Z., B. Ohana, Y. Allon, and J. Margalit. A mutant of *Bacillus thuringiensis* var. *israelensis* (BTI) resistant to antibiotics. *Applied Microbiology and Biotechnology.* 27, pp. 88-93, 1987.

Barloy, F., A. Delecluse, L. Nicolas, and M. M. Lecadet. Cloning and expression of the first anaerobic toxin gene from *Clostridium bifermentas* subsp. *malaysia*, encoding a new mosquitocidal protein with homologies to *Bacillus thuringiensis. Journal of Bacteriology.* 178, pp. 3099-3105, 1996.

Barloy, F., M. M. Lecadet, and A. Delecluse. Distribution of clostridial *cry*-like genes among *Bacillus thuringiensis* and *Clostridium* strains. *Current Microbiology.* 36, pp. 232-237, 1998.

Battista, J. R. Against all odds: The survival strategies of *Deinococcus radiodurans. Annual Review of Microbiology.* 51, pp. 203-224, 1997.

Baum, J. A. and T. Malvar. Regulation of insecticidal crystal protein production in *Bacillus thuringiensis. Molecular Microbiology.* 18, pp. 1-12, 1995.

Becker, N. Bacterial control of disease vectors — general strategy and further development, in *WHO, Informal Consultation on "Genetic manipulation with larvicidal bacteria for disease vector control,"* September 6-8, 1996, Cordoba, Spain.

Becker, N. and M. Ludwig. Mosquito control in West Germany. *Bulletin of the Society for Vector Ecology.* 8, pp. 85-93, 1983.

Becker, N. and M. Ludwig. Investigations on possible resistance in *Aedes vexans* field populations after a 10-years application of *Bacillus thuringiensis israelensis. Journal of the American Mosquito Control Association.* 9, pp. 221-224, 1994.

Becker, N., and Y. Margalith. Use of *Bacillus thuringiensis israelensis* against mosquitoes and blackflies, in *Bacillus thuringiensis, an Environmental Biopesticide: Theory and Practice,* Entwistle, P. F., J. S. Cory, M. J. Bailey, and S. R. Higgs, Eds., John Wiley & Sons, Chichester, U.K., pp. 147-170, 1993.

Becker, N. and P. Mercatoris. Efficacy of new tailor-made Bti formulations against mosquitoes, in *Abstracts of Second International Congress of Vector Ecology,* Orlando, FL, U.S., pp. 44-45, 1997.

Becker, N., S. Djakaria, A. Kaiser, O. Zulhasril, and H. W. Ludwig. Efficacy of a new tablet formulation of an asporogenous strain of *Bacillus thuringiensis israelensis* against larvae of *Aedes aegypti. Bulletin of the Society for Vector Ecology.* 16, pp. 176-182, 1991.

Becker, N., M. Zgomba, M. Ludwig, D. Petric, and F. Rettich. Factors influencing the efficacy of the microbial control agent *Bacillus thuringiensis israelensis. Journal of the American Mosquito Control Association.* 8, pp. 285-289, 1992.

Ben-Dov, E., V. Zalkinder, T. Shagan, Z. Barak, and A. Zaritsky. Spores of *Bacillus thuringiensis* var. *israelensis* as tracers for ingestion rates by *Tetrahymena pyriformis*. *Journal of Invertebrate Pathology*. 63, pp. 220-222, 1994.

Ben-Dov, E., S. Boussiba, and A. Zaritsky. Mosquito larvicidal activity of *Escherichia coli* with combinations of genes from *Bacillus thuringiensis* subsp. *israelensis*. *Journal of Bacteriology*. 177, pp. 2851-2857, 1995.

Ben-Dov, E., M. Einav, N. Peleg, S. Boussiba, and A. Zaritsky. Restriction map of the 125-kilobase of *Bacillus thuringiensis* subsp. *israelensis* carrying the genes that encode delta-endotoxins active against mosquito larvae. *Applied and Environmental Microbiology*. 62, pp. 3140-3145, 1996.

Ben-Dov, E., A. Zaritsky, E. Dahan, Z. Barak, R. Sinai, R. Manasherob, A. Khameraev, A. Troyetskaya, A. Dubitsky, N. Berezina, and Y. Margalith. Extended screening by PCR for seven *cry*-group genes from field-collected strains of *Bacillus thuringiensis*. *Applied and Environmental Microbiology*. 63, pp. 4883-4890, 1997.

Ben-Dov, E., E. Dahan, A. Zaritsky, Z. Barak, R. Sinai, R. Manasherob, A. Khameraev, A. Troyetskaya, A. Dubitsky, N. Berezina, and Y. Margalith. Novel *cry*-type genes detected by extended PCR screening from field-collected strains of *Bacillus thuringiensis*. *Israel Journal of Entomology*. 32, pp. 163-169, 1998.

Ben-Dov, E., G. Nissan, N. Pelleg, R. Manasherob, S. Boussiba, and A. Zaritsky. Refined, circular restriction map of the *Bacillus thuringiensis* subsp. *israelensis* plasmid carrying mosquito larvicidal genes. *Plasmid*. 42, 1999, in press.

Bhattacharya, P. R. Hyper-toxic mutant strains of *Bacillus thuringiensis* var. *israelensis*. *Indian Journal of Experimental Biology*. 33, pp. 801-802, 1995.

Bietlot, H. P. L., I. Vishnubhatla, P. R. Carey, M. Pozsgay, and H. Kaplan. Characterization of the cysteine residues and disulfide linkages in the protein crystal of *Bacillus thuringiensis*. *Biochemistry Journal*. 267, pp. 309-315, 1990.

Blaustein, L., and J. Margalith. Indirect effects of the fairy shrimp, *Branchipus schaefferi* and two ostracod species on *B. thuringiensis* var. *israelensis*-induced mortality in mosquito larvae. *Hydrobiologia*. 212, pp. 67-76, 1991.

Bourgouin, C., A. Klier, and G. Rapoport. Characterization of the genes encoding the haemolytic toxin and mosquitocidal δ-endotoxin of *Bacillus thuringiensis* var. *israelensis*. *Molecular and General Genetics*. 205, pp. 390-397, 1986.

Bourgouin, C., A. Delecluse, J. Ribier, A. Klier, and G. Rapoport. A *Bacillus thuringiensis* subsp. *israelensis* gene encoding a 125-kilodalton larvicidal polypeptide is associated with inverted repeat sequences. *Journal of Bacteriology*. 170, pp. 3575-3583, 1988.

Bourgouin, C., A. Delecluse, F. Torre, and J. Szulmajster. Transfer of the toxin protein genes of *Bacillus sphaericus* into *Bacillus thuringiensis* subsp. *israelensis* and their expression. *Applied and Environmental Microbiology*. 56, pp. 340-344, 1990.

Boussiba, S. Production of the nitrogen-fixing cyanobacterium *Anabaena siamensis* in a closed tubular reactor for rice farming. *Microbial Releases*. 2, pp. 35-39, 1993.

Boussiba, S., and A. Zaritsky. Mosquito biocontrol by the δ-endotoxin genes of *Bacillus thuringiensis* var. *israelensis* cloned in an ammonium-excreting mutant of a rice-field isolate of the nitrogen-fixing cyanobacterium *Anabaena siamensis*, in *Advances in Gene Technology: Feeding the World in the 21st Century*, Whelan, W.J., F. Ahmad, H. Bialy, S. Black, M.L. King, M.B. Rabin, L.P. Solomonson, and I.K. Vasil, Eds., Proc. Miami Bio/Technol. Winter Symp. IRL Press/Oxford Univ. Press, pp. 89, 1992.

Boussiba, S. and X. Q. Wu. Genetically engineered cyanobacteria as a BTI toxin genes delivery system: a biotechnological approach to the control of malaria mosquitoes, in *Proceedings of Combating Malaria*, Proc. UNESCO/WHO meeting of experts. UNESCO, Paris, pp. 49-64, 1995.

Boussiba, S., A. Zaritsky, and E. Ben-Dov. A Biocontrol Agent Containing an Endotoxin Gene. Israeli Patent Application No. 120,441, 1997.

Boussiba, S., A. Zaritsky, and E. Ben-Dov. A Biocontrol Agent Containing an Endotoxin Gene. International Patent Application No. PCT/IL98/00117, 1998.

Bravo, A., H. Agaisse, S. Salamitou, and D. Lereclus. Analysis of *crylAa* expression in *sigE* and *sigK* mutants of *Bacillus thuringiensis*. *Molecular and General Genetics*. 250, pp. 734-741, 1996.

Bulla, L. A., Jr., D. B. Bechtel, K. J. Kramer, Y. I. Shethna, A. I. Aronson, and P. C. Fitz-James. Ultrastructure physiology and biochemistry of *Bacillus thuringiensis*. *C.R.C. Critical Review Microbiology*. 8, pp. 147-204, 1980.

Burges, H. D. and K. A. Jones. Formulation of bacteria, viruses and protozoa to control insects, in *Formulation of Microbial Biopesticides*, H.D. Burges, Ed., Kluwer Academic Publishers, London, pp. 34-127, 1998.

Butko, P., F. Huang, M. Pusztai-Carey, and W. K. Surewicz. Membrane permeabilization induced by cytolytic δ-endotoxin CytA from *Bacillus thuringiensis* var. *israelensis*. *Biochemistry*. 35, pp. 11355-11360, 1996.

Butko, P., F. Huang, M. Pusztai-Carey, and W. K. Surewicz. Interaction of the delta-endotoxin CytA from *Bacillus thuringiensis* var. *israelensis* with lipid membranes. *Biochemistry*. 36, pp. 12862-12868, 1997.

Carbonneau, M. A., A. M. Melin, A. Perromat, and M. Clerc. The action of free radicals on *Deinococcus radiodurans* carotenoids. *Archives of Biochemistry and Biophysics*. 275, pp. 244-251, 1989.

Chandler, M. G. and R. H. Pritchard. The effect of gene concentration and relative gene dosage on gene output in *Escherichia coli*. *Molecular and General Genetics*. 138, pp. 127-141, 1975.

Chang, C., S-M. Dai, R. Frutos, B. A. Federici, and S. S. Gill. Properties of 72-kilodalton mosquitocidal protein from *Bacillus thuringiensis* subsp. *morrisoni* PG-14 expressed in *B. thuringiensis* subsp. *kurstaki* by using the shuttle vector pHT3101. *Applied and Environmental Microbiology*. 58, pp. 507-512, 1992.

Chang, C., Y.-M. Yu, S.-M. Dai, S. K. Law, and S. S. Gill. High-level *crylVD* and *cytA* gene expression in *Bacillus thuringiensis* does not require the 20-kilodalton protein, and the coexpressed gene products are synergistic in their toxicity to mosquitoes. *Applied and Environmental Microbiology*. 59, pp. 815-821, 1993.

Chen, X. J., A. Curtiss, E. Alcantara, and D. Dean. Mutations in Domain I of *Bacillus thuringiensis* δ-endotoxin Cry1A. Reduce the irreversible binding of toxin manduca sexta brush border membrane vesicles. *Journal of Biological Chemistry*. 270, pp. 6412-6419, 1995.

Cheong, H. and S. S. Gill. Cloning and characterization of a cytolytic and mosquitocidal δ-endotoxin from *Bacillus thuringiensis* subsp. *jegathesan*. *Applied and Environmental Microbiology*. 63, pp. 3254-3260, 1997.

Cheong, H., R. K. Dhesi, and S. S. Gill. Marginal cross-resistance to mosquitocidal *Bacillus thuringiensis* in Cry11A-resistant larvae: presence of Cry11A-like toxins in these strains. *FEMS Microbiology Letters*. 153, pp. 419-424, 1997.

Cheung, P. Y. K. and B. D. Hammock. Micro-lipid-droplet encapsulation of *Bacillus thuringiensis* subsp. *israelensis* δ-endotoxin for control of mosquito larvae. *Applied and Environmental Microbiology*. 50, pp. 984-988, 1985.

Chilcott,C. N., J. Kalamakoff, and J. S. Pillai. Characterization of proteolitic activity associated with *Bacillus thuringiensis* var. *israelensis* crystals. *FEMS Microbiology Letters*. 18, pp. 37-41, 1983.

Chungjatupornchai, W. Expression of the mosquitocidal protein genes of *Bacillus thuringiensis* subsp. *israelensis* and the herbicide-resistance gene *bar* in *Synechocystis* PCC6803. *Current Microbiology*. 21, pp. 283-288, 1990.

Chungiatupornchai, W., H. Höfte, J. Seurinck, C. Angsuthanasombat, and M. Vaeck. Common features of *Bacillus thuringiensis* toxins specific for Diptera and Lepidoptera. *European Journal of Biochemistry.* 173, pp. 9-16, 1988.

Cokmus, C. and Y. M. Elçin. Stability and controlled release properties of carboxymethylcellulose-encapsulated *Bacillus thuringiensis* var. *israelensis. Pesticide Science.* 45, pp. 351-355, 1995.

Couche, G. A., M. A. Pfannenstiel, and K. W. Nickerson. Structural disulfide bonds in the *Bacillus thuringiensis* subsp. *israelensis* protein crystal. *Journal of Bacteriology.* 169, pp. 3281-3288, 1987.

Crickmore, N., and D.J. Ellar. Involvement of a possible chaperonin in the efficient expression of a cloned CryII δ-endotoxin gene in *Bacillus thuringiensis. Molecular Microbiology.* 6, pp. 1533-1537, 1992.

Crickmore, N., E. J. Bone, J.A. Williams, and D.J. Ellar. Contribution of the individual components of the δ-endotoxin crystal to the mosquitocidal activity of *Bacillus thuringiensis* subsp. *israelensis FEMS Microbiology Letters.* 131, pp. 249-254, 1995.

Crickmore, N., D. R. Zeigler, J. Feitelson, E. Schnepf, J. Van Rie, D. Lereclus, J. Baum, and D. H. Dean. Revision of the nomenclature for the *Bacillus thuringiensis* pesticidal crystal proteins. *Microbiology and Molecular Biology Reviews.* 62, pp. 807-813, 1998.

Cucchi, A. and C. Sanchez de Rivas. SASP (small, acid-soluble spore proteins) and spore properties in *Bacillus thuringiensis israelensis* and *Bacillus sphaericus. Current Microbiology.* 36, pp. 220-225, 1998.

Cyanamide. Acobe® Biolarvicide: Technical Information, Field Trial Summary, American Cyanamide Company, Wayne, New Jersey, 1992.

Dai, S-M., and S. S. Gill. *In vitro* and *in vivo* proteolysis of the *Bacillus thuringiensis* subsp. *israelensis* CryIVD protein by *Culex quinquefasciatus* larval midgut proteases. *Insect Biochemistry and Molecular Biology.* 23, pp. 273-283, 1993.

Daly, M. J., L. Ouyang, P. Fuchs, and K. W. Minton. *In vivo* damage and recA-dependent repair of plasmid and chromosomal DNA in the radiation-resistant bacterium *Deinococcus radiodurans. Journal of Bacteriology.* 176, pp. 3508-3517, 1994.

Damgard, P. H., H. D. Larsen, B. M. Hansen, J. Bresciani, and K. Jorgensen. Enterotoxin-producing strains of *Bacillus thuringiensis* isolated from food. *Letters in Applied Microbiology.* 23, pp. 146-150, 1996.

de Barjac, H. A new subspecies of *Bacillus thuringiensis* very toxic for mosquitoes: *Bacillus thuringiensis* var. *israelensis* serotype 14 (in French). *C.R. Acad. Sci. (Paris).* 286D, pp. 797-800, 1978.

de Barjac, H. Bioassay procedure for samples of *Bacillus thuringiensis israelensis* using IPS-82 standard. WHO Report TDR/VED/SWG (5)(81.3), 1983.

de Barjac, H., and J. Larget. Toxicite de *Bacillus thuringiensis* var. *israelensis* serotype H-14 pour les larves de Phlebotomes, vecteurs de Leishmanoises. *Bulletin de la Societe de Pathology Exotic.*74, pp. 485-489, 1981.

Dean, D. H., F. Rajamohan, M. K. Lee, S.-J. Wu, X. J. Chen, E. Alcantara, and S. R. Hussain. Probing the mechanism of action of *Bacillus thuringiensis* insecticidal proteins by site-directed mutagenesis — a minireview. *Gene.* 179, pp. 111-117, 1996.

Delécluse, A., C. Bourgouin, A. Klier, and G. Rapoport. Specificity of action on mosquito larvae of *Bacillus thuringiensis* var. *israelensis* toxins encoded by two different genes. *Molecular and General Genetics.* 214, pp. 42-47, 1988.

Delécluse, A., C. Bourgouin, Klier, A., and Rapoport, G. Nucleotide sequence and characterization of a new insertion element, IS240, from *Bacillus thuringiensis israelensis. Plasmid.* 21, pp. 71-78, 1989.

Delécluse, A., J. F. Charles, A. Klier, and G. Rapoport. Deletion by *in vivo* recombination shows that the 28-kilodalton cytolytic polypeptide from *Bacillus thuringiensis* subsp. *israelensis* is not essential for mosquitocidal activity. *Journal of Bacteriology.* 173, pp. 3374-3381, 1991.

Delécluse, A., S. Poncet, A. Klier, and G. Rapoport. Expression of *cryIVA* and *cryIVB* genes, independently or in combination, in a crystal-negative strain of *Bacillus thuringiensis* subsp. *israelensis. Applied and Environmental Microbiology.* 59, pp. 3922-3927, 1993.

Delécluse, A., F. Barloy, and I. Thiery. Mosquitocidal toxins from various *Bacillus thuringiensis* and *Clostridium bifermentans*, in *Bacillus thuringiensis Biotechnology and Environmental Benefits*, vol. 1, Feng, T.-Y., K.-F. Chak, R. A. Smith, T. Yamamoto, Y. Margalith, C. Chilcott, and R. I. Rose, Eds., Hua Shiang Yuan Publishing Co., Taipei, Taiwan, pp. 125-142, 1995a.

Delécluse, A., M.-L. Rosso, and A. Ragni. Cloning and expression of a novel toxin gene from *Bacillus thuringiensis* subsp. *jegathesan*, encoding a highly mosquitocidal protein. *Applied and Environmental Microbiology.* 61, pp. 4230-4235, 1995b.

Dervyn, E., S. Poncet, A. Klier, and G. Rapoport. Transcriptional regulation of the *cryIVD* gene operon from *Bacillus thuringiensis* subsp. *israelensis. Journal of Bacteriology.* 177, pp. 2283-2291, 1995.

Dong, H., L. Nilsson, and C. G. Kurland. Gratuitous overexpression of genes in *Escherichia coli* leasds to growth inhibition and ribosome destruction. *Journal of Bacteriology.* 177, pp. 1497-1504, 1995.

Donovan, W. P., C. C. Dankocsik, and M. P. Gilbert. Molecular characterization of a gene encoding a 72-kilodalton mosquito-toxic crystal protein from *Bacillus thuringiensis* subsp. *israelensis. Journal of Bacteriology.* 170, pp. 4732-4738, 1988.

Donovan, W. P., Y. Tan, and A. C. Slaney. Cloning of the *nprA* gene for neutral protease A of *Bacillus thuringiensis* and effect of *in vivo* deletion of *nprA* on insecticidal crystal protein. *Applied and Environmental Microbiology.* 63, pp. 2311-2317, 1997.

Douek, J., M. Einav, and A. Zaritsky. Sensitivity to plating of *Escherichia coli* cells expressing the *cytA* gene from *Bacillus thuringiensis* var. *israelensis. Molecular and General Genetics.* 232, pp. 162-165, 1992.

Drobniewski, F. A. and D. J. Ellar. Purification and properties of a 28-kilodalton hemolytic and mosquitocidal protein toxin of *Bacillus thuringiensis* subsp. *darmstadiensis* 73-E10-2. *Journal of Bacteriology.* 171, pp. 3060-3065, 1989.

Earp, D. J. and D. J. Ellar. *Bacillus thuringiensis* var. *morrisoni* strain PG14: nucleotide sequence of a gene encoding a 27 kDa crystal protein. *Nucleic Acids Research.* 15, pp. 3619, 1987.

Elçin, Y. M. 1995. Control of mosquito larvae by encapsulated pathogen *Bacillus thuringiensis* var. *israelensis. Journal of Microencapsulation.* 12, pp. 515-523, 1987.

Elhai, J. Strong and regulated promoters in the cyanobacterium *Anabaena* PCC 7120. *FEMS Microbiology Letters.* 114, pp. 179-184, 1993.

Elhai, J. and C. P. Wolk. Conjugal transfer of DNA to cyanobacteria. *Methods in Enzymology.* 167, pp. 747-754, 1988.

Errington, J. *Bacillus subtilis* sporulation: regulation of gene expression and control of morphogenesis. *Microbiological Reviews.* 57, pp. 1-33, 1993.

Eskils, K. and A. Lovgren. Release of *Bacillus thuringiensis* subsp.*israelensis* in swedish soil. *FEMS Microbiology Ecology.* 23, pp. 229-237, 1997.

Federici, B. A. The future of microbial insecticides as vector control agents. *Journal of the American Mosquito Control Association.* 11, pp. 260-268, 1995.

Federici, B. A., and L.S. Bauer. Cyt1Aa protein of *Bacillus thuringiensis* is toxic to the cottonwood leaf beetle, *Crysomela scripta*, and suppresses high levels of resistance to Cry3Aa. *Applied and Environmental Microbiology.* 64, pp. 4368-4371, 1998.

Federici, B. A., P. Lüthy, and J. E. Ibarra. Parasporal body of *Bacillus thuringiensis israelensis*: structure, protein composition, and toxicity, in *Bacterial control of mosquitoes and black flies,* de Barjac, H. and D. J. Sutherland, Eds., Rutgers University Press, New Brunswick, NJ, pp. 16-44, 1990.

Feitelson, J. S., J. Payne, and L. Kim. *Bacillus thuringiensis*: insects and beyond. *Biotechnology.* 10, pp. 271-275, 1992.

Feldmann, F., A. Dullemans, and C. Waalwijk. Binding of the CryIVD toxin of *Bacillus thuringiensis* subsp. *israelensis* to larval dipteran midgut proteins. *Applied and Environmental Microbiology.* 61, pp. 2601-2605, 1995.

Ferre, J., M. D. Real, J. Van Rie, S. Jansens, and M. Peferoen. Resistance to the *Bacillus thuringiensis* bioinsecticide in a field population of *Plutella xylostella* is due to a change in a midgut membrane receptor. *Proceedings of National Academic Science USA.* 88, pp. 5119-5123, 1991.

Flores, H., X. Soberon, J. Sanchez, and A. Bravo. Isolated Domain II and III from *Bacillus thuringiensis* Cry1Ab delta-endotoxin binds to lepidopteran midgut membranes. *FEBS Letters.* 414, pp. 313-318. 1997.

Garguno, F., L. Thorne, A. M. Walfield, and T. J. Pollock. Structural relatedness between mosquitocidal endotoxins of *Bacillus thuringiensis* subsp. *israelensis. Applied and Environmental Microbiology.* 54, pp. 277-279, 1988.

Gazit, E. and Y. Shai. Structural characterization, membrane interaction, and specific assembly within phospholipid membranes of hydrophobic segments from *Bacillus thuringiensis* var. *israelensis* cytolityc toxin. *Biochemistry.* 32, pp. 12363-12371, 1993.

Gazit, E. and Y. Shai. The assembly and organization of the α5 and α7 helices from the pore-forming domain of *Bacillus thuringiensis* δ-endotoxin. *Journal of Biological Chemistry.* 270, pp. 2571-2578, 1995.

Gazit, E., P. La Rocca, M.S. Sanson, and Y. Shai. The structure and organization within the membrance of the helices composing the pore-forming domain of *Bacillus thuringiensis* δ-endotoxin are consistent with an "umbrella-like" structure of the pore. *Proceedings of National Academic Science USA.* 95, pp. 12289-12294, 1998.

Gazit, E., D. Bach, I. D. Kerr, M. S. P. Sansom, N. Chejanovsky, and Y. Shai. The α5 segment of *Bacillus thuringiensis* δ-endotoxin: *in vitro* activity, ion channel formation and molecular modelling. *Biochemistry Journal.* 304, pp. 895-902, 1994.

Gazit, E., N. Burshtein, D. J. Ellar, T. Sawyer, and Y. Shai. *Bacillus thuringiensis* cytolityc toxin associates specifically with its synthetic helices A and C in the membrane bound state. Implications for the assembly of oligomeric transmembrane pores. *Biochemistry.* 36, pp. 15546-15554, 1997.

Georghiou, G. P. Resistance potential to biopesticides and consideration of countermeasures, in *Pesticides and Alternatives,* Casida, J.E., Ed., Elsevier Science Publishers, New York, NY, pp. 409-420, 1990.

Georghiou, G. P. Mechanisms and microbial characteristics of invertebrate resistance to bacterial toxins, in *Proceedings of the VIth International Colloquium on Invertebrate Pathology and Microbial Control, IInd International Conference on Bacillus thuringiensis,* Montpellier, France, pp. 48-50, 1994.

Georghiou, G. P. and M. C. Wirth. Influence of exposure to single versus multiple toxins of *Bacillus thuringiensis* subsp. *israelensis* on development of resistance in the mosquito *Culex quinquefasciatus* (Diptera: Culicidae). *Applied and Environmental Microbiology.* 63, pp. 1095-1101, 1997.

Gill, S. S., G. J. P. Singh, and J. M. Hornung. Cell membrane interaction *Bacillus thuringiensis* subsp. *israelensis* cytolytic toxins. *Infectious Immunology.* 55, pp. 1300-1308, 1987.

Gill, S. S., E. A. Cowels, and P. V. Pietrantonio. The mode of action of *Bacillus thuringiensis* endotoxins. *Annual Review of Entomology.* 37, pp. 615-636, 1992.

Goldberg, L. J., and J. Margalith. A bacterial spore demonstrating rapid larvicidal activity against *Anopheles sergentii, Uranotaenia unguiculata, Culex univittatus, Aedes aegypti* and *Culex pipiens. Mosquito News.* 37, pp. 355-358, 1977.

Goldman, I.F., J. Arnold, and B.C. Carlton. Selection for resistance to *Bacillus thuringiensis* subsp. *israelensis* in field and laboratory populations of the mosquito *Aedes aegypti. Journal of Invertebrate Pathology.* 47, pp. 317-324, 1986.

Gonzalez, J. M., Jr. and B. C. Carlton. A large transmissible plasmid is required for crystal toxin production in *Bacillus thuringiensis* variety *israelensis. Plasmid.* 11, pp. 28-38, 1984.

Gould, F., A. Martinez-Ramirez, A. Anderson, J. Ferre, F. J. Silva, and W. J. Moar. Broad-spectrum resistance to *Bacillus thuringiensis* toxins in *Heliothis virescens. Proceedings of National Academic Science USA.* 89, pp. 7986-7990, 1992.

Grochulski, P., L. Masson, S. Borisova, M Puztai-Carey, J.-L. Schwartz, R. Brousseau, and M. Cygler. *Bacillus turingiensis* CryIA(a) insecticidal toxin: crystal structure and channel formation. *Journal of Molecular Biology.* 254, pp. 447-464, 1995.

Guerchicoff, A., C. P. Rubinstein, and R. A. Ugalde. Introduction and expression of an anti-dipteran toxin gene from *Bacillus thuringiensis* in nodulating rhizobia. *Cellular and Molecular Biology.* 42, pp. 729-735, 1996.

Guerchicoff, A., R. A. Ugalde, and C. P. Rubinstein. Identification and characterization of a previously undescribed *cyt* gene in *Bacillus thuringiensis* subsp. *israelensis. Applied and Environmental Microbiology.* 63, pp. 2716-2721, 1997.

Guillet, P., D. C. Kurtak, B. Philippon, and R. Meyer. Use of *Bacillus thuringiensis israelensis* for Onchocerciasis Control in West Africa, in *Bacterial Control of Mosquitoes and Black Flies*, de Barjac, H., and D. J. Sutherland, Eds., Rutgers University Press, New Brunswick, NJ, pp. 187-201, 1990.

Haldenwang, W. G. The sigma factors of *Bacillus subtilis. Microbiological Reviews.* 59, pp. 1-30, 1995.

Harris, M. A., R. D. Oetting, and W. A. Gardner. Use of entomopathogenic nematodes and a new monitoring technique for control of fungus gnats, *Bradysia coprophila* (Diptera: Sciaridae), in floriculture. *Biological Control.* 5, pp. 412-418, 1995.

Held, G. A., C. Y. Kawanishi, and Y. S. Huang. Characterization of the parasporal inclusion of *Bacillus thuringiensis* subsp. *kyushuensis. Journal of Bacteriology.* 172, pp. 481-483, 1990.

Höfte, H. and H. R. Whiteley. Insecticidal crystal proteins of *Bacillus thuringiensis. Microbiological Reviews.* 53, pp. 242-255, 1989.

Horak, P., J. Weiser, L. Mikes, and L. Kolarova. The effect of *Bacillus thuringiensis* M-exotoxin on trematode cercariae. *Journal of Invertebrate Pathology.* 68, pp. 41-49, 1996.

Hotha, S., and R.-M. Banik. Production of alkaline protease by *Bacillus thuringiensis* H-14 in aqueous two-phase systems. *Journal of Chemical Technology and Biotechnology.* 69, pp. 5-10, 1997.

Hoti, S. L. and K. Balaraman. Formation of melanin pigment by a mutant of *Bacillus thuringiensis* H-14. *Journal of General Microbiology.* 139, pp. 2365-2369, 1993.

Ignoffo, C. M., C. Garcia, and M. J. Kroha. Laboratory tests to evaluate the potential efficacy of *Bacillus thuringiensis* var. *israelensis* for use against mosquitoes. *Mosquito News.* 41, pp. 85-93, 1981.

Ihara, H., E. Kuroda, A. Wadano, and M. Himino. Specific toxicity of δ-endotoxins from *Bacillus thuringiensis* to *Bombyx mori. Bioscience Biotechnology Biochemistry.* 57, pp. 200-204, 1993.

Ikemura, T. Correlation between abundance of *Escherichia coli* tRNAs and the occurrence of the respective codons in its protein genes. *Journal of Molecular Biology.* 146, pp. 1-21, 1981.

Ishii. T. and M. Ohba. Investigation of mosquito-specific larvicidal activity of a soil isolate of *Bacillus thuringiensis* serovar *canadensis. Current Microbiology.* 35, pp. 40-43, 1997.

Jensen, G. B., A. Wilcks, S. S. Petersen, J. Damgaard, J. A. Baum, and L. Andrup. The genetic basis of the aggregation system in *Bacillus thuringiensis* subsp. *israelensis* is located on the large conjugative plasmid pXO16. *Journal of Bacteriology.* 177, pp. 2914-2917, 1995.

Jensen, G. B., L. Andrup, A. Wilcks, L. Smidt, and O. M. Poulsen. The aggregation-mediated conjugation system of *Bacillus thuringiensis* subsp. *israelensis:* host range and kinetics of transfer. *Current Microbiology.* 33, pp. 228-236, 1996.

Kase, L. E. and P. L. Branton. US Patent 4-631-857, 1986.

Kawalek, M. D., S. Benjamin, H.-K. Lee, and S. S. Gill. Isolation and identification of novel toxins from a new mosquitocidal isolate from Malaysia, *Bacillus thuringiensis* subsp. *jegathesan. Applied and Environmental Microbiology.* 61, pp. 2965-2969, 1995.

Keller, B., N. Becker, and G.-A. Langenbruch. The use of microbial control agents in pest control programs in Europe, in *Proceedings of the VIth International Colloquium on Invertebrate Patholology and Microbial Control, IInd International Conference on Bacillus thuringiensis,* Montpellier, France, pp. 264-265, 1994.

Khawaled, K., Z. Barak, and A. Zaritsky. Feeding behavior of *Aedes aegypti* larvae and toxicity of dispersed and of naturally encapsulated *Bacillus thuringiensis* var. *israelensis. Journal of Invertebrate Pathology.* 52, pp. 419-426, 1988.

Khawaled, K., E. Ben-Dov, A. Zaritsky, and Z. Barak. The fate of *Bacillus thuringiensis* var. *israelensis* in *B. thuringiensis* var. *israelensis*-killed pupae. *Journal of Invertebrate Pathology.* 56, pp. 312-316, 1990.

Kim, M.-K., and B.-K. Ahn. Growth and production of insecticidal crystal proteins of *Bacillus thuringiensis* as affected by carbon sources. *Agricultural Chemistry and Biotechnology.* 39, pp. 177-182, 1996.

Knowles, B. and D. J. Ellar. Colloid-osmotic lysis is a general feature of the mechanism of action of *Bacillus thuringiensis* δ-endotoxins with different insect specificities. *Biochimica et Biophysica Acta.* 924, pp. 509-518, 1987.

Knowles, B. H., M. R. Blatt, M. Tester, J. M. Horsnell, J. Carroll, G. Menestrina, and D. J. Ellar. A cytolytic δ-endotoxin from *Bacillus thuringiensis* var. *israelensis* forms cation-selective channels in planar lipid bilayers. *FEBS Letters.* 244, pp. 259-262, 1989.

Knutti, H. J. and W. R. Beck. The control of blackfly larvae with Teknar, in Black fly Ecology, Population Management, An annotated world list, K, C. Kim, and R. W. Merritt, Eds., Pennsylvania State University, University Park, PA, pp. 409-418, 1987.

Kondo, S., M. Fujiwara, M. Ohba, and T. Ishii. Comparative larvicidal activities of the four *Bacillus thuringiensis* serovars against chironomid midge, *Paratanytarsus grinmii* (Diptera: Chironomidae). *Microbiological Research.* 150, pp. 425-428, 1995a.

Kondo, S., M. Ohba, and T. Ishii. Comparative susceptibility of chironomid larvae (Dipt., Chironomidae) to *Bacillus thuringiensis* serovar *israelensis* with special reference to altered susceptibility due to food difference. *Journal of Applied Entomology.* 119, pp. 123-125, 1995b.

Koni, P. A. and D. J. Ellar. Cloning and characterization of novel *Bacillus thuringiensis* cytolitic delta-endotoxin. *Journal of Molecular Biology.* 229, pp. 319-327, 1993.

Kraemer-Schafhalter, A., and A. Moser. Kinetic study of *Bacillus thuringiensis* var. *israelensis* in lab-scale batch process. *Bioprocess Engineering*. 14, pp. 139-144, 1996.

Kumar, P. A., R. P. Sharma, and V. S. Malik. The insecticidal proteins of *Bacillus thuringiensis*. *Advances in Applied Microbiology*. 42, pp. 1-43, 1996.

Kurtak, D. Insecticide resistance in the onchocerciasis control programme. *Parasitology Today*, 2, pp. 20-21, 1986.

Lacey, L. A. *Bacillus thuringiensis* serotype H-14, in *Biological Control of Mosquitoes*, H. C. Chapman, Ed., American Mosquito Control Association, Fresno, pp. 132-158, 1985.

Lacey, L. A. and M. S. Goettel. Current developments in microbial control of insect pests and prospects for the early 21st century. *Entomophaga*. 40, pp. 3-27, 1995.

Lacey, L. A. and C. M. Lacey. The medical importance of riceland mosquitoes and their control using alternatives to chemical insecticides. *Journal of the American Mosquito Control Association*. 6, pp. 1-93, 1990.

Lacey, L. A., H. Escaffre, B. Philippon, A. Sékétéli, and P. Guillet. Large river treatment with *Bacillus thuringiensis* (H-14) for the control of *Simulium damnosum s.l.* in the Onchocerciasis Control Programme. *Tropical and Medical Parasitology*. 33, pp. 97-101, 1982.

Lahkim-Tsror, L., C. Pascar-Glusman, Y. Margalith, and Z. Barak. Larvicidal activity of *Bacillus thuringiensis* subsp. *israelensis* serotype H-14 in *Aedes aegypti*: Histopathological studies. *Journal of Invertebrate Pathology*. 41, pp. 104-116, 1983.

Lee, H.-K. and S. S. Gill. Molecular cloning and characterization of a novel mosquitocidal protein gene from *Bacillus thuringiensis* subsp. *fukuokaensis*. *Applied and Environmental Microbiology*. 63, pp. 4664-4670, 1997.

Lee, H. L., E. R. Gregorio Jr., M.S. Khadri and P. Seleena. Ultralow volume application of *Bacillus thuringiensis* susp. *israelensis* for the control of mosquitoes. *Journal of the American Mosquito Control Association*. 12, pp. 651-655, 1996.

Lee, S. G., W. Eckblad, and A. Bulla, Jr. Diversity of protein inclusion bodies and identification of mosquitocidal protein in *Bacillus thuringiensis* subsp. *israelensis*. *Biochemical and Biophysical Research Communications*. 126, pp. 953-960, 1985.

Lereclus, D., A. Delecluse, and M. M. Lecadet, Diversity of *Bacillus thuringiensis* toxins and genes, in *Bacillus thuringiensis, an Environmental Biopesticide: Theory and Practice*, Entwistle, P. F., J. S. Cory, M. J. Bailey, and S. R. Higgs, Eds., John Wiley & Sons, Chichester, U.K., pp. 37-69, 1993.

Levy, R. Controlled release of mosquito larvicides and pupicides from a crosslinked polyacrylamide Culigel® SP superabsorbent polymer matrix, in *Proceedings of 16th International Symposium on Controlled Release of Bioactive Materials*, Pearlman, R. and J. A. Miller, Eds., pp.437-438. 1989.

Li, J., J. Carroll, and D. J. Ellar. Crystal structure of insecticidal δ-endotoxin from *Bacillus thuringiensis* at 2.5 A resolution. *Nature*. 353, pp. 815-821, 1991.

Li, J., P. A. Koni, and D. J. Ellar. Structure of the mosquitocidal δ-endotoxin CytB from *Bacillus thuringiensis* sp. *kyushuensis* and implications for membrane pore formation. *Journal of Molecular Biology*. 257, pp. 129-152, 1996.

Lim, L. H. and S. Poorani. Mass scale microdroplet application of mosquitocidal *Bacillus thuringiensis* for the control of vectors. *Second International Symposium on Biopesticides*, Wuhan, China. October 26-30, 1998.

Liu, J.-W., W. H., Yap, T. Thanabalu, and A. G. Porter. Efficient synthesis of mosquitocidal toxins in *Asticcacaulis excentricus* demonstrates potential of cram-negative bacteria in mosquito control. *Nature Biothechnology*. 14, pp. 343-347, 1996.

Liu, Y.-B., and B. E. Tabashnik. Inheritance of resistance to the *Bacillus thuringiensis* toxin Cry1C in the Diamondback moth. *Applied and Environmental Microbiology*. 63, pp. 2218-2223, 1997.

Liu, Y.-B., B. E. Tabashnik, W. J. Moar, and R. A. Smith. Synergism between *Bacillus thuringiensis* spores and toxins against resistant and Diamondback moths (*Plutella xylostella*). *Applied and Environmental Microbiology*. 64, pp. 1385-1389, 1998.

Liu, Y.-T., M-J. Sui, D-D. Ji, I-H. Wu, C-C. Chou, and C-C. Chen. Protection from ultraviolet iradiation by melanin of mosquitocidal activity of *Bacillus thuringiensis* var. *israelensis*. *Journal of Invertebrate Pathology*. 62, pp. 131-136, 1993.

Lopez-Meza, J. E., B. A. Federici, W. J. Poehner, A. M. Martinez-Castillo, and J. E. Ibarra. Highly mosquitocidal isolates of *Bacillus thuringiensis* subspecies *kenyae* and *entomocidus* from Mexico. *Biochemical Systematics and Ecology* 23, pp. 461-468, 1995.

Luo, K, B. E. Tabashnik, and M. J. Adang. Binding of *Bacillus thuringiensis* Cry1Ac toxin to aminopeptidase in susceptible and resistant Diamondback moth (*Plutella xylostella*). *Applied and Environmental Microbiology*. 63, pp. 1024-1027, 1997.

Mahillon, J., R. Rezsöhazy, B. Hallet, and J. Delcour. IS231 and other *Bacillus thuringiensis* transposable elements: a review. *Genetica*. 93, pp. 13-26, 1994.

Manasherob, R., E. Ben-Dov, A. Zaritsky and Z. Barak. Protozoan-enhanced toxicity of *Bacillus thuringiensis* var. *israelensis* δ-endotoxin against *Aedes aegypti* larvae. *Journal of Invertebrate Pathology*. 63, pp. 244-248, 1994.

Manasherob, R., E. Ben-Dov, Z. Liu, and A. Zaritsky. The 19 kDa protein of *Bacillus thuringiensis* subsp. *israelensis* did not protect *Escherichia coli* cells from lethal effect of CytA, in *Abstracts of 29th Annual Meeting of the Society for Invertebrate Pathology and IIIrd International Colloquium on Bacillus thuringiensis,* Cordoba, Spain, pp. 51, 1996a.

Manasherob, R., E. Ben-Dov, Y. Margalith, A. Zaritsky, and Z. Barak. Raising activity of *Bacillus thuringiensis* var. *israelensis* against *Anopheles stephensi* larvae by encapsulation in *Tetrahymena pyriformis* (Hymenostomatida: Tetrahymenidae). *Journal of the American Mosquito Control Association*. 12, pp. 627-631, 1996b.

Manasherob, R., E. Ben-Dov, Z. Liu, Z. Barak, and A. Zaritsky. Characterization of P19 expressed in acrystalliferous *Bacillus thuringiensis* subsp. *israelensis* and in *Escherichia coli,* in *Abstracts of 30th Annual Meeting of the Society for Invertebrate Pathology,* Banff, Canada, pp. 45, 1997a.

Manasherob, R., M. Myasnik, E. Ben-Dov, Z. Barak, and A. Zaritsky. Introduction of *Bacillus thuringiensis* subsp. *israelensis* δ-endotoxin genes to the extremely photoresistant bacterium *Deinoccocus Radiodurans* R1, in *Abstracts of 30th Annual Meeting of the Society for Invertebrate Pathology,* Banff, Canada, pp. 45, 1997b.

Manasherob, R., S. Boussiba, E. Ben-Dov, X.Wu, and A. Zaritsky. Protection of heterologous *Bacillus thuringiensis* subsp. *israelensis* toxin from UV-B in *Anabaena* PCC 7120, in *Abstracts of XIth European Meeting of the Society for Vector Ecology, pp.* 41, Lisbon, Portugal, 1998a.

Manasherob, R., E. Ben-Dov, A. Zaritsky, and Z. Barak. Germination, growth and sporulation of *Bacillus thuringiensis* subsp. *israelensis* in excreted food vacuoles of the protozoan *Tetrahymena pyriformis*. *Applied and Environmental Microbiology*. 64, pp. 1750-1758, 1998b.

Margalith, Y. Biological control by *Bacillus thuringiensis* subsp. *israelensis* (Bti); history and present status. *Israel Journal of Entomology*. 23, pp. 3-8, 1989.

Margalith, Y. Discovery of *Bacillus thuringiensis israelensis,* in *Bacterial Control of Mosquitoes and Black Flies,* de Barjac, H. and D. J. Sutherland, Eds., Rutgers University Press, New Brunswick, NJ, pp. 3-9, 1990.

Margalith, Y. and H. Bobroglo. The effect of organic materials and solids in water on the persistence of *Bacillus thuringiensis* var. *israelensis* serotype H14. *Zeitschrift fuer Angewandte Entomologie* 97, pp. 516-520, 1984.

Margalith, Y., E. Zomer, Z. Erel and Z. Barak. Development and application of *Bacillus thuringiensis* var. *israelensis* H-14 as an effective biological control agent against mosquitoes in Israel. *Biotechnology.* 1, pp. 74-75, 1983.

Margalith, Y., A. Markus, and Z. Pelah. Effect of encapsulation on the persistence of *Bacillus thuringiensis* var. *israelensis* (H-14). *Applied Microbiology and Biotechnology.* 19, pp. 382-383, 1984.

Margalith, Y., L. Lahkim-Tsror, C. Gascar-Gluzman, H. Bobroglio, and Z. Barak. Biological Control of mosquitoes in Israel, in *Integrated Mosquito Control Methodologies,* vol. 2, Laird, M. and J. W. Miles, Ed., Academic Press, London, Orlando, and New York, pp. 361-374, 1985.

Margalith, Y., N. Becker, C. Back, and A. Zaritsky, *Bacillus thuringiensis* subsp. *israelensis* as a biological control agent of mosquitoes and black flies, in *Bacillus thuringiensis Biotechnology and Environmental Benefits,* vol. 1, Feng, T.-Y., K.-F. Chak, R. A. Smith, T. Yamamoto, Y. Margalith, C. Chilcott, and R. I. Rose, Eds., Hua Shiang Yuan Publishing Co., Taipei, Taiwan, pp. 521-556, 1995.

Marrone, P. G., and S. C. MacIntosh, Resistance to *Bacillus thuringiensis* and resistance management, in *Bacillus thuringiensis, an Environmental Biopesticide: Theory and Practice,* Entwistle, P. F., J. S. Cory, M. J. Bailey, and S. R. Higgs, Eds., John Wiley & Sons, Chichester, U.K., pp. 221-235, 1993.

Mattimore, V. and J. R. Battista. Radioresistance of *Deinococcus radiodurans*: Function necessary to survive ionizing radiation are also necessary to survive prolonged desiccation. *Journal of Bacteriology.* 178, pp. 633-637, 1995.

McClintock, T., C. R. Schaffer, and R. D. Sjoblad. A comparative review of the mammalian toxicity of *Bacillus thuringiensis*-based pesticides. *Pesticide Science.* 45, pp. 95-105, 1995.

McGaughey, W. H. Isect resistance to the biological incecticide *Bacillus thuringiensis. Science.* 229, pp. 193-195, 1985.

McGaughey, W. H., and M. E. Whalon. Managing insect resistance to *Bacillus thuringiensis* toxins. *Science.* 258, pp. 1451-1455, 1992.

McLean, K. M. and H. R. Whiteley. Expression in *Escherichia coli* of a cloned crystal protein gene of *Bacillus thuringiensis* subsp. *israelensis. Journal of Bacteriology.* 169, pp. 1017-1023, 1987.

Merritt, R.W., R.H. Dadd, and E.D. Walker. Feeding behavior, natural food, and nutritional relationships of larval mosquitoes. *Annual Review of Entomology.* 37, pp. 349-376, 1992.

Molloy, D. P. Progress in the biological control of blackflies with *Bacillus thuringiensis,* with emphasis on temperate climates, in *Bacterial Control of Mosquitoes and Black Flies,* de Barjac, H., and D. J. Sutherland, Eds., Rutgers University Press, New Brunswick, NJ, pp. 161-186, 1990.

Molloy, D. P., S. P. Wright, B. Kaplan, J. Gerardi, and P. Peterson. Laboratory evaluation of commercial formulations of *Bacillus thuringiensis israelensis* against mosquito and blackfly larvae. *Journal of Agricultural Entomology.* 1, pp. 161-168, 1984.

Moseley, B. E. B. Photobiology and Radiobiology of *Micrococcus (Deinococcus) radiodurans. Photochemical and Photobiological Reviews* 7, pp. 223-274, 1983.

Mouchès, C., M. Magnin, J. B. Bergè, M. de Silvestri, V. Beyssat, N. Pasteur, and G. P. Georghiou. Overproduction of detoxifying asterases in organophosphate-resistant *Culex* mosquitoes and their presence in other insects. *Proceedings of National Academic Science USA.* 84, pp. 2113-2116, 1987.

Mulla, M. S. Activity, field efficacy, and use of *Bacillus thuringiensis israelensis* against mosquitoes, in *Bacterial Control of Mosquitoes and Black Flies,* de Barjac, H. and D. J. Sutherland, Eds., Rutgers University Press, New Brunswick, NJ, pp. 134-160. 1990.

Mulla, M. S., B. A. Federici, and H. A. Darwazeh. Larvicidal efficacy of *Bacillus thuringiensis* ser. H-14 against stagnant water mosquitoes and its effect on non-target organisms. *Environmental Entomology* 11, pp. 788-795, 1982.

Mulla, M. S., J. D. Chaney, and J. Rodcharoen. Control of nuisance aquatic midges *(Diptera: Chironomidae)* with the microbial larvicide *Bacillus thuringiensis* var. *israelensis* in a man-made lake in Southern Calfornia. *Bulletin of the Society for Vector Ecology.* 15, pp. 176-184, 1990.

Mulligan, F. S., III, C. H. Schaefer, and W. H. Wilder. Efficacy and persistence of *Bacillus sphaericus* and *Bacillus thuringiensis* H-14 against mosquitoes under laboratory and field conditions. *Journal of Economic Entomology* 73, pp. 684-688, 1980.

Murphy, R. C., and S. E. Stevens, Jr. Cloning and expression of the *crylVD* gene of *Bacillus thuringiensis* subsp. *israelensis* in the cyanobacterium *Agmenellum quadruplicatum* PR-6 and its resulting larvicidal activity. *Applied and Environmental Microbiology.* 58, pp. 1650-1655, 1992.

Murthy, P. S. R. Mucous membrane irritancy study of *Bacillus sphaericus* (1593) and *B. thuringiensis* (H-14) formulations in rabbit. *Biological Memoirs.* 23, pp.11-13, 1997.

Nambiar, P. T. C., S.-W. Ma, and V. N. Iyer. Limiting an insect infestation of nitrogen-fixing root nodules of the pigeon pea *(Cajanus cajun)* by engineering the expression of an entomocidal gene in its root nodules. *Applied and Environmental Microbiology.* 56, pp. 2866-2869, 1990.

Nielsen-LeRoux, and J. F. Charles. Binding of *Bacillus sphaericus* binary toxin to a specific receptor on midgut brush-border membranes from mosquito larvae. *European Journal of Biochemistry.* 210, pp. 585-590, 1992.

Nielsen-LeRoux, C., J. F. Charles, I. Thiery, and G. P. Georghiou. Resistance in a laboratory population of *Culex quinquefasciatus* (Diptera: Culicidae) to *Bacillus sphaericus* binary toxin is due to a change in the receptor on midgut brush-border membranes. *European Journal of Biochemistry.* 228, pp. 206-210, 1995.

Nielsen-LeRoux, C., F. Pasquier, J. F. Charles, G. Sinegre, B. Gaven, and N. Pasteur. Resistance to *Bacillus sphaericus* involves different mechanisms in *Culex pipiens* (Diptera: Culicidae) larvae. *Journal of Medical Entomology.* 34, pp. 321-327, 1997.

Nishimoto, T., H. Yoshisue, K. Ihara, H. Sakai, and T. Komano. Functional analysis of block 5, one of the highly conserved amino acid sequences in the 130-kDa CryIVA protein produced by Bacillus turingiensis subsp. israelensis. *FEBS Letters.* 348, pp. 249-254, 1994.

Nordström, K. Control of plasmid replication: theoretical considerations and practical solutions, in *Plasmids in bacteria,* Helinski, D. R., S. N. Cohen, D. B. Clewell, D. A. Jackson, and A. Hollaender, Eds., Plenum Publ., New York, pp. 189-214, 1985.

Ohana, B., Y. Margalith, and Z. Barak. Fate of *Bacillus thuringiensis* subsp. *israelensis* under simulated field conditions. *Applied and Environmental Microbiology.* 53, pp. 828-831, 1987.

Ohba, M., H. Saitoh, K. Miyamoto, K. Higuchi, and E. Mizuki. *Bacillus thuringiensis* serovar *higo* (flagellar serotype 44), a new serogroup with a larvicidal activity preferential for the anopheline mosquito. *Letters in Applied Microbiology.* 21, pp. 316-318, 1995.

Olliaro, P. and P. I. Trigg. Status of antimalarial drugs under development. *Bulletin of the World Health Organization.* 73, pp. 565-571, 1995.

Oppert, B., K. J. Kramer, D. E. Johnson, S. C. MacIntosh, and W. H. McGaughey. Altered protoxin activation by midgut enzymes from a *Bacillus thuringiensis* resistant strain of *Plodia interpunctella. Biochemical and Biophysical Research Communications.* 198, pp. 940-947, 1994.

Oppert, B., K. J. Kramer, R. W. Beeman, D. Johnson, and W. H. McGaughey. Proteinase-mediated insect resistance to *Bacillus thuringiensis* toxins. *Journal of Biological Chemistry.* 272, pp. 23473-23476, 1997.

Orduz, S., T. Diaz, N. Restrepo, M. M. Patiño, and M. C. Tamayo. Biochemical immunological and toxicological characteristics of the crystal proteins of *Bacillus thuringiensis* subsp. *medellin. Memorials do Instituto Oswaldo Cruz.* 91, pp. 231-237, 1996.

Orduz, S., M. Realpe, R. Arango, L.A. Murillo, and A. Delécluse. Sequence of the *cryIIBbI* gene from *Bacillus thuringinesis* subsp. *medellin* and toxicity analysis of its encoded protein. *Biochimica et Biophysica Acta.* 1388, pp. 267-272, 1998.

Orlova, M. V., T. A. Smirnova, L. A. Ganushkina, V. Y. Yacubovich, and R. R. Azizbekyan. Insecticidal activity of *Bacillus laterosporus. Applied and Environmental Microbiology.* 64, pp. 2723-2725, 1998.

Painter, M. K., K. J. Tennessen, and T. D. Richardson. Effects of repeated applications of *Bacillus thuringiensis israelensis* on the mosquito predator *Erythemis simplicicollis* (Odonata: Libellulidae) from hatching to final instar. *Environmental Entomology.* 25, pp. 184-191, 1996.

Pang, Y., R. Frutos, and B. A. Federici. Synthesis and toxicity of full length and truncated bacterial CryIVD mosquitocidal proteins expressed in lepidopteran cells using a baculovirus vector. *Journal of General Virology.* 73, pp. 89-101, 1992.

Panjaisee, S., S. Charoenpornwatana, S. Pantuwatana, A. Bhumiratana, and W. Panbangred. High production of CryIVB protein does not lead to formation of crystalline inclusions in *Bacillus thuringiensis* subsp. *israelensis. World Journal of Microbiology and Biotechnology.* 13, pp. 319-327, 1997.

Pao-intara, M., C. Angsuthanasombat, and S. Panyim. The mosquito larvicidal activity of 130 kDa delta-endotoxin of *Bacillus thuringiensis* var. *israelensis* resides in the 72 kDa amino-terminal fragment. *Biochemical and Biophysical Research Communications.* 153, pp. 294-300, 1988.

Park, H. W., H. S. Kim, D. W. Lee, Y. M. Yu, B. R. Jin, and S. K. Kang. Expression and synergistic effect of three types of crystal protein genes in *Bacillus thuringiensis. Biochemical and Biophysical Research Communications.* 214, pp. 602-607, 1995.

Poncet, S., A. Delecluse, G. Anello, A. Klier, and G. Rappaport. Transfer and expression of the *cryIVB* and *cryIVD* genes of *Bacillus thuringiensis* subsp. *israelensis* in *Bacillus sphaericus* 2297. *FEMS Microbiology Letters.* 117, pp. 91-96, 1994.

Poncet, S., A. Delecluse, A. Klier, and G. Rapoport. Evaluation of synergistic interactions among the CryIVA, CryIVB, and CryIVD toxic components of *B. thuringiensis* subsp. *israelensis* crystals. *Journal of Invertebrate Pathology.* 66, pp. 131-135, 1995.

Poncet, S., E. Dervyn, A. Klier, and G. Rapoport. Spo0A represses transcription of the *cry* toxin genes in *Bacillus thuringiensis. Microbiology.* 143, pp. 2743-2751, 1997a.

Poncet, S., C. Bernard, E. Dervyn, J. Cayley, A. Klier, and G. Rappaport. Improvement of *Bacillus sphaericus* toxicity against dipteran larvae by integration, via homologous recombination, of the Cry11A toxin gene from *Bacillus thuringiensis* subsp. *israelensis. Applied and Environmental Microbiology.* 63, pp. 4413-4420, 1997b.

Porter, A. G., E. W. Davidson, and J.-W. Liu. Mosquitocidal toxins of bacilli and their genetic manipulation for effective biological control of mosquitoes. *Microbiological Reviews.* 57, pp. 838-861, 1993.

Prieto-Samsonov, D. L., R. I. Vazquez-Padron, C. Ayra-Pardo, J. Gonzalez-Cabrera, and G. A. Riva. *Bacillus thuringiensis*: from biodiversity to biotechnology. *Journal of Industrial Microbiology and Biotechnology.* 19, pp. 202-219, 1997.

Purcell, M. and D. J. Ellar. The identification and characterisation of novel proteinacious components of the *Bacillus thuringiensis* subsp. *israelensis* parasporal inclusion, in *Abstracts of the 30th Annual Meeting of the Society for Invertebrate Pathology*, Banff, Canada, pp. 53, 1997.

Ragni, A., I. Thiery, and A. Delecluse. Characterization of six highly mosquitocidal *Bacillus thuringiensis* strains that do not belong to H-14 serotype. *Current Microbiology*. 32, pp. 48-54, 1996.

Rajamohan, F., E. Alcantara, M. K. Lee, X. J. Chen, A. Curtiss, and D. H. Dean. Single amino acid changes in Domain II of *Bacillus thuringiensis* Cry1Ab δendotoxin affect irreversible binding to *Manduca sexta* midgut membrane vesicles. *Journal of Bacteriology*. 177, pp. 2276-2282, 1995.

Rang, C., M. Bes, V. Lullien-Pellerin, D. Wu, B. Federici, and R. Frutos. Influence of the 20-kDa protein from *Bacillus thuringiensis* ssp. *israelensis* on the rate of production of truncated Cry1C proteins. *FEMS Microbiology Letters*. 141, pp. 261-264, 1996.

Rashed, S. S. and M. S. Mulla. Factors influencing ingestion of particulate materials by mosquito larvae (Diptera: Culicidae). *Journal of Medical Entomology*. 26, pp. 210-216, 1989.

Ravoahangimalala, O., and J.-F. Charles. *In vitro* binding of *Bacillus thuringiensis* var. *israelensis* individual toxins to midgut cells of *Anopheles gambiae* (Diptera: Culicidae). *FEBS Letters*. 362, pp. 111-115, 1995.

Ravoahangimalala, O., J.-F. Charles, and J. Schoeller-Raccaud. Immunological localization of *Bacillus thuringiensis* serovar *israelensis* toxins in midgut cells of intoxicated *Anopheles gambiae* larvae Diptera: Culicidae. *Research in Microbiology*. 144, pp. 271-278, 1993.

Ravoahangimalala, O., I. Thiery, and G. Sinegre. Rice field efficacy of deltamethrin and *Bacillus thuringiensis israelensis* formulations on *Anopheles gambiae* S. S. in the Anjiro region of Madagascar. *Bulletin of the Society for Vector Ecology*. 19, pp. 169-174, 1994.

Reddy, S. T., N. S. Kumar, and G. Venkateswerlu. Comparative analysis of intracellular proteases in sporulated *Bacillus thuringiensis* strains. *Biotechnology Letters*. 20, pp. 279-281, 1998

Reeve, J., L. H. Kligman, and R. Anderson. Are natural lipids UV-screening agents? *Applied Microbiology and Biotechnology*. 33, pp. 161-166, 1990.

Restrepo, N., D. Gutierrez, M. M. Patino, I. Thiery, A. Delecluse, and S. Orduz. Cloning, expression and toxicity of a mosquitocidal toxin gene of *Bacillus thuringiensis* subsp. *medellin*. *Memorials do Instituto Oswaldo Cruz*. 92, pp. 257-262, 1997.

Robacker, D. C., A. J. Martinez, J. A. Garcia, M. Diaz, and C. Romero. Toxicity of *Bacillus thuringiensis* to Mexican fruit fly (Diptera: Tephritidae). *Journal of Economic Entomology*. 89, pp. 104-110, 1996.

Rodcharoen, J. and M. S. Mulla. Cross-resistance to *Bacillus sphaericus* strains in *Culex quinquefasciatus*. *Journal of the American Mosquito Control Association*. 12, pp. 247-250, 1996.

Roh, J. Y., H. W. Park, Y. H. Je, D. W. Lee, B. R. Jin, H. W. Oh, S. S. Gill, and S. K. Kang. Expression of mosquitocidal crystal protein genes in non-insecticidal *Bacillus thuringiensis* subsp. *israelensis*. *Letters in Applied Microbiology*. 24, pp. 451-454, 1997.

Rosso, M. L. and A. Delecluse. Contribution of the 65-kilodalton protein encoded by the cloned gene *cry19A* to the mosquitocidal activity of *Bacillus thuringiensis* subsp. *jegathesan*. *Applied and Environmental Microbiology*. 63, pp. 4449-4455, 1997a.

Rosso, M. L. and A. Delecluse. Distribution of the insertion element IS240 among *Bacillus thuringiensis* strains. *Current Microbiology*. 34, pp. 348-353, 1997b.

Saitoh, H., K. Higuchi, E. Mizuki, S. H. Hwang, and M. Ohba. Characterization of mosquito larvicidal parasporal inclusions of a *Bacillus thuringiensis* serovar *higo* strain. *Journal of Applied Microbiology*. 84, pp. 883-888, 1998.

Samba, E. World Health Organization Onchocerciasis Control Program, WHO/OCP/CTD/94.1. pp. 1-25, 1994.

Saraswathi, A. and L. S. Ranganathan. Larvicidal effect of *Bacillus thuringiensis* var.*israelensis* on *Tabanus triceps* (Thunberg) (Diptera: Tabanidae). *Indian Journal Experimental Biology*. 34, pp. 1155-1157, 1996.

Schnell, D. J., M. A. Pfannenstiel, and K. W. Nickerson. Bioassay of solubilized *Bacillus thuringiensis* subsp.*israelensis* crystals by attachment to latex beads. *Science (Washington)*. 223, pp. 1191-1193, 1984.

Schwartz, J. L., M. Juteau, P. Grochulski, M. Cygler, G. Prefontaine, R. Brousseau, and L. Masson. Restriction of intramolecular movements within the CryIAa toxin molecule of *Bacillus thuringiensis* through disulfide bond engineering. *FEBS Letters*. 410, pp.397-402, 1997.

Sekar, V. Genetics of *Bacillus thuringiensis israelensis*, in *Bacterial Control of Mosquitoes and Black Flies*, de Barjac, H., and D. J. Sutherland, Eds., Rutgers University Press, New Brunswick, NJ, pp. 66-77, 1990.

Sekar, V. and B. C. Carlton. Molecular cloning of the delta-endotoxin gene of *Bacillus thuringiensis* var. *israelensis*. *Gene*. 33, pp. 151-158, 1985.

Seleena, P., H. L. Lee, and M. M. Lecadet. A novel insecticidal serotype of *Clostridium bifermentans*. *Journal of the American Mosquito Control Association*. 13, pp. 395-397, 1997.

Sen, K., G. Honda, N. Koyama, M. Nishida, A. Neki, H. Sakai, M. Himeno, and T. Komano. Cloning and nucleotide sequences of the two 130 kDa insecticidal protein genes of *Bacillus thuringiensis* var. *israelensis*. *Agricultural and Biological Chemistry*. 52, pp. 873-878, 1988.

Servant, P., M.-L. Rosso, S. Hamon, S. Poncet, A. Delécluse, and G. Rappaport. Production of CryllA and CryllBa toxins in *Bacillus sphaericus* confers toxicity towards *Aedes aegypti* and resistant *Culex* populations. *Applied and Environmental Microbiology*. 65, pp. 3021-3026, 1999.

Sharma, R. D. *Bacillus thuringiensis*: A biocontrol agent of *Meloidogyne incognita* on barley. *Nematologia Brasileira*. 18, pp. 79-84, 1994.

Shestakov, S. V. and N. T. Khyen. Evidence for genetic transformation in blue-green alga *Anacystis nidulns*. *Molecular and General Genetics*. 107, pp. 372-375, 1970.

Siegel, J. P. and J. A. Shadduck. Mammalian safety of *Bacillus thuringiensis israelensis*, in *Bacterial Control of Mosquitoes and Black Flies*. de Barjac, H. and D. J. Sutherland, Eds., Rutgers University Press, New Brunswick, NJ, pp. 202-217, 1990.

Silva-Filha, M.-H., L. Regis, C. Nielsen-LeRoux, and J.-F. Charles. Low-level resistance to *Bacillus sphaericus* in a field-treated population of *Culex quinquefasciatus* (Diptera: Culicidae). *Journal of Economic Entomology*. 88, pp. 525-530, 1995.

Silva-Filha, M.-H., C. Nielsen-LeRoux, and J.-F. Charles. Binding kinetics of *Bacillus sphaericus* binary toxin to on midgut brush-border membranes of Anopheles and Culex sp. mosquito larvae. *European Journal of Biochemistry*. 247, pp. 754-761, 1997.

Smith, G. P., J. D. Merrick, E. J. Bone, and D. J. Ellar. Mosquitocidal activity of the CryIC δ-endotoxin from *Bacillus thuringiensis* subsp. *aizawai*. *Applied and Environmental Microbiology*. 62. pp. 680-684, 1996.

Smits, P. H. and H. J. Vlug. Control of Tipulid larvae with *Bacillus thuringiensis* var. *israelensis*. *Proceedings of the Vth International Colloquium on Invertebrate Patholology and Microbial Control*, Adelaide, pp. 343, 1990.

Soltes-Rak, E., D. J. Kushner, D. D. Williams, and J. R. Coleman. Effect of promoter modification on mosquitocidal *crylVB* gene exoression in *Synechococcus* sp. strain PCC 7942. *Applied and Environmental Microbiology*. 59, pp. 2404-2410, 1993.

Soltes-Rak, E., D. J. Kushner, D. D. Williams, and J. R. Coleman. Factors regulating *crylVB* expression in the cyanobacterium *Synechococcus* PCC 7942. *Molecular and General Genetics*. 246, pp. 301-308, 1995.

Stevens, S.E., Jr., R.C. Murphy, W.J. Lamoreaux, and L.B. Coons. A genetically engineered mosquitocidal cyanobacterium. *Journal of Applied Phycology*. 6, pp. 187-197, 1994.

Studier, F. W. and B. A. Moffatt. Use of bacteriophage T7 RNA polymerase to direct selective high-level expression of cloned genes. *Journal of Molecular Biology*. 189, pp. 113-130, 1986.

Tabashnik, B. E. Evaluation of synergism among *Bacillus thuringiensis* toxins. *Applied and Environmental Microbiology*. 58, pp. 3343-3346, 1992.

Tabashnik, B. E. Evolution of resistance to *Bacillus thuringiensis*. *Annual Review of Entomology*. 39, pp. 47-79, 1994.

Tabashnik, B. E., N. Finson, M. W. Jonson, and W. J. Moar. Resistance to toxins from *Bacillus thuringiensis* subsp. *kurstaki* causes minimal cross-resistance to *Bacillus thuringiensis* subsp. *aizawai* in the Diamondback moth (Lepidoptera: Plutellidae). *Applied and Environmental Microbiology*. 59, pp. 1332-1335, 1993.

Tabashnik, B. E., N. Finson, F. R. Groeters, W. J. Moar, M. W. Johnson, K. Luo, and M. J. Adang. Reversal resistance to *Bacillus thuringiensis* in *Plutella xylostella*. *Proceedings of National Academic Science USA*. 91, pp. 4120-4124, 1994.

Tailor, R., J. Tippett, G. Gibb, S. Pells, D. Pike, L. Jordan, and S. Ely. Identification and characterisation of a novel *Bacillus thuringiensis* δ-endotoxin entomocidal to coleopteran and lepidopteran larvae. *Molecular Microbiology*. 6, pp. 1211-1217, 1992.

Tandeau de Marsac, N., F. de la Torre, and J. Szulmajster. Expression of the larvicidal gene of *Bacillus sphaericus* 159M in the cyanobacterium *Anacystis nidulans*. *Molecular and General Genetics*. 209, pp. 396-398, 1987.

Tang, J. D., S. Gilboa, R. T. Roush, and A. M. Shelton. Inheritance, stability, and lack-of-fithess costs of field-selected resistance to *Bacillus thuringiensis* in diamondback moth (Lepidoptera: *Plutellidae*) from Florida. *Journal of Economic Entomology*. 90, pp. 732-741, 1997.

Thanabalu, T., J. Hihdley, S. Brenner, C. Oei, and C. Berry. Expression of the mosquitocidal toxins of *Bacillus sphaericus* and *Bacillus thuringiensis* subsp. *israelensis* by recombinant *Caulobacter crescentus*, a vehicle for biological control of aquatic insect larvae. *Applied and Environmental Microbiology*. 58, pp. 905-910, 1992.

Thiery, I., A. Delecluse, M. C. Tamayo, and S. Orduz. Identification of a gene for CytIA-like hemolysin from *Bacillus thuringiensis* subsp. *medellin* and expression in a crystal-negative *B. thuringiensis* strain. *Applied and Environmental Microbiology*. 63, pp. 468-473, 1997.

Thiery, I., S. Hamon, A. Delecluse, and S. Orduz. The introduction into *Bacillus sphaericus* of the *Bacillus thuringiensis* subsp. *medellin cytIAb1* gene results in higher susceptibility of resistant mosquito larva populations to *B. sphaericus*. *Applied and Environmental Microbiology*. 64, pp. 3910-3916, 1998.

Thomas, W. E. and D. J. Ellar. *Bacillus thuringiensis* var. *israelensis* crystal δ-endotoxin: effects on insect and mammalian cells *in vitro* and *in vivo*. *Journal of Cell Science*. 60, pp. 181-197, 1983a.

Thomas, W. E. and D. J. Ellar. Mechanism of action of *Bacillus thuringiensis* var. *israelensis* insecticidal δ-endotoxin. *FEBS Letters*. 154, pp. 362-368, 1983b.

Thorne, L., F. Garguno, T. Thompson, D. Decker, M. Zounes, M. Wild, A. M. Walfieldand, and T. J. Pollock. Structural similarity between the Lepidoptera- and Diptera-specific isecticidal endotoxins genes of *Bacillus thuringiensis* subsp. *kurstaki* and *israelensis*. *Journal of Bacteriology.* 166, pp. 801-811, 1986.

Tidwell, M. A., D. C. Williams, T. A. Gwinn, C. J. Pena, S. H. Tedders, G. E. Gonzalvez, and Y. Mekuria. Emergency control of *Aedes aegypti* in the Dominican Republic using Scorpion 20 ULV forced-air generator. *Journal of the American Mosquito Control Association.* 10, pp. 403-406, 1994.

Trissicook, M., S. Pantuwatana, A. Bhumiratana, and W. Panbangred. Molecular cloning of the 130-kilodalton mosquitocidal δ-endotoxin gene of *Bacillus thuringiensis* subsp. *israelensis* in *Bacillus sphaericus*. *Applied and Environmental Microbiology.* 56, pp. 1710-1716, 1990.

Uawithya, P., T. Tuntitippawan, G. Katzenmeier, S. Panyim, and C. Angsuthanasombat. Effects on larvicidal activity of single proline substitutions in α3 or α4 of the *Bacillus thuringiensis* Cry4B toxin. *Biochemistry and Molecular Biology International.* 44, pp.825-832, 1998.

Vaishnav, D. D. and R. L. Anderson. Uptake and loss of *Bacillus thuringiensis* var. *israelensis* by *Daphnia magna* in laboratory exposure. *Environmental Toxicology and Chemistry.* 14, pp. 763-766, 1995.

Van Frankenhuyzen, K. The challenge of *Bacillus thuringiensis*, in *Bacillus thuringiensis, an Environmental Biopesticide: Theory and Practice*, Entwistle, P. F., J. S. Cory, M. J. Bailey, and S. R. Higgs, Ed., John Wiley & Sons, Chichester, U.K., pp. 1-35, 1993.

Van Rie, J., W. H. McGaughey, D. E. Johnson, B. D. Barnett, and H. Van Mellaert. Mechanism of insect resistance to microbial *Bacillus thuringiensis*. *Science.* 247, pp. 72-74, 1990.

Vellanoweth, R. L. and J. C. Rabinowitz. The influence of ribosome-binding-site elements on translational efficacy in *Bacillus subtilis* and *Escherichia coli in vivo*. *Molecular Microbiology.* 6, pp. 1105-1114, 1992.

Visick, J. E. and H. R. Whiteley. Effect of a 20-kilodalton protein from *Bacillus thuringiensis* subsp. *israelensis* on production of the CytA protein by *Escherichia coli*. *Journal of Bacteriology.* 173, pp. 1748-1756, 1991.

Waalwijck, C., A. M. Dullemans, M. E. S. van Workum, and B. Visser. Molecular cloning and the nucleotide sequence of the Mr 28000 protein gene of *Bacillus thuringiensis* subsp. *israelensis*. *Nucleic Acids Research.* 13, pp. 8206-8217, 1985.

Ward, E. S. and D. Ellar. Nucleotide sequence of a *Bacillus thuringiensis* var. *israelensis* gene encoding a 130-kDa delta-endotoxin. *Nucleic Acids Research.* 15, pp. 7195, 1987.

Ward, E. S. and D. J. Ellar. Cloning and expression of two homologous genes of *Bacillus thuringiensis* subsp. *israelensis* which encode 130-kilodalton mosquitocidal proteins. *Journal of Bacteriology.* 170, pp. 727-735, 1988.

Ward, E. S., A. R. Ridley, D. J. Ellar, and J. A. Todd. *Bacillus thuringiensis* var. *israelensis* δ-endotoxin: cloning and expression of the toxin in sporogenic and asporogenic strains of *Bacillus subtilis*. *Journal of Molecular Biology.* 191, pp. 13-22, 1986.

Ward, E. S., D. J. Ellar, and C. N. Chilcot. Single amino acid changes in the *Bacillus thuringiensis* var. *israelensis* δ-endotoxin affect the toxicity and expression of the protein. *Journal of Molecular Biology.* 202, pp. 527-535, 1988.

Wassmer, D. Confusion in Margalitaville. *Wing Beats, American Mosquito Control Association.* 6, pp. 16-18, 1995.

Widner, W. R. and H. R. Whiteley. Two highly related insecticidal crystal proteins of *Bacillus thuringiensis* subsp. *kurstaki* possess different host range specificities. *Journal of Bacteriology.* 171, pp. 965-974, 1989.

Wirth, M. C. and G. P. Georghiou. Cross-resistance among CryIV toxins of *Bacillus thuringiensis* subsp. *israelensis* in *Culex quinquefasciatus* (Diptera: Culicidae). *Journal of Economic Entomology.* 90, pp. 1471-1477, 1997.

Wirth, M. C., M. Marquine, G. P. Georghiou, and N. Pasteur. Esterases A2 and B2 in *Culex quinquefasciatus* (Diptera: Culicidae) role in organophosphate resistance and linkage. *Journal of Medical Entomology.* 27, pp. 202-206, 1990.

Wirth, M. C., G. P. Georghiou, and B.A. Federici. CytA enables CryIV endotoxins of *Bacillus thuringiensis* to overcome high levels of CryIV resistance in the mosquito, *Culex quinquefasciatus. Proceedings of National Academic Science USA.* 94, pp. 10536-10540, 1997.

Wirth, M. C., A. Delecluse, B.A. Federici, and W.E. Walton. Variable cross-resistance to Cry 11B from *Bacillus thuringiensis* subsp. *jegathesan* in *Culex quinquefasciatus* (Diptera: Culicidae) resistant to single or multiple toxins of *Bacillus thuringiensis* subsp. *israelensis. Applied Environmental Microbiology.* 64, pp. 4174-4179, 1998.

Wolfersberger, M. G., X. J. Chen, and D. H. Dean. Site-directed mutations in the third domain of *Bacillus thuringiensis* δ-endotoxin CryIAa affect its ability to increase the permeability of *Bombyx mori* midgut brush border membrane vesicles. *Applied and Environmental Microbiology.* 62, pp. 279-282, 1996.

Wolk, C. P., A. Vonshak, P. Kehoe, and J. Elhai. Construction of shuttle vectors capable of conjugative transfer from *Escherichia coli* to nitrogen-fixing filamentous cyanobacteria. *Proceedings of National Academic Science USA.* 81, pp. 1561-1565, 1984.

WHO (World Health Organization). Tropical disease research: progress 1995-1996. Thirteenth Programme report of the UNDP/World Bank/WHO Special Programme for Research and Training in Tropical Diseases.World Health Organization, Geneva, 1997.

Wu, D. and F. N. Chang. Synergism in mosquitocidal activity of 26 and 65 kDa proteins from *Bacillus thuringiensis* subsp. *israelensis* crystal. *FEBS Letters.* 190, pp. 232-236, 1985.

Wu, D. and B. A. Federici. A 20-kilodalton protein preserves cell viability and promotes CytA crystal formation during sporulation in *Bacillus thuringiensis. Journal of Bacteriology.* 175, pp. 5276-5280, 1993.

Wu, D. and B. A. Federici. Improved production of the insecticidal CryIVD protein in *Bacillus thuringiensis* using *crylA(c)* promoters to express the gene for an associated 20-kDa protein. *Applied Microbiology and Biotechnology.* 45, pp. 697-702, 1995.

Wu, D., J. J. Johnson, and B. A. Federici. Synergism of mosquitocidal toxicity between CytA and CryIVD proteins using inclusions produced from cloned genes of *Bacillus thuringiensis. Molecular Microbiology.* 13, pp. 965-972, 1994.

Wu, X., S.J. Vennison, H. Liu, E. Ben-Dov, A. Zaritsky, and S. Boussiba. Mosquito larvicidal activity of transgenic *Anabaena* strain PCC 7120 expressing combinations of genes from *Bacillus thuringiensis* subsp. *israelensis. Applied and Environmental Microbiology.* 63, pp. 1533-1537, 1997.

Xudong, X., K. Renqiu, and H. Yuxiang. High larvicidal activity of intact cyanobacterium *Anabaena* sp. PCC 7120 expressing gene 51 and gene 42 of *Bacillus sphaericus* sp. 2297. *FEMS Microbiology Letters.* 107, pp. 247-250, 1993.

Yap, W. H., T. Thanabalu, and A. G. Porter. Influence of transcriptional and translational control sequences on the expression of foreign genes in *Caulobacter crescentus. Journal of Bacteriology.* 176, pp. 2603-2610, 1994a.

Yap, W. H., T. Thanabalu, and A. G. Porter. Expression of mosquitocidal toxin genes in a gas-vacuolated strain of *Ancylobacter aquaticus. Applied and Environmental Microbiology.* 60, pp. 4199-4202, 1994b.

Yomamoto, T. and R. E. McLaughlin. Isolation of a protein from the parasporal crystal of *Bacillus thuringiensis* var. *kurstaki* toxic to the mosquito larva, *Aedes taeniorhynchus. Biochemical and Biophysical Research Communications.* 103, pp. 414-421, 1981.

Yoshida, K.-I., Y. Matsushima, K. Sen, H. Sakai, and T. Komano. Insecticidal activity of a peptide containing the 30th to 695th amino acid residues of the 130-kDa protein of *Bacillus thuringiensis* var. *israelensis*. *Agricultural and Biological Chemistry*. 53, pp. 2121-2127, 1989a.

Yoshida, K.-I., K. Sen, H. Sakai, and T. Komano. Expression of two 130-kDa protein genes of *Bacillus thuringiensis* var. *israelensis* in *Bacillus subtilis*. *Agricultural and Biological Chemistry*. 53, pp. 3033-3035, 1989b.

Yoshisue, H., K.-I. Yoshida, K. Sen, H. Sakai, and T. Komano. Effects of *Bacillus thuringiensis* var. *israelensis* 20-kDa prorein on production of the *Bti* 130-kDa crystal protein in *Escherichia coli*. *Bioscience Biotechnology Biochemistry*. 56, pp. 1429-1433, 1992.

Yoshisue, H., T. Fukada, K-I. Yoshida, K. Sen, S.-I. Kurosawa, H. Sakai, and T. Komano. Transcriptional regulation of *Bacillus thuringiensis* subsp. *israelensis* mosquito larvicidal crystal protein gene *cryIVA*. *Journal of Bacteriology*. 175, pp. 2750-2753, 1993a.

Yoshisue, H., T. Nishimoto, H. Sakai, and T. Komano. Identification of a promoter for the crystal protein-encoding gene *cryIVB* from *Bacillus thuringiensis* subsp. *israelensis*. *Gene*. 137, pp. 247-251, 1993b.

Yoshisue, H., K. Ihara, T. Nishimoto, H. Sakai, and T. Komano. Expression of the genes for insecticidal crystal proteins in *Bacillus thuringiensis*: *cryIVA*, not *cryIVB*, is transcribed by RNA polymerase containing σ^H and that containing σ^E. *FEMS Microbiology Letters*. 127, pp. 65-72, 1995.

Yoshisue, H., H. Sakai, K. Sen, M. Yamagiwa, and T. Komano. Identification of a second transcriptional start site for the insecticidal protein gene *cryIVA* of *Bacillus thuringiensis* subsp. *israelensis*. *Gene*. 185, pp. 251-255, 1997.

Yu, Y.-M., M. Ohba, and S. S. Gill. Characterization of mosquitocidal activity of *Bacillus thuringiensis* subsp. *fukuokaensis* crystal proteins. *Applied and Environmental Microbiology*. 57, pp. 1075-1081, 1991.

Zaritsky, A. Biotechnology in the service of pesticide production and use for integrated control of mosquitoes, in *Proceeding of Combating Malaria*, Proc. UNESCO/WHO meeting of experts. UNESCO, Paris, pp. 31-48, 1995.

Zaritsky, A. and K. Khawaled. Toxicity in carcasses of *Bacillus thuringiensis* var. *israelensis*-killed *Aedes aegypti* larvae against scavenging larvae: Implications to bioassay. *Journal of the American Mosquito Control Association*. 2, pp. 555-559, 1986.

Zaritsky, A., V. Zalkinder, E. Ben-Dov, and Z. Barak. Bioencapsulation and delivery to mosquito larvae of *Bacillus thuringiensis* H-14 toxicity by *Tetrahymena pyriformis*. *Journal of Invertebrate Pathology*. 58, pp. 455-457, 1991.

Zaritsky, A., E. Ben-Dov, Z. Barak, and V. Zalkinder. Digestibility by and pathogenicity of the protozoan *Tetrahymena pyriformis* to larvae of *Aedes aegypti*. *Journal of Invertebrate Pathology*. 59, pp. 332-334, 1992.

Zhang, J., P.C. Fitz-James, and A. Aronson. Cloning and characterization of a cluster of genes encoding polypeptides present in the insoluble fraction of the spore coat of *Bacillus subtilis*. *Journal of Bacteriology*. 175, pp. 3757-3766, 1993.

SECTION VI

Regulatory Aspects

CHAPTER **9**

Quarantines and Regulations, Pest Risk Analysis, and International Trade

Robert P. Kahn, Gary L. Cave, John K. Greifer, and Edwin Imai

CONTENTS

1-56670-478-2/00/$0.00+$.50
© 2000 by CRC Press LLC

9.1   INTRODUCTION*

The movement of articles in international trade is subject to a wide spectrum of rules, regulations, policies, and/or guidelines implemented by parliaments or government agencies of various nations. These agencies include customs, public health, marketing, food and drug, narcotic enforcement, environmental protection, and plant health/plant quarantine. The principal objective of plant quarantine activities is the exclusion of exotic, quarantine significant plant pests and pathogens along man-made pathways. Exclusion is an environmentally sound method of control because if the entry of these organisms is prevented, the need for pesticides, biological control agents, or other integrated pest management activities for eradication, suppression, or containment would not be required.

The objectives of this chapter are to discuss:

---

* By Robert P. Kahn.

- regulatory, quarantine, and exclusion concepts and activities in general and their application to international trade of fruits and vegetables,
- pest risk analysis (PRA) upon which these activities may be based, and
- the interaction of quarantine, international trade, and PRA.

Plant health or plant quarantine programs in most countries including the U.S. usually have three components:

- Exclusion of pests and pathogens of quarantine and economic importance that might be moved *along man-made pathways* when articles are imported, or the reduction of the risk to an acceptable level in moving such hazardous organisms.
- Containment, suppression, and/or eradication of pests and pathogens that may have recently entered *along man-made or natural pathways*, and
- Assistance to exporters of plant products, such as fruits and vegetables, to meet the regulatory requirements of the importing countries.

International trade can be considered a two-way street; the exports of one country are the imports of another country. Both countries are concerned with excluding pests and pathogens. Pest risk analysis is often used as a biological tool for determining risk levels leading to promulgation of rules and regulations and deploying resources to protect a country's agriculture against the entry of exotic, quarantine significant pests and pathogens. The importing country promulgates regulations, develops operating procedures at points or ports of entry, establishes inspection stations at major ports, deploys inspectors at these and other ports of entry, and when resources are limited allocates them to higher risk situations.

In essence, regulatory activities and international trade are closely interwoven with pest risk analysis serving as an interface. As discussed later in this chapter, international standards for PRA are in place that will provide the biological justification for regulatory activities, including the entry status of imported agricultural commodities.

9.2 PEST AND PATHOGEN EXCLUSION*

9.2.1 Regulatory Concepts

The legal, geographical, biological, political, social, administrative, and economic aspects of plant quarantine in general and plant pest and pathogen exclusion in particular have been reviewed in detail (Kahn, 1979, 1991 and references cited therein). In this section, legal, biological, and geographical aspects are discussed in general as they relate to pest risk analysis and international trade in agricultural product.

9.2.1.1 Legal Basis

Legislation enacted by Congress and approved by the President gives the Secretary of Agriculture broad authority to protect U. S. agriculture from pests and

---

* By Robert P. Kahn.

diseases. Among the legislative acts passed by Congress, which are the legal basis for Animal and Plant Health and Inspection Service (APHIS) programs for *agricultural imports*, are the following:

- *Federal Seed Act (FSA) of 1939, as amended* restricts the entry of agriculture and vegetable seed to insure seed purity and freedom from seeds of certain named weed species.
- *The Federal Plant Pest Act (FPPA) of 1957, as amended* regulates the movement (along man-made pathways) of plant pests. The FPPA is the basis for authority of inspectors to board ships and restrict entry or require treatment of any cargo that may be infected or infested with certain pests or disease agents. Authority is also given for emergency action and promulgation of regulations necessary to prevent spread.
- *Federal Noxious Weed Act of 1973, as amended* restricts the entry of seed deemed to be harmful to agricultural crops, livestock, fish and wildlife resources, public health, irrigation, and navigation. Only seeds of weeds named in the Act are regulated, but some additional weeds are regulated under the FSA.
- *Plant Quarantine Act (PQA) of 1912, as amended* authorizes the Secretary of Agriculture to establish quarantines which restrict or prohibit the entry of host plants, plant parts (such as seeds or fruits) and other products or articles in order to protect U.S. crops from specific pests or disease agents.

The regulations that relate to fruit and vegetable imports are found in the Code of Federal Regulations (Anon., 1998), 7CFR 319, Foreign Quarantine Notice, Sub-part 319.56 which

- restricts or prohibits fresh fruits and vegetables including herbs from all countries,
- restricts frozen fruits and vegetables, and
- includes fresh-cut flowers imported for decoration if fresh fruit is attached.

Section 9.2 of this chapter is concerned primarily with 319.56, although some of these products are also covered by some other Sub-parts, such as: 319.15 Sugarcane; 319.24 Corn Diseases; 319.28 Citrus fruits; 319.37 Nursery Stock, Plants, Roots, Bulbs, Seed, and Other Plant Products (including fruits and vegetables capable of propagation, such as potatoes and whole coconuts); and 319.41 Indian Corn or Maize, Broomcorn and Related plants. Of these, 319.56 is used later in this section to illustrate how quarantine affects the importation of agricultural products in general and fruits and vegetables in particular.

In the U.S. and 105 other countries, the legal basis for regulatory activities related to *agricultural exports* is the *International Plant Protection Convention (IPPC) of 1951, as amended* (Anon., 1952; also see Section 9.4), and administered by the Food and Agriculture Organization (FAO) of the United Nations. Another legal basis for the international trade in agricultural commodities is the General Agreement on Tariffs and Trades (GATT) and its Sanitary and Phytosanitary Measures (SPS) component. Under the SPS portion, FAO has been given the responsibility to develop international standards for pest risk analysis. Member countries (including the U.S.) must be able to biologically justify their regulations and quarantine operations, particularly when they interfere with international trade (see Section 9.4).

9.2.1.2 Biological Aspects

The biology of a pest or pathogen, i.e., its life cycle, is an essential component of PRA. This in turn is the foundation of biologically justified regulations and regulatory procedures that affect international trade in agricultural products. The life cycle is the sequence of events that takes place from the appearance of one stage of an organism to its reappearance in the next generation. Some disease agents or pests have stages or life forms that enable them to survive the pesticide, as well as biological and/or physical stresses which occur along the various man-made pathways, and to become established in new areas.

Exclusion, as a control strategy, is based on the identity of the organism and knowledge of its life cycle as influenced by environmental stress, including climate, weather, farm practices, natural and man-made pathways, host range, and other factors (Kahn, 1982, 1988, 1991).

9.2.1.3 Geographical Basis

The distribution of exotic pests and pathogens in various continents, regions, or countries is the geographic basis of quarantine actions that often impact international trade in agriculture and other commodities. The geographic distribution of pests and pathogens is determined by the ecological range of the organism and its host(s). In addition, due to the rapid transportation of plants and other agricultural products along man-made pathways, the distances most organisms can be moved are much longer and the time required much shorter that the movement along natural pathways.

The following are examples of natural pathways:

- Vectors, such as fungi, insects, mites, slugs, snails, birds, rodents, and other animals,
- Propagules transported passively by storms, air currents, wind, and jet streams,
- Propagules transported passively by ocean currents or land surface drainage,
- Rain-generated aerosols (splashing may release fungus spores or bacterial cells in an aerosol moved passively by air currents),
- Self-locomotion of spores and cells (short-distance spread by bacteria or fungus zoospores', or the movement of nematodes in water in the soil, often in combination with natural dispersal,
- Natural dispersal of infected seed, sometimes assisted by active dispersal of seeds into air currents, followed by passive spread by air currents, wind, or animals.

Among the man-made pathways are the following:

- Mail containing articles that can harbor pathogens (e.g., infected propagative materials, live cultures of pathogens).
- Baggage containing fruits or vegetables for consumption, plant propagative material, or other articles in relatively small volumes.
- Agricultural cargo (e.g., agricultural raw materials, plant propagative materials, commodities in large volumes).
- Non-agricultural cargo (contaminated with pests and pathogens in packing materials, soil, sand, ballast, etc.).

9.2.3 Quarantine 7CFR319.56, Fruits and Vegetables

Quarantine 319.56 and other regulations enable PPQ to establish phytosanitary measures, regulatory actions, and other safeguards to reduce to an acceptable or tolerable level the probability or chances that an exotic pest will enter and become established as a result of the commercial importation of approved fruits and vegetables. These measures may be stipulated on the import permit or by standard operating procedures at ports (air or maritime) or points (land borders) of entry. In this section, some of the safeguards employed as a means of pest exclusion are discussed. Most of these are environmentally sound procedures for pest control by exclusion or regulatory control.

9.2.3.1 Approval of Permit Applications

During the period between 1955 and May 1998, about 4000 permits were issued to allow the importation of fruits and vegetables. In 1997, almost 22% of all permits issued under all plant quarantine regulations were for fruits and vegetables. A survey of PRAs for applications under 319.56 show that the following were among the mitigation measures those required (PPQ, APHIS, 1998).

- Inspection only upon arrival.
- A mandatory treatment is prescribed.
- The commodity must enter from an approved pest-free area as determined by survey and other criteria approved by PPQ for specified pest(s), but the commodity is prohibited from other areas of the country.
- A mandatory treatment is required in conjunction with a pest-free area in a given country.
- In a country with an approved pest-free area, a mandatory treatment is required if the commodity originates from area in that country that is not pest-free.
- A mandatory treatment is required in conjunction with preclearance (Section 9.2.3.3.1).
- A mandatory treatment is required in conjunction with a work plan which details how the commodity is to be processed and under what prescribed safeguards.
- The commodity must be certified to be free of a specific pest or pathogen.

9.2.3.2 Disapproval of Permit Applications

As a result of PRA, some applications to import specific fruits or vegetables from specific countries were disapproved. Among the reasons for disapproval were the following:

- No approved treatment is available.
- Not from a pest-free area in a country where pest free areas have been approved (Section 9.4.3.5.3).
- A PPQ-approved work plan is required as a condition of entry, but pest risk documentation does not show that one was submitted and approved.
- The fruit or vegetable was prohibited by

- the Food and Drug regulations. (e.g., *Piper betel*, betel nut);
- 7CFR 319.37. (e.g., *Berberis* spp. barberry);
- 7CFR 319.24.(e.g., *Zea mays*, corn).

9.2.3.3 Inspection

Inspection is a mitigation procedure that takes place when a commodity arrives at a port or point of entry in the U.S., including its territories and possessions. Inspection is conducted by a visual examination of a consignment by trained plant quarantine officers. The method of sampling and sampling size is prescribed in operational manuals. Officers look for signs (insect eggs, fungus fruiting bodies, bacterial oozes, etc.) or symptoms (the response of plants to infection or infestation).

Phytosanitary Certificates — The phytosanitary certificate (PC) is a safeguard document that not only certifies plant health based on inspection but also certifies the identity of the fruits or vegetables and their geographic origin. The PC is issued by the quarantine service of the exporting country and addressed to the quarantine service of the importing country. The PC should conform to the IPPC model certificate. The weight given to the PC by the importing country depends on its past experience with PCs from a given exporting country. Consequently, in recent years most exporting countries strive to uphold the credibility, accuracy, biological soundness, and conformity to international standards of their PCs, but this was not always so in the past. Inspections are made by authorized certifying officials (PPQ officers, state or county cooperators). The cooperators must meet specified standards for education, training, experience, and the ability to identify pests and pathogens.

The PC certifies that the products have been inspected according to appropriate procedures, and they are considered to be free from quarantine pests, practically free from other injurious pests, and conform to the phytosanitary regulations of the importing country. Some countries also accept state phytosanitary certificates which may certify origin, treatments, active growth field inspections, virus indexing, or other special conditions. The certificate must contain the quantity and name of the product as well as the botanical name, the number and description of the packages, distinguishing marks, and certified origin.

Preclearance — Preclearance inspection is a safeguard taken at the origin of the commodity, in conjunction with several other safeguards agreed to by the exporting country and the U.S. PPQ officers work alongside their counterparts in the exporting country. A work plan is developed that spells out the responsibilities of the U.S. and the exporting country. Among the safeguards that may be specified in a preclearance program are one or more the following: pest-free areas, insect trapping, surveys for insects and disease agents, mandatory treatments, and inspection. PPQ may monitor with local inspections during the growing seasons, packing, storing, and shipping of approved fruits and vegetables.

Treatments — Among the safeguards for imports are treatments which consist of a chemical or physical processes, or modification of the environment (around the commodity) to eradicate a pest. Fortunately, many of the PPQ treatments prescribed are environmentally sound since they involve physical or environment-control procedures such as hot water, warm soapy water plus brushing, vapor heat, high temperature forced air, dry heat, low temperature or refrigeration during transit of fruits, and quick freezing. Some of the low temperature treatments may be combined with a chemical treatment such as fumigation. Chemical procedures, mostly fumigations (e.g., methyl bromide) are used when other methods are not effective.

According to the Montreal Convention, the use of methyl bromide is being phased out, but research in replacement treatments is ongoing. Chemical treatments must meet requirements against target pests — conditions, dosages, concentrations, and time. These treatments are conducted by persons trained in pesticide application and safety.

9.3   PEST RISK ANALYSIS*

9.3.1   Introduction

9.3.1.1   International Trade and Plant Pest Risk

The growth in international travel and trade poses many problems for those who protect the agricultural, natural, and human resources of the U.S. Both air and sea transport are faster than ever before, and the volume of people and commodities crossing U.S. borders is growing faster than resources necessary to regulate them. This brings with it a certain probability (risks) that harmful nonindigenous organisms could be introduced into the U.S., resulting in negative consequences for producers, consumers, and those involved in other aspects of agribusiness.

The process for estimating the possibility of these negative consequences is risk analysis. Risk analysis is a formalized process consisting of the identification, assessment, and management of nonindigenous organisms which may move with imported commodities. In the past, with reference to certain organisms, agricultural import decisions were based on a stated policy of "zero risk." This concept, made with the scientific tools and concepts of the time, resulted in perhaps overly restrictive quarantine measures. This approach was taken because there were few alternatives, technology was not as advanced as it is today, and the mechanisms of pest and disease transmission were not as well understood as they are now. Thus, quarantine officials attempted to avoid any and all risks. This arrival at less stringent phytosanitary management is largely derived through experience. It is interesting to note that those countries with only recent exposure to pest risk assessment also opt for the most restrictive measures. However, with the increasing importance of modern risk analysis, regulators can now assemble and analyze pest information in a thorough, consistent, and transparent manner. As a consequence, decisions regarding the denial of imports, or the mitigation measures under which commodities may be imported, can now be made on sound scientific and technical bases.

* By Gary L. Cave

9.3.1.2 International Standards

The Agreement on Sanitary and Phytosanitary (SPS) Measures of the General Agreement on Tariffs and Trade (GATT) requires members to base their animal, plant, and human health requirements related to trade on international standards. These plant health measures are to be based on standards developed under the auspices of the International Plant Protection Convention (IPPC), implemented by the United Nations' Food and Agriculture Organization (FAO). Similarly, the North American Free Trade Agreement (NAFTA) requires the U.S., Canada and Mexico to base their regulations on guidelines and standards developed by the North American Plant Protection Organization (NAPPO) (both the IPPC and NAPPO have been formally recognized by GATT and NAFTA, respectively, as the appropriate bodies for establishing international and regional SPS measures). The SPS Agreement also requires that members make their risk analysis procedures transparent and available to other interested members (SPS Annex B).

Plant Protection and Quarantine (PPQ), an organizational unit within the Animal and Plant Health Inspection Service (APHIS) of the U.S. Department of Agriculture, has delegated authority for the protection of the plant resources of the U.S. In this capacity, PPQ has been involved internationally and regionally in the development of International Standards for Phytosanitary Measures (ISPMs). PPQ has participated in each of the IPPC Working Groups organized to draft ISPMs, and has provided review comments on all redrafted standards. Among the ISPMs drafted to date are the four Supplemental Standards for Pest Risk Analysis (IPPC) and the Pest Risk Analysis Guidelines (NAPPO).

9.3.1.3 APHIS and Risk Assessment as Defined by the SPS Agreement

Risk analysis, as defined by APHIS, is equivalent to risk assessment as defined in the SPS Agreement. Annex A of the SPS Agreement defines risk assessment as "the evaluation of the likelihood of entry, establishment or spread of a pest or disease within the territory of an importing Member according to the sanitary or phytosanitary measures which might be applied, and of the associated potential biological and economic consequences; or the evaluation of the potential for adverse effects on human or animal health arising from the presence of additives, contaminants, toxins, or disease-causing organisms in food, feedstuffs and beverages." In addition to this, APHIS recognizes risk analysis as a process comprising *risk assessment* (the scientific evaluation of the biological risks and potential consequences), *risk management* (a process of determining appropriate mitigation measures to reduce risk), and *risk communication* (the sharing of risk information) (USDA, 1996). The results of APHIS risk analyses provide well-supported recommendations to APHIS decision makers to help ensure safe trade.

9.3.2   APHIS Risk Analysis Principles

APHIS recognizes that there are various approaches to risk analysis. The selection of the approach depends on the particular circumstances associated with the commodity and the current pest or disease information. Regardless of the approach, APHIS believes that a credible risk analysis process must embody the following principles (USDA, 1996):

- GATT consistent
- Science based
- Well documented
- Flexible
- Open to review

9.3.2.1   GATT Consistent

APHIS risk analysts understand and comply with GATT SPS terms and principles, and produce agency recommendations that can withstand GATT/World Trade Organization (WTO) challenges. GATT explicitly states that regulations based on "international standards, guidelines and recommendations developed under the auspices of the Secretariat of the IPPC, in cooperation with regional organizations operating within the framework of the International Plant Protection Convention," should not be challenged. Compliance with the SPS Agreement also means that APHIS is committed to using relevant standards of the IPPC, the International Office of Epizootics, or other relevant international or regional organizations recognized by WTO. Alternatives to the standards may be used when supported by objective risk analyses.

9.3.2.2   Science Based

Data used in APHIS risk analyses are collected and evaluated using the best available scientific methods. APHIS analysts recognize the importance of describing uncertainty and identifying data gaps, and actively solicit input and review from the scientific community to the extent necessary to confirm the scientific integrity of the analyses.

9.3.2.3   Well Documented

Data used in the risk analyses are organized, evaluated and referenced in a systematic manner and in sufficient detail in order to allow interested parties to understand the process.

9.3.2.4   Flexible

Because of the variety of pest and disease situations evaluated using risk analysis, methods that apply to one situation may be irrelevant or misleading in evaluating another. While acknowledging that various methods can be used, APHIS analysts are able to articulate the rationale for the choice of a method. Flexibility also means that the risk analysis process is dynamic and able to accommodate new information and technology.

9.3.2.5 Open to Review

APHIS acknowledges its responsibility to document the risk analysis process and allow interested parties to provide relevant scientific information and comments on the process and results.

9.3.3 Components of the APHIS Risk Analysis Process

When a risk analysis is initiated due to a proposed action such as a commodity importation or other relevant event, APHIS analysts will identify and record background information and situation-specific details such as the source of the request, the origin, proposed destination, and intended use for a commodity. The analysis then follows the general process outlined below.

9.3.3.1 Risk Assessment

APHIS defines risk assessments as the evaluation of the likelihood and the biological and economic consequences of entry, establishment, or spread of a pest or disease agent within the territory of an importing country. Risk assessments also consider the degree of uncertainty associated with a proposed action. The degree of uncertainty depends upon the availability and quality of pest/disease data. An agent poorly known biologically cannot be assessed as precisely as one for which more precise biological information is available. This high degree of biological uncertainty may justify conservative estimates. However, APHIS also recognizes the importance of updating risk assessments as additional scientific information becomes available (SPS Article 5.7).

A risk assessment evaluates the unmitigated pest or disease risk to determine if the risk is sufficient to warrant mitigation. The focus is on establishing the existence of biological and economic consequences and the likelihood of their occurrence. In many cases, broad agreement concerning this risk negates the need for formal risk assessments.

9.3.3.2 Overview of the Risk Assessment Process

PPQ has been working to develop a science-based, quantitative risk assessment process (USDA, 1995) harmonized with guidelines provided by NAPPO (Anon., 1993) and FAO (Anon., 1996a). Pest risk assessment is one of the three stages of a pest risk analysis (Anon., 1993; Anon., 1996a):

Stage 1 — Initiating the process for analyzing pest risk (identifying pests or pathways for which the pest risk analysis is needed) (Figure 9.1)..

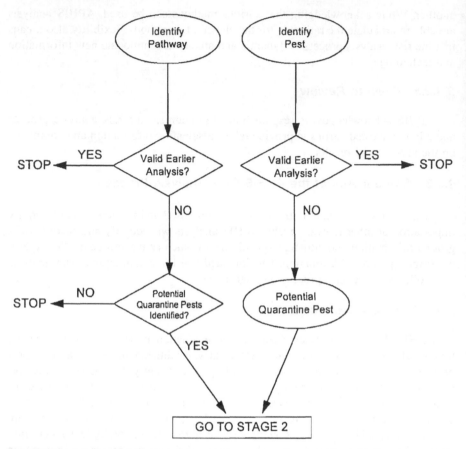

Figure 9.1   A flow chart for pest risk analysis process. Stage 1: The initiation phase of the analysis (Anon., 1996a).

Stage 2 — Assessing pest risk (determining pest status, characterized in terms of geographic distribution, biology, likelihood of entry, establishment and spread, and economic importance) (Figure 9.2).

Stage 3 — Managing pest risk (developing, evaluating, comparing and selecting options for mitigating the risk) (Figure 9.3)

There are two general initiating points for a pest risk analysis: identification of a pathway or identification of a pest (Anon., 1993; Anon., 1996a). Pathway-initiated risk analyses arise from, but are not limited to, the following situations:

- An international trade is initiated in a new commodity.
- A pathway other than commodity import is identified.
- A policy decision is taken to establish or revise phytosanitary regulations or requirements.

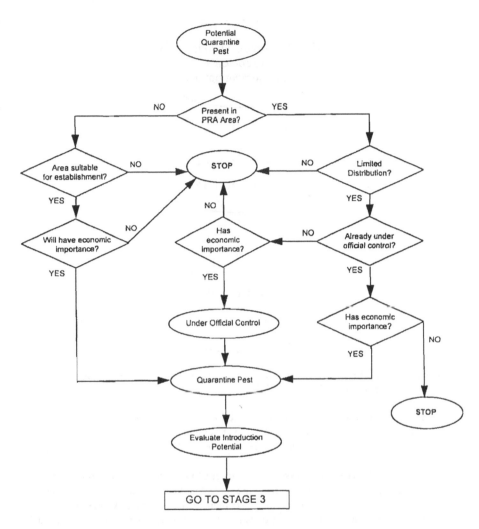

Figure 9.2   A flow chart for pest risk analysis process. Stage 2: The pest risk assessment phase of the analysis (Anon., 1996a).

- Pest-initiated analyses include, but are not limited to, the following: a quarantine pest is discovered in a new area and/or a pest is repeatedly intercepted at ports of entry; a new pest is identified by scientific research.
- PPQ pathway-initiated pest risk assessments are conducted at two levels: qualitative and quantitative. Both types follow the same framework, the difference being that in quantitative assessments, quarantine pests are examined in greater detail and provide a quantitative assessment of the likelihood of introduction.

There are nine steps in APHIS pathway-initiated pest risk assessments:

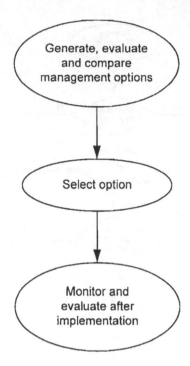

Figure 9.3   A flow chart for pest risk analysis process. Stage 3: The risk management phase of the analysis (Anon., 1996a).

Step 1 — Initiating Event: Proposed Action — Introductory remarks are made concerning the events initiating this risk assessment.

Step 2 — Weediness Potential — The weediness potential of the imported species is assessed.

Step 3 — Previous Risk Assessments, Current Status, and Pest Interceptions — Previous pest risk assessments from the same country or region and the same host, commodity, or relative are identified. If there is an existing risk assessment that adequately assesses the risks in question, the risk assessment stops. Appropriate current importations, e.g., same commodity from other countries, or other commodities from the country in question are also described, and pertinent pest interceptions at U.S. ports of entry are reported.

Step 4 — Pest List — A list of potential quarantine pests associated with the plant species to be imported is compiled (regardless of plant part requested for importation).

Step 5 — Identify Quarantine Pests — Quarantine pests (*sensu* Anon., 1996b; Anon., 1996) are identified. A determination is then made which of the quarantine pests must be analyzed further. Only these latter pests are subjected to Steps 6 through 9.

Step 6 — Quarantine Pests Likely To Follow Pathway — Only the pests that may be reasonably expected to remain with the commodity are listed here.

Step 7 — Consequences Of Introduction: Economic/Environmental Importance — The consequences of introduction refer to an estimation of the severity of negative impacts that might result from the introduction of the quarantine pests. These negative impacts are described in the context of five Risk Elements (REs):.

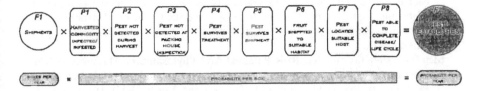

Figure 9.4   Scenario analysis: a linear event tree illustrating the introduction and establishment of pests through commodity importation.

RE #1 — Climate — Host Interaction
RE #2 — Host Range
RE #3 — Dispersal Potential
RE #4 — Economic Impact
RE #5 — Environmental Impact

The criteria for estimating risks based on these elements are qualitative, but numerical values (0 to 3 points) are assigned for each element.

Step 8 — Likelihood of Introduction — In a qualitative risk assessment, two REs are used to estimate the likelihood of introduction:

RE # 6 — Quantity of Commodity Imported

RE # 7 — Pest Opportunity (Survival and Access to Suitable Habitats and Hosts)

As in the previous REs, the criteria for estimating risks based on these elements are qualitative, but again, numerical values (0 to 3 points) are assigned for each element. The total of the seven risk ratings provides a numerical estimate of pest risk potential for each pest. The assessor may comment briefly on risk management options at this point in qualitative assessments.

In a quantitative risk assessment, a probabilistic risk assessment method is used rather than the REs. This probabilistic method is composed of three components: scenario analysis, input value estimation, and Monte Carlo simulation.

Scenario Analysis — A linear event tree (Figure 9.4) is used to depict the events (nodes) that would have to occur before pests could be introduced, and to estimate the pest outbreak frequency. The frequency of pest outbreaks is calculated by multiplying the number of shipments per year (F_1) by the probability of each event ($P_1...P_n$). The resultant frequency of pest outbreaks (F_2) is expressed on an annual basis

Estimation Of The Input Probabilities — In a typical risk assessment, the input values for the scenario analysis are usually unknown, and must be estimated. The source of this uncertainty may arise from a number of factors — natural spatial and temporal variation, environmental variation, data gaps, unconfirmed data, relationships among multiple components in a node, etc. Probabilistic methods can be used to estimate the values for the input parameters. This is accomplished by expressing estimates as probability distributions, e.g., a normal distribution, rather than point estimates (point estimates do not have this flexibility in accounting for uncertainty). By expressing the probabilities as distribution functions, a Monte Carlo sampling technique can be applied to account for the uncertainty of estimated probabilities (Kaplan, 1992).

Monte Carlo Simulations — In a typical Monte Carlo simulation, input values for the calculations are drawn from the specified input probability distributions; for each iteration, a computer program randomly selects a value from each of the input probability distributions. After performing a specified number of iterations, the computer generates a probability distribution of estimates for the frequency of pest introductions.

Step 9 — Conclusion: Pest Risk Potential And Phytosanitary Measures — The processes of risk assessment and management are interrelated. However, in the PPQ process, the scope is limited to Stages 1 and 2 of the FAO risk assessment process. Nevertheless, brief comments may be made on mitigation measures for the quarantine pests identified in the assessment.

9.3.4 Philosophical Aspects of Pest Risk Assessment

9.3.4.1 What Risk Assessments Can And Cannot Achieve

Pest risk assessment is comprised of three elements: the probability of an unwanted, non-indigenous organism being moved; the probability of this organism becoming established in a new area; and the status of this biotic agent as a pest. These three concepts should be framed in the context of international business, and must not be used as impediments to trade.

An achievable objective of the risk assessment process is to provide a reasonable estimation of the overall risk presented by specific non-indigenous organisms or with non-indigenous organisms associated with specific pathways, e.g., conveyances such as railroad cars; or specific commodities, e.g., fruits, vegetables, and nursery stock. The assessment should strive for theoretical accuracy while remaining comprehensible and manageable, and the scientific data should be collected, organized and recorded in a formal and systematic manner (Betterly et al., 1983). In addition, assessments should communicate the relative amount of uncertainty involved and may provide recommendations for the mitigation measures that reduce the risk.

Uncertainty, as previously illustrated, is inherent in any risk assessment, and can be divided into three subsets: methodological or process uncertainty; assessor uncertainty; and biological or organismal uncertainty. Each one brings its own impedimenta. However, a good risk analysis can help in identifying important sources of uncertainty and representing that uncertainty in a quantitative way.

Methodological uncertainty can be reduced by modification of the process when procedural errors are detected or when new risk methodologies are developed. Assessor uncertainty can be managed by having qualified and conscientious persons conduct the assessments. However, the quality of the risk analysis will, to some extent, always reflect the quality of the assessor(s). Biological uncertainty is the most difficult to manage of the three. It is the biological uncertainty that dictates the need for development of the pest risk process. Although it is not possible to identify which non-indigenous organisms will create problems, it can be assumed that some will (Elton, 1958; Sailer, 1978). By necessity, pest risk assessments focus on those organisms for which biological information is available. By developing detailed assessments for known pests that inhabit a variety of niches on the parent species, e.g., foliage, within the bark/wood, etc., effective mitigation measures can be developed to eliminate unknown organisms.

Objectives not achievable in a risk assessment include the determination of quarantine security or the appropriate level of protection (this lies more in the realm of risk management). Quarantine security is a defined level of pest risk, below which additional mitigation is not required. Quarantine security does not have a fixed value, but is dynamic and subjective in nature, as it is tied to the changing values of acceptable (tolerable) risk and the uncertainty surrounding the conclusions of risk assessments. However, PPQ decisions regarding the appropriate mitigation measure(s) are based on the conclusions of the risk assessment process; extant, proven mitigation options; and input by relevant stakeholders.

A risk assessment cannot give a precise determination of the timing of establishment of an exotic organism, or the specific ecological (Crawley, 1987; Drake et al., 1993; Kogan, 1990) or agricultural (Betterley et al., 1983; McGregor, 1973; Sailer, 1978) impacts caused by an introduced organism. The establishment potential of a non-indigenous organism involves the idiosyncrasies of the organism and the exigencies of the new environment. These cannot be predicted *a priori* by general statements based solely on the biology or economics of the organism (Crawley, 1987; Kogan, 1990; Sailer, 1978). Even if extensive biological information exists about an organism, many scientists believe that the turbulence of the ecological dynamics precludes any accurate predictions of the future (Drake et al., 1993; May, 1987; Porter et al., 1988).

9.3.4.2 The Philosophical Balance Between What Risk Assessments Can and Cannot Achieve

When conducting risk assessments for regulatory purposes, the most serious obstacles to overcome are historical precedence, and the limitations presented by legal, operational, and political parameters. In order to focus on pest risk, assessments need to be completed in an atmosphere as free as possible of regulatory and political influences (NRC, 1983). The assessment must demonstrate risk in order to initiate regulatory action – estimations of risk must be made in order to restrict or prohibit specific pathways. But the assessment must start with the premise that everything is risk-neutral until determined otherwise.

9.3.5 Trade Risk Analysis and the Future

As a discipline, agricultural risk analysis is still evolving. The manner in which risk analyses are applied may vary depending on the commodities involved, and there is uncertainty in some of the measurements used. However, all risk analyses share the common bonds of commitment to scientific principles, consistency, and transparency. Additionally, risk analyses identify gaps in knowledge and highlight research needs, thus aiding policy makers in refocusing resources to fill in these data and research needs.

Even with the advances in trade risk analysis, there may still be difficulties. Import decisions, even when based on the most rigorous risk analyses, are not always accepted by trading partners. This relates to the uncertainty inherent in the analyses, and the agendas of the stakeholders — some will focus on the benefits of the trade, while others will place their accent on the potential costs. Decision-makers must be well informed, and the decisions should involve consultation and even compromise. Even though trade is mutually beneficial to both parties, it must be remembered that there is at least minimal pest risk associated with the trade movement of any material. Trade is, by nature, a risky business.

9.4   THE INTERACTION OF PLANT QUARANTINE AND INTERNATIONAL TRADE*

9.4.1   Introduction

Today, the need for effective national safeguarding systems is balanced by the increasing reliance among producers for an open trade system where they can freely export and sell their products in global markets. Increasingly, U.S. plant quarantine officials, like their regulatory counterparts in other countries, face new challenges in balancing the need for vigorous national plant protection programs with the need for an outward-oriented trade policy which increases the export opportunities for domestic producers.

To a great extent, the 1994 North American Free Trade Agreement (NAFTA) (Anon., 1993) and the 1995 GATT Uruguay Round Agreement provide a framework for balancing the need for health-related protection measures in trade with the need for disciplines and rules to prevent the use of health protection measures as illegal or unjustified barriers to trade. Both the NAFTA and Uruguay Round Agreements contain elaborate international rules which govern the use of sanitary (human and animal health) and phytosanitary (plant health) regulations in trade. The rules are found in the Uruguay Round "Agreement on the Application of Sanitary and Phytosanitary Measures" (herein referred to as the SPS Agreement).

Generally, the SPS Agreement requires governments to adopt sanitary and phytosanitary (SPS) measures — including import restrictions, treatments, and other border control measures — which affect trade in an open, non-discriminatory, and scientific fashion. Concepts such as risk assessment, regionalization, equivalence, and transparency — new concepts for many governments around the world — have suddenly risen to the level of an international obligation. Today, plant health authorities worldwide face a collective challenge in developing and adopting risk assessment and risk management systems and practices that are consistent with their obligations under the SPS Agreement.

9.4.2   Background

This section examines the conditions which led to the Uruguay Round agreement in agriculture. This is followed with a discussion of key provisions of the SPS Agreement, particularly their application in the phytosanitary area.

---

* By John K. Greifer and Edwin Imai.

9.4.2.1 Agricultural Trade Relations Prior to the Uruguay Round

Shortly after World War II, a small group of industrialized nations, including the U.S., negotiated the first significant trade agreement, known as the General Agreement on Tariffs and Trade (GATT). Under this treaty, countries agreed to reduce commercial barriers which affected trade in certain manufactured products. These countries shared the view that protectionist policies and practices (e.g., high tariffs, quotas, and other trade distorting practices), which had proliferated among countries following the Depression of the 1930s, resulted in high costs to consumers who were prevented from choosing potentially cheaper foreign goods, stifled economic growth, and frustrated the export aspirations of domestic industries interested in selling their products overseas.

Agricultural trade had been excluded from previous rounds of GATT negotiations due to the generally held view that domestic food production was a national security concern. A general policy goal for many countries, following two world wars, was to be self-reliant in food production. Many developed countries, including the U.S., maintained a variety of programs to protect farm income from erratic production and commodity price fluctuations. Consequently, countries were reluctant, for many years, to alter a variety of domestic agricultural policies and programs which were aimed at protecting and supporting their farm sector through the use of high tariffs, restrictive import quotas, and domestic income support programs, including export subsidies.

By the 1970s it became increasingly evident that the agricultural policies of the 1950s and 1960s were effectively ensuring high food production in the industrialized countries. However, such policies, combined with modern production technologies, were eventually creating large supplies which could not be absorbed by domestic markets. This became the source of increasing trade tensions, especially between large agricultural producing and subsidizing nations which were competing for many of the same foreign markets with the same products. The U.S. and the European Community (today known as the European Union) repeatedly found themselves in clashes over trade.

9.4.2.2 Tokyo Round, GATT

When countries embarked in the Tokyo Round of GATT negotiations (1973–79), the U.S. took an aggressive position in advocating significant reductions in government support programs, especially export subsidies and other domestic support programs which distorted agricultural trade and created false competitive advantages based on government support rather than farmers' productivity. However, U.S. efforts to include agricultural trade reform in the Tokyo Round failed largely because of opposition from the European Community and insufficient support from other large agricultural producing nations.

By the 1980s, other large agricultural producing nations, such as Argentina, Australia, Brazil, and New Zealand, became increasingly concerned with agricultural trade conditions. These countries increasingly shared the view that agricultural trade had become distorted by a variety of trade practices and domestic support programs which limited their own comparative advantages and trade potential. The increasing popularity of market-oriented policies during the 1980s, along with the growing frustration with global agricultural trade conditions, produced sufficient support for agricultural trade reform. The opportunity arose for such reform in 1986 when countries agreed to launch a new GATT Round which became known as the Uruguay Round (1986–94).

9.4.2.3   Uruguay Round, GATT

The Uruguay Round was the most comprehensive GATT round ever conducted, covering new sectors of trade such as services, and first-time negotiations on agricultural trade. Large agricultural producing nations such as the U.S., Australia, New Zealand, Argentina, Brazil, and Mexico were vigorous advocates of significant agricultural trade reform, including the elimination of tariffs and export subsidies. Even the European Union, which remained strongly committed to maintaining its agricultural policies regardless of the trade distortions they created, eventually joined the consensus in addressing agricultural trade in the Uruguay Round. The costs of administering government supports in agriculture had begun to get excessive, even for the European Community.

9.4.2.4   North American Free Trade Agreement (NAFTA)

Bringing closure to the Uruguay Round negotiations was nearly unattainable. At times it appeared there would be no Uruguay Round agreement. In this environment of uncertainty regarding the Uruguay Round negotiations — along with the emergence of regional trade blocks in Europe and Asia — the U.S. agreed to negotiate a regional free trade agreement with Mexico and Canada. NAFTA negotiations were initiated in 1990, completed in 1993 (Anon.,1993), and went into effect in January 1994. The U.S. objective in NAFTA negotiations was to lock in and stimulate further gains in regional trade and to hedge against the possible failure of a Uruguay Round agreement.

Like the Uruguay Round negotiations, the NAFTA includes an agreement on agricultural trade as well as a code of rules governing the use of sanitary and phytosanitary measures. The SPS provisions of the NAFTA and WTO SPS Agreement are essentially identical. Therefore, comments on the SPS Agreement can be assumed to apply to the NAFTA SPS code as well. Generally, regional trade, including the use of dispute settlement procedures, between the U.S. and Mexico (or Canada) is guided by the NAFTA.

9.4.3   Sanitary and Phytosanitary (SPS) Measures, GATT

9.4.3.1   Emergence of an SPS Regime

It was in the context of the GATT Uruguay Round negotiations (1986–94) that countries negotiated the reduction of agricultural tariffs as well as rules to control the use of sanitary and phytosanitary regulations in trade. These issues came together because negotiators realized that as tariffs came down governments could, unless rules were in place, resort to the use of sanitary or phytosanitary measures as a new form of trade protectionism.

The original 1947 GATT treaty contained a general exemption allowing signatory countries to maintain measures to protect the health of consumers, animals, and plants as long as such measures were not applied in an unjustifiably discriminatory fashion or as a disguised restriction on trade (Anon., 1994, Article XXb). Efforts were later undertaken during the Tokyo Round (1973–79) to improve this particular GATT article through the "Tokyo Round Agreement on Technical Barriers to Trade" (commonly referred to as the Standards Code). The Standards Code, which covered the use of technical standards (i.e., product requirements) in trade, including sanitary and phytosanitary measures, provided for harmonization of standards through international bodies, notification requirements, and a specific process for settling disputes over standards.

The Standards Code, however, proved to be deficient in several respects. First, only 37 countries agreed to be bound to the Standards Code. Over the years, more and more countries joined the GATT, but were not required to be bound to each GATT-related agreement, including the Standards Code. Second, it did not adequately cover the use of health-related standards. This became evident when GATT panels were formed to settle a number of sanitary and phytosanitary-related disputes and were unable to do so because the Standards Code did not provide sufficient legal guidance by which to settle these SPS-related conflicts. The dispute between the U.S. and the EC over the EC's ban on beef from cattle treated with growth-promoting hormones was probably the most visible trade dispute involving a sanitary measure which went unresolved by a GATT panel because the Standards Code lacked sufficient legal scope. For these and other reasons, a consensus emerged over the years leading up to the GATT Uruguay Round that "the GATT and the Standards Code had failed to stem disruptions of trade in international markets caused by proliferating technical restrictions, including SPS measures" (Roberts, 1998).

In addition to establishing new rules to govern the use of sanitary and phytosanitary requirements in trade, the Uruguay Round also resulted in the establishment of the World Trade Organization (WTO, 1994), including several Committees within the WTO structure to monitor implementation of the various agreements (including the SPS Agreement) and to administer the dispute settlement procedures. Under the dispute settlement procedures, negotiated during the Uruguay Round, WTO member countries have the right to challenge other WTO members whose import measures or policies are inconsistent with the SPS Agreement.

9.4.3.2 Basic Rights

The SPS Agreement recognizes the fundamental right of countries to protect the health and life of their consumers, animals, and plants against pests, diseases, and other threats to health. Furthermore, the SPS Agreement explicitly recognizes the sovereign right of countries to set their level of protection for pest or disease threats at the level they deem appropriate. However, the basic right to protect against harmful pests and disease is tempered by several rules aimed at preventing the use of health

measures in an unjustified, arbitrary, or discriminatory fashion. The primary obligation is that SPS protection measures must be based on either a relevant international
standard that is established *either* by the international standards body recognized by
the SPS Agreement *or* a scientific risk assessment. For plant quarantine authorities,
this means being able to demonstrate the threat of a particular pest or disease of
concern that makes a particular phytosanitary regulation (i.e., import requirements
or border controls) necessary.

9.4.3.3   Harmonization

The SPS Agreement encourages *but does not require* countries to harmonize
their SPS measures, to the greatest extent possible, by basing their health measures
on relevant international standards (Anon., 1994, Article 3.1: Harmonization is
intended to reduce unnecessary variances between countries' technical standards —
differences which can often be the source of trade friction.

The SPS Agreement defines harmonization as: "the establishment, recognition,
and application of common sanitary and phytosanitary measures by different countries" (Anon., 1994, Annex A) The SPS Agreement recognizes three international
standard-setting bodies as the official entities for developing health-related standards,
guidelines, and recommendations.

These international bodies include:

- Codex Alimentarious for food safety standards,
- International Plant Protection Convention (IPPC) for plant health standards, and
- Office of International Epizootics (OIE) for animal health standards.

Under the SPS Agreement, a phytosanitary measure which is based on or con-
forms to an existing and relevant international standard is presumed to be in compliance with all aspects of the SPS Agreement (Anon., Article 3.2). International
standards are referred to as "safe harbor standards" in the sense that their use makes
that measure immune to challenge. In these instances, a risk assessment is unnecessary. However, if a country chooses not to use an existing international standard,
that country is required to base its measure on a risk assessment and be prepared to
notify the reasons for their deviating from the relevant international standard (Anon.,
Article 3.3 and Annex B.5).

9.4.3.4   Relation to the International Plant Protection
        Convention (IPPC)

The IPPC is a treaty, dating back to 1952, aimed at promoting international
cooperation to control and prevent the spread of harmful plant pests associated with
the movement of people and commodities. The IPPC is deposited with the Food and
Agricultural Organization (FAO) of the United Nations; currently, 106 countries are
signatories. The IPPC calls upon member countries to:

- establish national plant protection organizations to carry out activities to monitor and report emerging pest problems, inspect imports, and certify exports to meet importing countries phytosanitary requirements, and
- form regional plant protection organizations in order to work together on a regional basis to combat pests in the region and prevent their spread across national boundaries.

As a result of changes in international trade, including the emergence of new plant quarantine concepts and the signing of the WTO SPS Agreement, IPPC member countries undertook to modernize and update the Convention in 1995. A primary objective of the revision was to ensure that the Convention adequately addressed the role envisioned for it under the SPS Agreement, particularly its role in setting global phytosanitary standards.

There are three important aspects of the revised Convention. First, the revised IPPC makes provisions for the establishment of a Commission for Phytosanitary Matters. This Commission will serve as the body for setting priorities and for developing and adopting global phytosanitary standards. The existing IPPC did not include provisions for standard-setting activities. An interim standard-setting procedure was adopted in 1993. However, this procedure proved to be ad hoc and inefficient. The new Commission for Phytosanitary measures are expected to formalize, clarify, and accelerate the standards-setting process. More importantly, the adoption of standards will be done by plant quarantine authorities from the member countries rather than political delegations, as is the practice under the existing interim procedure.

Second, the revised IPPC expands the scope of the Convention to allow the use of phytosanitary measures for quarantine and regulated non-quarantine pests. Essentially, the revised Convention allows countries to adopt phytosanitary measures to guard against "regulated pests." Regulated pests include "quarantine" pests and "regulated non-quarantine" pests. Quarantine pests, which are not a new concept, are defined as: "a pest of potential economic importance to the area endangered thereby and not yet present there, or present but not widely distributed and being officially controlled." A regulated non-quarantine pest is defined as: "a non-quarantine pest whose presence in plants for planting affects the intended use of those plants with an economically unacceptable impact and which is therefore regulated within the territory of the importing contracting party." The focus on propagative material reflects the general view among quarantine officials that pests associated with plant material destined for planting represents a higher level of pest risk and spread than plant commodities destined for consumption. In either case, plant health officials must be prepared to provide the scientific evidence and basis (i.e., pest risk assessment) for their phytosanitary measures. The new scope covering regulated non-quarantine pests is an improvement over the existing Convention because it clarifies and disciplines the phytosanitary measures countries can adopt for pests which do not meet the strict definition of a quarantine pest, but which are, nonetheless, of regulatory concern to the importing country.

Third, like the SPS Agreement, the revised Convention clearly recognizes a country's right to protect its plant resources, including cultivated and non-commercial (wild flora) plants, from potentially harmful pests, but clarifies and emphasizes that phytosanitary measures must be "technically justified." Technically justified is defined in the revised Convention to mean "justified on the basis of conclusions reached by using an appropriate pest risk analysis or, where applicable, another comparable examination and evaluation of available scientific information."(Anon., 1977b, Article II).

This emphasis on the need to provide technical justification and pest risk assessment for phytosanitary measures is an improvement over the existing Convention language because it heightens the importance of scientific evidence in evaluating pest risks and steers member countries away from using phytosanitary regulations in an arbitrary fashion to block trade for commercial protection purposes.

9.4.3.5   Components

Risk Assessment — The SPS Agreement emphasizes the use of scientific principles as a basis for health-related protection measures in trade, which means basing phytosanitary measures on a risk assessment (or some comparable evaluation of scientific evidence). The SPS Agreement defines some basic terms related to risk assessment, including the following:

> **Risk Assessment** — The SPS Agreement defines risk assessment as *"the evaluation of the likelihood of entry, establishment or spread of a pest or disease within the territory of an importing Member according to the sanitary or phytosanitary measures which might be applied, and of the associated potential biological and economic consequences; or the evaluation of the potential for adverse effects on human or animal health arising from the presence of additives, contaminants, toxins, or disease-causing organisms in food, feedstuffs and beverages"* (Anon., 1994, Annex A).
>
> **Risk Assessment factors** — The SPS Agreement identifies a number of factors which countries must take into account when conducting a risk assessment: *"relevant processes and production methods, relevant inspection, sampling and testing methods; prevalence of specific diseases or pests; existence of pest- or disease-free areas; relevant ecological and environmental conditions; and quarantine or other treatments"* (Anon., 1994, Article 5.2).
>
> **Economic Consequences** — The SPS Agreement requires countries to take the following economic factors into account when evaluating risks to plant or animal health: *"potential damage in terms of loss of production or sales in the event of the entry, establishment or spread of a pest or disease; the costs of control or eradication in the territory of the importing Member; and the relative cost-effectiveness of alternative approaches to limiting risk"* (Anon., 1994, Article 5.3). This does not preclude the consideration of other relevant consequences associated with pest introductions, including non-quantitative impacts on the environment (e.g., harm to wild flora and forests).

The SPS Agreement does not include other basic terms from the field of risk, including risk analysis, risk management, and risk communication. It is expected that the usage and definitions of these terms are to be guided by the appropriate standard-setting bodies (i.e., Codex, OIE, and IPPC). In this regard, the SPS Agreement encourages countries to take into account risk assessment standards and techniques of the relevant standard setting body (Anon. 1994, Article 5.1).

Setting the Appropriate Level of Protection — The SPS Agreement recognizes and maintains the right of countries to determine and set an "appropriate level of protection" (ALP) for all pest or disease threats. While the SPS Agreement maintains the right of countries to determine their own APL, the SPS Agreement contains several disciplines to prevent countries from setting their ALP in an arbitrary or discriminatory fashion. The SPS Agreement defines the ALP as: "the level of protection deemed appropriate by the member establishing a sanitary or phytosanitary measure to protect human, animal, or plant life or health" (Anon., 1994, Annex A). A note is included indicating that "many members otherwise refer to this concept as the *acceptable* level of risk." The SPS Agreement treats the ALP and "acceptable level of risk" as synonymous terms.

In setting the ALP, the SPS Agreement requires countries to *"avoid arbitrary or unjustifiable distinctions in the levels of protection it considers to be appropriate in different situations, if such distinctions result in discrimination or a disguised restriction on international trade"* (Anon. 1994, Article 5.5). The objective is to prevent arbitrary behavior when it comes to setting the ALP in different, but comparable, risk situations. Different levels of protection may exist for different commodities for justifiable reasons. However, countries should be prepared to provide a science-based rationale for such differences.

Also, under the SPS Agreement, countries must ensure that their SPS measures are not more trade restrictive than necessary to achieve its ALP (Anon. 1994, Article 5.6). A measure is considered more trade-restrictive than required when there is another reasonable measure available that provides the appropriate level of protection sought by the importing country and which is significantly less restrictive to trade (Anon. 1994, Article 5.6, footnote).

Countries are required to provide information regarding their risk assessment procedures (including the factors that were taken into consideration) as well as information on how and why it selected a particular level of protection (Anon. 1994, Annex B.3(c)). The emphasis on making regulatory decisions and actions transparent is intended to curb the ability of countries to set arbitrary and non-science based SPS measures.

To further the goal of consistency in risk management decision making, the WTO SPS Committee, which is a committee consisting of representatives from all WTO member countries, is mandated to develop guidelines to promote consistency in the levels of protection applied for similar, identical, or comparable risks. This effort is now underway in the SPS Committee.

Regionalization — Under the SPS Agreement, countries are committed to adapting their import requirements to the health conditions of the specific area or region where a plant or animal commodity originates. This is the concept of regionalization, or the idea of recognizing areas or regions which present a low pest or disease risk and allowing trade in animal or plant commodities from those areas. Plant quarantine officials generally do not use the term "regionalization." Instead, the concept of "pest-free areas," which has the same meaning, is used. Under the SPS Agreement, a region (or pest-free area) may be all of a country, part of a country, or all or parts of several countries.

The concept of regionalization recognizes that pest and disease conditions may vary across a country as a result of ecological, environmental, and quarantine differences. This concept follows the basic premise that regulatory measures must be based on scientific principles. Hence, countries must be prepared to consider scientific evidence which may demonstrate the existence of a pest or disease-free area within an otherwise infested country. The burden of demonstrating a pest or disease-free area rests with the exporting country. The importing country's obligation is to be clear about the administrative and risk assessment procedures which would be used to evaluate free area requests.

The concept of regionalization contained in the SPS Agreement as applied to trade in plant commodities is not new. In the U.S., some examples of pest-free zones that existed prior to the negotiation of the SPS Agreement include currently recognized exotic-fruit-fly-free zones in Mexico, Chile, Brazil, and Australia. Other pest-free area requests are currently under review.

The U.S. general requirements for recognizing a pest free area are: 1) that within the past 12 months, the plant protection service of the exporting country must demonstrate the absence of pest infestations in the "free area" based on surveys approved by the U.S.; 2) the exporting country must adopt and enforce regulations to prevent the introduction of injurious pests into the alleged free area, and these regulations must be deemed by the U.S. to be at least equivalent to U.S. import requirements designed to prevent the introduction into the U.S. and interstate spread of injurious insects; and 3) the plant protection service of the exporting country must submit to the U.S. its detailed survey procedures and how it will enforce its quarantine requirements around the free area to prevent future introductions of injurious insects into this protected zone (Anon., 1998).

Millions of dollars' worth of U.S. fruit exports depend to a great extent on other countries recognizing U.S. quarantines (e.g., fruit fly in parts of Florida) and U.S. assertions that its products originate in production areas outside the quarantine zones. Maintenance of free areas, in turn, depends on the continued ability to demonstrate that effective controls, surveillance, monitoring, and emergency response programs are in place.

Equivalence — Under the SPS Agreement, countries are required to recognize another country's SPS measure as equivalent to their own when the exporting country demonstrates that its treatments or pest control procedures provide the importing country's desired level of quarantine security.

Equivalence encourages countries to recognize that different procedures (e.g., inspection, certification, testing, surveying, trapping, fumigation, and other treatments or practices) can be used to achieve the level of protection demanded by the importing country. The burden is on the exporting country to objectively demonstrate that its system or practices, while different from the importing country's measures, still achieves the importing country's plant quarantine security goals. This concept of evaluating alternative quarantine security measures is not new in trade, but it is not always applied consistently. For example, the United States has typically sought and obtained, on a bilateral basis, foreign recognition of a particular quarantine procedure (e.g., inspection, trapping, or some other practice aimed at eliminating a particular pest risk) used in the United States in order to facilitate the export of fruit or vegetable commodities.

Transparency — Under the SPS Agreement, countries are required to make their rule-making process transparent by:

- providing advance notification to WTO members of new proposed phytosanitary measures which may affect trade;
- making available the scientific basis for specific phytosanitary regulations to interested parties upon request, including documentation on the appropriate level of protection selected; and
- providing countries an opportunity to comment on its proposed rules before they are implemented (60-day comment period expected).

An exception to this advance notification rule exists for emergency disease or pest situations.

Like other provisions in the SPS Agreement, the transparency rule is intended to compel regulatory officials to ensure that their SPS measures are based on relevant scientific evidence and are consistent with previous risk management decisions. Advance notification of rules is intended to give affected parties, both domestic and foreign, an opportunity to provide relevant information on proposed rule changes and to anticipate and adjust to any regulatory actions which may affect trade. Notice and comment rule making is already used by several member nations, and in the U.S., is required by the Administrative Procedures Act.

Emergency Provisions — Provisions exist in both the SPS Agreement, WTO, and in the NAFTA to allow countries to adopt emergency SPS measures without prior notification and publication for comment in cases where an urgent problem of health exists (Anon. 1994, Annex B.6). These provisions are consistent with U.S. legislation which authorizes U.S. plant health authorities to publish interim rules when an emergency quarantine is needed to prevent the entry and/or spread of plant pests. Interim rules are effective immediately. A comment period exists, but the rule is already in effect. After the comment period closes (usually 60 days), U.S. plant health authorities publish a follow-up document in the *Federal Register*. If there are no changes based on comments received, the follow-up document is an affirmation of the interim rule. If there are changes based on comments received, a final rule is published that explains any revisions.

The SPS Agreement does not define an "urgent problem of health." Until more detailed language is developed and accepted, the U.S. would consider a new or changed pest or disease situation in a foreign country that presents a potential threat to U.S. animal or plant health to be a legitimate basis for taking emergency action. As information and assurances are provided regarding the status of a newly reported pest or disease situation, the emergency or interim action would be modified accordingly.

Dispute Settlement — At the WTO, a Dispute Settlement Body (DSB) was established to administer the dispute settlement system. Similarly, NAFTA established the Free Trade Commission to oversee implementation of the NAFTA obligations and to address specific issues, including disputes, which may arise between Mexico, the U.S., and Canada.

While existing as separate systems, the GATT and NAFTA dispute settlement procedures share some common features. For example, in both systems, disputes begin with bilateral consultations. If these discussions fail to resolve the issue, a complaining party may request formation of a panel. Panels, consisting of individuals drawn from an established list of recognized experts in the field of international law and agreed to by the parties, may seek recommendations and advice from the relevant international standard-setting organizations (i.e., OIE, Codex, IPPC, or their regional subsidiary organizations), individual experts, or appoint a board of experts to evaluate the technical aspects of a given issue.

A key question panels will ask when reviewing a phytosanitary measure which may be subject to dispute is whether that measure is based on a relevant international standard. If not, the next test is whether the measure was based on a risk assessment. Recent panel reviews, such as the one formed to evaluate the U.S.-EU hormone dispute, highlight the important role international standards and risk assessment play in justifying and defending an SPS measure.

If a panel issues an opinion that a measure is in violation of the SPS Agreement, the offending government has the option of either changing the WTO-inconsistent measure or keeping it and compensating the complaining party for the value of impaired trade. If compensation is not provided, the complaining party would be permitted to suspend some trade concession of equivalent value to lost trade.

9.4.4 Summary

By the 1980s many governments were turning more and more to free market principles as the engine of economic growth. The farm economy, like other business sectors, was to be increasingly exposed to market forces rather than managed and supported through government programs. To ensure that domestic producers could make a viable transition from farm support to farm competition it was a necessary precondition that markets, including foreign markets, were open and fair. This newfound interest in agricultural trade reform and promoting free market principles in agriculture was reflected in the Uruguay Round negotiations. The result was comprehensive reductions in tariffs and the establishment of global trade rules for agriculture, including rules to prevent the use of phytosanitary regulations as disguised barriers to trade.

While the traditional GATT framework, prior to the Uruguay Round, recognized the right of countries to protect themselves from pest and disease risks, it was generally recognized that this exemption from free trade rules was too broad and undisciplined to curb the use of health restrictions as disguised barriers to trade. The expectation was that health-related barriers would proliferate as countries negotiated away traditional forms of commercial protection, such as tariffs and quotas, for their domestic industries. Consequently, the Uruguay Round negotiations resulted in an agreement that requires countries to document and make available the scientific underpinnings of their sanitary and phytosanitary regulations. Thus far, the most effective elements of the SPS Agreement appear to be those which require governments to make their regulatory processes open to inquiry and the obligations requiring governments to demonstrate the scientific basis for their sanitary and phytosanitary regulations.

The SPS Agreement put countries on a path toward harmonizing their SPS measures, using international standards and effectively reducing arbitrariness in technical regulations by requiring countries to base their measures on scientific evidence and principles. A more difficult goal will be to harmonize country's "appropriate levels of protection" and ensuring that countries are not arbitrarily inconsistent about where they set their levels of protection for comparable or identical pest risks. In this regard, discussions are currently underway both within and among member nations to better understand the risk management process and how to ensure consistency in risk management decisions.

LITERATURE CITED

Anon. *International Plant Protection Convention of 1951*. FAO, Rome, 1952

Anon. *NAPPO Standards for Pest Risk Analysis*. North American Plant Protection Organization, Ottawa, pp. 1-9, 1993.

Anon. *International Standards for Phytosanitary Measures. Part 1. Import Regulations. Guidelines for Pest Risk Analysis*. Publication No. 2, Secretariat of the International Plant Protection Convention, FAO, UN, Rome, 1996a.

Anon. *International Standards for Phytosanitary Measures. Glossary of Phytosanitary Terms*. Publication No. 5, Secretariat of the International Plant Protection Convention, FAO, UN, Rome, 1996b

Anon. *NAPPO Compendium of Phytosanitary Terms*. B.E. Hopper (Ed.) 1996c.

Anon. *North American Free Trade Agreement Between the Government of the United States of America, The Government of Canada and the Government of the United Mexican States*, Volume I, U.S. Government Printing Office, Washington, D.C. 1993.

Anon. *Agreement on the Application of Sanitary and Phytosanitary Measures*, General Agreement on Trade and Tariffs. pp. 69-73. 1994.

Anon. *Guideline for Plant Pest Risk Analysis of Imported Commodities*, Version 1.7, Plant Protection and Quarantine, APHIS, USDA, Riverdale, MD, 277 pp., 1997a.

Anon. *International Plant Protection Convention, New Revised Text*, FAO Conference at its 29th Session, November, 1997, FAO, Rome., 1977b.

Anon. *Code of Federal Regulations, Title 7, Agriculture, Subtitle B, Regulations of the Department of Agriculture; Chapter III, Animal and Plant Health Inspection Service, Department of Agriculture*: Parts 300-399, January 1, 1998, Office of the Federal Register, National Archives and Records Administration, Washington D.C. 1998.

Betterley, S. C., E. Miller, R. R. Ormiston, M. J. Shannon, and L. D. Fuell. *Feasibility Study on Pest Risk Analysis*. USDA. (unpublished), 1983.

Crawley, M. J. What Makes a Community Invisible, in: *Colonization, Succession, and Stability*, Gray, A. J., M. J. Crasley, and P. J. Edwards, Eds., Blackwell Scientific Publishers, Oxford, 1987.

Drake, J. A., T. E. Flum, G. J. Wittman, T. Voskuil, A. M. Hoylman, C. Cresson, D. A. Kenney, G. R. Huxel, C. S. Larue, and J. R. Duncan. The Construction and Assembly of an Ecological Landscape. *J. Anim. Ecol.* 62: 117-130, 1993.

Elton, C. S. *The Ecology of Invasions by Animals and Plants.* Methuen and Co., Ltd., London, 1958.

Hopper, B, E, Ed., *NAPPO Compendium of Phytosanitary Terms*, NAPPO, Nepean, Ontario, Canada, 1996.

Kahn, R. P. A Concept of Pest Risk Analysis. *EPPO Bull.* 9:119-130, 1979.

Kahn, R. P. The Host as a Vector: Exclusion as a Control, in *Pathogens, Vectors, and Plant Disease,* Harris, K. F. and K. Maramorosch, Eds., Academic Press, NY, 1982.

Kahn, R. P. *Plant Protection and Quarantine, Vol. 1, Biological Concepts,* CRC Press, Boca Raton, 226 pp., 1988.

Kahn, R. P. Exclusion as a Disease Control Strategy, in *Annual Review of Phytopathology.*

Cook, R, J., G. A. Zentmeyer, E. B. Cowling, Eds., Annual Reviews, Inc., Palo Alto, CA, 1991.

Kahn, R P. Biological Concepts, in *Containment Facilities and Safeguards for Exotic Pests and Pathogens,* Kahn, R. P. and S. B. Mathur, Eds., American Phytopathological Society Press, St. Paul, MN, 1998.

Kaplan, S. "Expert information" versus "expert opinions." Another approach to the problem of eliciting/combining/using expert knowledge in PRA. *Reliability Engineering and System Safety* 35: 61-72, 1992.

Kogan, M. *Biological aspects of exotic pest exclusion.* Paper presented at the Annual Meeting of the Entomological Society of America, December 3, 1990. New Orleans, LA, 1990.

May, R. M. Nonlinearities and complex behavior in simple ecological and epidemiological models. *Ann. New York Acad. Sci.* 504: 1-15, 1987.

McGregor, R. C. *The Emigrant Pests.* USDA (unpublished), 1973.

NRC (National Research Council). *Risk Assessment in the Federal Government: Managing the Process.* National Academy Press, Washington, D.C. 1983.

Porter, S. D., B. V. Eimerern, and L. E. Gilbert. Invasion of Red Imported Fire Ants (Hymenoptera: Formicidae): microgeography of competitive replacement. *Ann. Entomol. Soc. Amer.* 81: 913-18, 1988.

PPQ, APHIS, Unpublished data, 1998.

Roberts, D. The WTO SPS Agreement: Overview and Implications for Animal Health and Animal Product Import Decision Making, Riverdale, MD., Conference Draft, *USDA Workshop on the Use of Economic Analysis in Evaluating Animal and Animal Product Import Requests*, March, 1988.

Sailer, R. S. Our immigrant insect fauna. *Bull. Entomol. Soc. Amer.* 24: 3-11, 1978.

USDA. Pathway-Initiated Pest Risk Assessment: Guidelines for Qualitative Assessments, ver 4.0 (unpublished), 1995.

USDA. APHIS Trade Risk Analysis Approach (unpublished), 1996.

WTO (World Trade Organization). The Results of the Uruguay Round of Multilateral Trade Negotiations, The Legal Texts (Geneva, Switzerland: GATT Secretariat), 1994.

BIBLIOGRAPHY

Hoekman, B. and M. Kostecki, *Political Economy of The World Trading System from GATT to WTO.* Oxford University Press, Oxford, 1995.

Josling, T., S. Tangerman, and T. K. Warley, *Agriculture in the GATT.* St. Martin's Press, New York, 1996.

CHAPTER **10**

Regulation and its Economic Consequences

Michael Ollinger and Jorge Fernandez-Cornejo

CONTENTS

1-56670-478-2/00/$0.00+$.50
© 2000 by CRC Press LLC

10.1   PESTICIDE REGULATION AND ITS ECONOMIC CONSEQUENCES

Together with fertilizer and improved seeds, pesticides have played a major role in increasing farm productivity. For example, corn yields rose three-fold over the past 40 years. Despite this positive effect, evidenced by the willingness of U.S. farmers to spend $8.3 billion on pesticides in 1996 (USDA, 1997), there is concern about contamination of ground and surface water, the effects on wildlife, and the potential health risks to farm workers and, because of residues, to consumers (Harper and Zilberman, 1989). These potential side effects have prompted the U.S. Government to strictly regulate pesticides.

Regulatory stringency exists along a continuum. At one extreme are pesticides with use prohibitions, such as DDT, that are believed to have a high likelihood of causing harm to human health or of having a severe impact on fish and wildlife. At the other extreme is naturally occurring or self contained pesticides, such as pheromones, that have almost no use restrictions.

Legislative history suggests that pesticide regulation has gradually become more stringent. The very first federal legislation, *The Federal Insecticide Act of 1910*, dealt with correctly identifying container contents through a label requirement. The first real attempt to control pesticide toxicity came with the Federal Food, Drug, and Cosmetic Act of 1938 (FFDC) and the Federal Insecticide, Fungicide, and Rodenticide Act of 1948 (FIFRA). This legislation established procedures for setting tolerances, prohibited a high immediate toxicity to humans, and required handling instructions. However, regulators at the United States Department of Agriculture (USDA) and the Food and Drug Administration (FDA) primarily responded to the compliance requirement concerns of registrants and the verification of pesticide efficacy.

Throughout the 1950s and 1960s, publications, such as *Silent Spring* by Rachel Carson, caused growing consumer concern about the harmful side-effects of pesticides and other chemicals. Congress responded by enacting legislation that increased regulatory restrictiveness. However, it was not until 1972 with passage of the amendments to FIFRA and FFDC that regulation made a transition toward more aggressive protection of human health and safety and the environment.* This legislation imposed a strict burden of proof for verifying pesticide safety, and gave the Environmental Protection Agency (EPA), which was now the primary enforcer of pesticide regulation, the power to cancel and suspend pesticide registrations. Although testing requirements for new pesticides grew substantially after 1972, existing pesticides that had been approved in the less demanding regulatory period before 1972 remained in use, and, except for chemicals with the most serious side-effects, were not required to meet existing regulatory standards until the 1988 FIFRA amendment. The recently enacted Food Quality Protection Act (FQPA) of 1996 promises to increase stringency for dietary risks for both new and existing pesticides.

The first section of this chapter discusses the powers that the FFDC and FIFRA amendments and other legislation, such as the Safe Drinking Water Act, conferred on regulators. The second section deals with regulatory issues, such as pesticide registration requirements. The third section contains a discussion of research on the economic consequences of pesticide regulation.

10.2 PESTICIDE LEGISLATION

The FFDCA of 1938 and FIFRA in 1948 form the basis of modern pesticide regulation. Under the FFDCA, Congress gave the FDA the authority to establish procedures for setting tolerances. With FIFRA, Congress assigned USDA the tasks of registering all pesticides sold in interstate commerce against manufacturers claims of effectiveness, and ensuring that the pesticide label states whether the pesticide is toxic to humans and fish and wildlife.

10.2.1 Major Amendments to Pesticide Legislation

Congress amended FIFRA in 1959, 1964, and 1967 and FFDCA in 1954, 1958, and 1960. The FIFRA amendments granted the USDA the authority to regulate all pesticides, closed a loophole that enabled companies to register pesticides when regulators felt more data were required, and made it necessary for pesticides to meet tolerances to gain registration. The FFDCA amendments required producers to provide regulators with data about the acute, intermediate, subchronic (short-term), chronic (long-term), and other health effects of pesticides. They also included the Delaney Clause, which defined any pesticide that concentrates in processed food as an additive and prohibited the use of additives that "induce cancer" in experimental

---

* Unless otherwise noted, we will refer to amendments to FIFRA and FFDC as FIFRA amendments because FIFRA is the primary legislative vehicle through which EPA regulation is enforced.

animals or man. Under this standard, FDA set pesticide tolerances in processed foods at zero. USDA, on the other hand, set tolerances for raw commodities at a level necessary to protect human health.

Hatch (1982) asserts that growing public awareness of the carcinogenic and environmental effects of pesticides led to the transfer of administration of FIFRA and the setting of tolerances for FFDCA from USDA and FDA to the Environmental Protection Agency (EPA). The subsequent 1972 FIFRA amendments greatly increased regulatory stringency by requiring the EPA to examine the effects of pesticides on fish and wildlife safety; demanding more extensive testing of the chronic and acute toxicity effects; and, for the first time, granting the EPA the authority to cancel or suspend pesticides that may pose unreasonably adverse health or environmental effects. FIFRA also gave states substantial latitude in regulating the distribution, sale, and use of pesticides, and mandated that all existing pesticides be reregistered within four years, with pesticides that failed to meet existing standards subject to cancellation if estimated risks outweighed benefits.

It took several years for the EPA to establish the precise regulatory standards and guidelines (rules) necessary to implement the 1972 FIFRA amendment. The first formal rules, published in the *Federal Register* in 1978 (USEPA, 1978), mainly dealt with testing requirements for proving the safety of chemical compounds to human health. A second set of rules in 1982 (USEPA, 1982) dealt mainly with environmental testing requirements. A third set of more informal rules in 1994 (USEPA, 1994) had only modest requirements for safety to both human health and the environment.

Several critical aspects of the 1972 amendment were ambiguous, making implementation through formal rules difficult. Concern centered on the costs of registering pesticides with low measurable environmental risks; the development of pesticides for low acreage crops, such as most fruits and vegetables (minor-use pesticides); and the reregistration of existing pesticides. A major debate centered on the use of existing test data to register new pesticides. This situation arose when a second manufacturer wanted to market a product similar to one already on the market. Overall testing costs and the costs to the new manufacturer of product development would have been lower if the new manufacturer could have used existing data. However, the original pesticide developer wanted to prevent the new manufacturer from using the data because such use would have facilitated the market entry of a competitor. Hence, a conflict of incentives existed that could be settled only with new legislation.

The 1978 amendment eased data requirements for pesticides that pose low environmental risks and gave the EPA the right to reduce data requirements for minor-use pesticides. This legislation also strengthened state enforcement of pesticide use and allowed the registration of pesticides for specific local needs. Additionally, the amendment permitted certain crop uses that were not inconsistent with the label and simplified the registration process. Finally, the 1978 amendment gave new manufacturers the right to use original producer data but required the new manufacturers to compensate the original developers. The amount of compensation was to be decided through arbitration. This aspect of legislation encouraged producers to begin the registration process for many minor crop pesticides that they formerly had withheld from the market.

The 1978 amendment eliminated the pesticide registration deadline established in the 1972 FIFRA amendment. Instead, it required an expeditious process, and allowed the EPA to evaluate the 600 active ingredients used in product formulations rather than the 35,000 registered pesticide products. However, by 1986, only one active ingredient had completed its final reassessment and only 130 had completed their preliminary reassessments. In response, Congress demanded completion of reregistration by 1997 through the 1988 FIFRA amendment. Reregistration priority was based on: (i) potential for post-harvest residues on human food or animal feed; (ii) toxicity effects; (iii) outstanding data requirements; and (iv) likelihood of worker exposure.

The amendment covered all pesticides registered before 1984 and affected 1153 active ingredients grouped by EPA into 611 similar chemicals. Funding for this reregistration program came from a one-time reregistration fee for each active ingredient and an annual maintenance fee for each registered product. Exempt from fees were minor use pesticides and antimicrobial pesticides with annual production of less than million pounds and registered for nonfarm uses.

The 1988 FIFRA amendment also expedited the registration process for "me-too" pesticides and expanded EPA's authority to regulate the storage, transportation, and disposal of pesticides, containers, rinses, and contaminated materials. "Me-too" pesticides have the same active ingredient, similar use patterns, and comparable use directions as a previously registered pesticide.

The Delaney Clause became controversial in 1987 when the National Academy of Science (NAS) issued a report entitled "Regulating Pesticides in Food: the Delaney Paradox." NAS suggested that all pesticides should be regulated based on consistent standards; total dietary exposure to oncogenic pesticides could be reduced if a negligible risk standard was applied consistently; and EPA should focus on pesticides in the most commonly consumed crops/foods/feeds and improve pesticide-residual testing methodologies. In 1988, EPA adopted a negligible risk standard of one cancer incident in one million people. However, the Ninth Circuit Court ruled in 1992 that this risk standard was inconsistent with the Delaney Clause, compelling EPA to threaten cancellation of over 30 pesticides.

A more recent NAS (1993) study, "Pesticides in the Diets of Infants and Children," highlighted the sensitivity of children to the health risks of pesticide residues. The controversy surrounding this publication and the 1988 NAS report encouraged Congress to enact the 1996 FQPA. This legislation reduced registration stringency by eliminating the paradox of holding raw and processed foods to different standards; prohibiting states from setting residue standards that differ from federal standards; and facilitating the registration of minor-use and reduced-risk pesticides. However, FQPA greatly increased restrictiveness by mandating that all pesticides have supporting test data updated every 15 years; requiring an aggregate pesticide tolerance in which exposure to air, water, and other nonfood sources (excluding occupational exposure) and all food sources is considered when establishing a tolerance for a new pesticide use; mandating the use of up to a ten-fold safety factor in setting tolerances for infants and children; substantially reducing the cases where economic benefits of pesticide use were considered relative to dietary risk for raw commodities; and ordering tests for estrogenic and other endocrine effects.

10.2.2 Other Legislation that Influences Pesticides

10.2.2.1 Endangered Species Act

The Endangered Species Act was passed to protect animal and plant species from extinction and to conserve the ecosystem on which endangered species depend. Its provisions restrict pesticide use thought to affect fish and wildlife by establishing a threshold (lowest) application rate that "may affect" endangered species. In 1991, 31 pesticides had "may affect" determinations, requiring EPA to consult with the Fish and Wildlife Service (FWS) over a course of action. The course of action of FWS has been to define areas where a pesticide use was banned or restricted — made part of the label.

10.2.2.2 Clean Air and Clean Water Acts

The Clean Air Act dictates allowable concentrations of specified pollutant emissions based on state and federal limits. Although this legislation primarily affects manufacturing, it has led to phase-out of methyl-bromide, a chemical that can deplete stratospheric ozone.

The Clean Water Act limits discharges of toxic pesticides from manufacturing plants. Several pesticides, such as aldrin/dieldrin, DDT, and endrin, have either highly limited or no permissible discharge from production plants. The Clean Water Act also required states to establish management plans containing management practices for best controlling pesticides, fertilizers, manure, and other nonpoint pollution sources (NPS) on a watershed basis.

10.2.2.3 Coastal Zone Management Act

The Coastal Zone Management Act (CZMA) requires states with coastal zones to submit a program that will control NPS pollution in coastal waters. With management plans not required until 1999 and only small penalties for failing to implement them once they are in place, this legislation will likely have only a small effect on pesticide use.

10.2.2.4 Safe Drinking Water Act

The Safe Drinking Water Act (SDWA) is far more relevant to pesticide use than the CZMA. Established in 1986, SDWA requires states to protect public water wells from contamination from all sources of pollution, including pesticides. The SDWA requires that Public Water Systems ensure that their water supplies do not exceed the Maximum Contaminant Level for four insecticides and nine herbicides. States are the primary enforcers and have brought more than 9000 pesticide abuse cases against farmers and other pesticide users (Ribaudo and Gadsby, 1998).

As pointed out by Ribaudo and Gadsby (1998), six states — Arizona, California, Iowa, Kansas, Montana, and Wisconsin — are addressing pesticides in water with regulations. The Arizona plan establishes a list of pesticides — currently about 152 chemicals — that have groundwater leaching potential. If the pesticide or any toxic breakdown is found in groundwater, or in the soil below the root zone, the pesticide is subject to regulation. If the pesticide is carcinogenic, mutagenic, teratogenic, or toxic to humans, and the label cannot be modified such that crop or geographic uses or application rates do not threaten groundwater, the registration is canceled. For pesticides that do not pose risks to human health, continued use is allowed under certain restrictions.

California's law is slightly more complex, requiring the state to set numeric values for six chemical characteristics that define a chemical's ability to leach into groundwater. During registration, the manufacturer must submit information on these characteristics. If the value for one characteristic exceeds the prescribed numeric value, the chemical is placed on the groundwater protection list. If any of the chemical characteristics are either found in groundwater, eight feet below the surface, or below the root zone or the zone of microbial activity, the pesticide can be restricted or canceled. Restrictions, as in many other states, may apply to only some pesticide management zones because leachability is affected by application rates, soil type, and topographical features. Cancellation can occur statewide if all pesticide management zones show similar signs of leaching.

Iowa and Kansas monitor groundwater for pesticides and promote the use of voluntary best management practices if a contamination trend is detected. If voluntary measures fail to control pesticide leaching, then the states impose mandatory restrictions. Montana and Wisconsin have established leaching triggers that, when reached, require the state to take remedial action. This action includes state restrictions on pesticide use according to soil type and application timing and prohibitions on pesticide application through irrigation systems. If these and further measures fail, the chemical is prohibited in the area overlaying the aquifer.

10.2.2.5 State Legislation

States have the primary responsibility for enforcing federal pesticide regulations. Under this mandate, states encourage proper pesticide use and record common pesticide use practices, and determine whether pesticide use is in accordance with label directions. States also have the power to regulate the sale and use of pesticides, and, until enactment of the FQPA, could establish more stringent tolerances than those required by EPA. The power to regulate the sale and use of pesticides is exerted through state registration programs that, in many cases, are based on federal standards. States may also prohibit or restrict federally registered pesticides and demand environmental and safety data in addition to the data submitted to EPA, but cannot impose labeling or packaging requirements that differ from those mandated under FIFRA. States can also facilitate the registration of pesticides for special needs, and, after consultation with the EPA, register pesticides for which federal uses have been denied, disapproved, suspended, or canceled.

The diversity of state laws makes them difficult to summarize. Instead, we consider California, which has one of the most active regulatory programs. Established in 1901 and made stricter in 1969, California's regulatory program includes registration and data requirements; licensing of pest control operators, dealers, and advisors; assessment of pesticide taxes; worker safety restrictions; and limits on pesticide use and permit availability. It also requires companies to submit all test data developed to meet EPA standards and all data required uniquely by California to the California Environmental Protection Agency (CEPA). The unique California data includes dermal absorption tests and mixer/loader, applicator, and reentry data. Additionally, state law requires all pesticides to meet current standards. If a company does not adhere to this requirement, CEPA generates the data itself and charges the company for its service. For more detailed discussion of state regulatory requirements, see *Pesticide Regulation Handbook* (1991).

The FQPA could have a major impact on state regulation. This legislation prohibits states from setting tolerances that differ from those imposed by EPA unless the state petitions the EPA for an exemption. EPA would likely approve the petition only if a state can justify an exemption by citing local conditions, and then only if it does not violate federal law. States can also regulate pesticide use in other ways. For example, California has numerous area-based restrictions.

10.2.2.6   1996 Farm Bill

The Farm Bills passed every five years also have provisions related to pesticide legislation. The 1996 Farm Bill requires all applicators, including farmers, to maintain records on restricted use pesticides; establishes a revolving fund to help register minor use pesticides; requires EPA to create a program to coordinate minor use activities; and mandates that USDA provide educational, research, and technical assistance programs for users and dealers of agrichemicals.

10.3   REGULATION ISSUES AND AGENCIES INVOLVED

All pesticides must be registered with the EPA prior to use by passing through a series of demanding field tests. With registration comes the right and the requirement to print labels that clearly indicate acute, chronic, and other harmful human effects, adverse environmental impacts, and handling and application instructions. Pesticide registrations can be canceled if future tests reveal adverse side-effects. EPA also grants permits and exemptions of pesticide restrictions for emergencies and for pesticide testing purposes. Below, we consider pesticide registration, tolerances, data and testing requirements, label requirements, emergency use permits, and cancellations and suspensions. We also consider the Applicator and Worker Protection Standard, applicator certification, and the regulation of transgenic crops and biological pesticides. For more detailed discussions of regulation issues, see *Pesticide Regulation Handbook* (1991), *Farm Chemicals Handbook '95*, and NAS (1987).

10.3.1   Pesticide Registration

A firm must register a pesticide for use by showing that the pesticide has a "reasonable certainty" of no harm to human health, as defined under the FDCA, and will not cause "unreasonable adverse effects" on human health or the environment, as defined under FIFRA. To make its case, the firm submits data on pesticide field tests and characteristics to the EPA, which either approves, asks for more data, or rejects the registration application. This process is repeated for each crop use. The number of tests and testing costs drop dramatically for crop uses within crop classes, such as kale and spinach, but less so for more dissimilar crops, such as lettuce and grains. Additionally, the type of data and the cost of generating it varies by soil conditions, and climatic and topographical features. The EPA can reject a pesticide for all uses or for use on some crops or in some areas. For example, EPA may completely reject a pesticide use for clover but not corn, or a use for corn in Florida but not Iowa.

10.3.2 Pesticide Tolerances

Established by EPA and enforced by USDA for meat and poultry and by FDA for fruits, vegetables, and grains, tolerances are the legally allowed amounts of pesticide residues permitted in foods sold in interstate commerce. Since pesticides must have tolerances to be registered, granting a tolerance is tantamount to granting a pesticide registration.

Until 1996, one tolerance could exist for residues in a commodity as a raw food item and another tolerance for residues in the same commodity as a processed food, e.g., apples for direct consumption and apples used in apple juice. Under Section 408 of the Federal Food, Drug, and Cosmetic Act of 1954, tolerances for residues in or on raw commodities were set at levels necessary to protect public health, while taking into account the need for an adequate, wholesome, and economical food supply. Under Section 409, which controls tolerances for pesticide residues in processed foods, residues had to be proven safe. This section was interpreted to mean that there must exist "reasonable certainty" that "no harm" to consumers would result when the pesticide is put to its intended use. Moreover, the Delaney Clauses prohibited the use of a food additive found to cause cancer. A pesticide was considered an additive if it concentrated in the processed food.

Combined Sections 408 and 409 suggest that economic benefits could be considered and carcinogenic compounds could exist at levels not believed to cause cancer in the setting of tolerances for raw commodities. However, economic benefits could not be used, and carcinogenic additives that concentrate were prohibited in processed foods. The EPA enforced this clause by denying tolerances for both raw and processed commodities if residue concentration occurred and the Delaney Clause applied. EPA used the raw commodity tolerance for processed foods if the pesticide was used on raw commodities and the Delaney Clause was not an issue. According to the EPA, pesticide concentration existed if there was a greater quantity of the pesticide in the processed food than on the raw commodity.

Raw agricultural commodity tolerances are based on the results of field trials designed to achieve the highest residues likely under normal agricultural practices. These agricultural practices include the highest recommended application rates and weather and climatic conditions that permit the highest residue levels on crops. Average residue levels on raw commodities can be lower than the established tolerance because climactic conditions dissipate residues more rapidly in some areas (NAS, 1987). Since average pesticide residue levels can be lower than the raw commodity tolerance, some pesticide concentration in processed foods was possible.

The EPA establishes a tolerance by first determining whether the pesticide causes a significant increase in tumors in animals. If it does, residues are not permitted beyond those considered "safe" for human consumption. If there is no significant increase in tumors, then EPA calculates an Acceptable Daily Intake (ADI) for the substance in question. EPA sets an ADI level equal to the threshold, i.e., the highest dose at which no adverse human health effects are observed, divided by a safety factor of 100. Pesticide exposure is calculated by summing residues from all food sources. If residues from all previously approved tolerances and the new tolerance is less than the ADI level, then the pesticide use is approved. EPA assumes that the pesticide residues are present at full tolerance on all harvested acres for which a tolerance exists and is proposed.

Under the Delaney Clause prior to FQPA, the issue of pesticide concentration was extremely important in tolerance setting. The EPA set tolerances for pesticides that were found to cause tumors but that did not concentrate in processed foods by generating an upper-bound estimate of human cancer risks based on animal bioassays. Using an assumption of lifetime exposure, EPA then extrapolated the pesticide effects of high pesticide doses on animals to the much lower doses ordinarily consumed by humans. Then, EPA predicted what would likely happen to humans under actual exposure conditions and approved tolerances if there was an upper bound risk of less than one in one million of contracting cancer from normal food consumption.

Under FFDCA, food — either raw or processed — is adulterated unless a pesticide residue is within the tolerance limit. The tolerance level is based on the sum of risks from all food uses; all other exposures, such as drinking water and lawn care products; and exposure from pesticides with a common mechanism of toxicity. The tolerance for each food use must be set so that this aggregate standard is met (i.e., all food and other uses of the pesticide). Additionally, consistent with the previous practice, pesticides with threshold effects are to have tolerance set 100 times below the maximum level determined not to cause dietary risks, including cancer.

The FQPA establishes a uniform tolerance standard for raw commodities and processed foods. Pesticides with no threshold effects are to have tolerances set at a level at which there will be no greater than a negligible increase in cancer. Also, to ensure with reasonable certainty that no harm would result to infants and children, the EPA will establish an additional ten-fold margin of safety and would likely restrict pesticide uses for some crops heavily consumed by children and infants.

10.3.3   Data and Testing Requirements

The pesticide data submitted to EPA for pesticide registration includes a description of the pesticide chemistry; identity, description, and quantity of residues expected to be present in food from the crop; analytical procedures used in obtaining the residue data (procedures must permit replication); and acute, chronic, and subchronic toxicity tests and mutagenicity and pesticide metabolism studies from residues on crops, in animal feed derived from crop by-products, or from forages and resulting residues in meat, poultry, milk, fish, or eggs. Data requirements also include an assessment of human toxicity from residues remaining after application; a determination of human health hazards due to reentry; product efficacy; and evaluations of spray drift and impact on nontarget organisms, such as fish and birds.

The purpose of the test data is to permit estimation and tracking of the principal food residues that are likely to result from commercial use of a pesticide under variable climactic and soil conditions. This generally requires extrapolation from data on a limited number of trials done in different parts of the country. EPA generally uses the highest residue levels reported in such tests (NAS, 1987).

As of 1992, there were 70 different types of laboratory and field tests, some of which cost millions of dollars and take several years to complete. The tests examine the chemistry, toxicology, environmental fate, and ecological effects of both the parent compound and its major impurities, degradation products, and metabolites; and include two-generation reproduction and teratogenicity studies, a mutagenicity study, delayed neuropathy, plant and animal metabolism, and toxicology studies, consisting of acute (immediate effect), subchronic (up to 90 days), and chronic (long-term) tests, as well as oncogenicity and chronic feeding studies. Other tests are used to evaluate the effects of pesticides on aquatic systems and wildlife, farmworker health, and the environment. Additionally, the FQPA requires a screening program for estrogenic and other endocrine effects.

10.3.4 Pesticide Labeling

The registration process concludes with the approval of a product label. Label information includes producer name, weight, and ingredient statements; precautionary and classification statements (if restricted); and legally binding instructions on how, when, or where the pesticide may be applied or not applied, and on which crops the pesticide can be used. Pesticide classification is based primarily on the acute toxicity of the product and the site at which it is used. Examples of use restrictions can be for a high probability of pesticide leaching and pesticide migration due to weather conditions, such as high winds. The directions for pesticide use specify the method of application required to minimize farm worker exposure and impact on the environment, and also concern over reentry intervals, rotational crop restrictions, and specific warnings to prevent crop injury. The label also provides warnings of pesticide toxicity and health complications arising from ingestion or exposure; identifies environmental or other farm crop effects; gives the acute toxicity rating; and states whether the pesticide is carcinogenic, has other chronic effects, or is harmful to human health. Disclosure is also required for inert ingredients because these compounds may contribute to acute overall product hazard.

10.3.5 Pesticide Exemptions and Experimental Use Permits

EPA grants exemptions from regulation to many types of biological control agents and other pesticides regulated by other agencies, and substances that contain or are treated with a pesticide to protect the article or substance itself. In addition,

EPA grants tolerance exemptions when it appears that no hazard to public health will result from pesticide residues. Tolerance exemptions must be obtained for all inert ingredients in formulations applied to raw agricultural commodities.

The EPA may exempt state governments or federal agencies from FIFRA if use will not cause unreasonable harm to the environment and a pesticide is needed to control unexpected pest outbreaks, prevent the spread of disease, or reduce a significant risk to an endangered or threatened species, beneficial organisms, or the environment. Exemptions are not permitted for pesticides that were canceled, unless EPA modifies the terms of the cancellation or suspension. Moreover, these exemptions apply to specific uses of a pesticide and may be for limited time periods. Other types of exemptions include quarantine, public health, and crisis exemptions. Additionally, to permit pesticide testing under actual field conditions, the EPA issues Experimental Use Permits (EUP). Products subject to EUP may be distributed only under the conditions outlined in the EUP.

10.3.6   Pesticide Cancellation and Suspension

Congress gave EPA authority to cancel pesticide registrations that have unreasonable adverse effects on man or the environment. Cancellation requires a Special Review Process (SRP) that is initiated if the pesticide may pose a serious risk of acute injury or chronic human effects; could be a threat to the existence of any endangered or threatened species; or could otherwise pose a risk to humans or the environment. Since determining unreasonable effects include consideration of both costs and benefits, the EPA weighs the economic benefits against the environmental costs.

After the SRP, the EPA typically makes label changes requiring warnings, use precautions, and other handling and use practices, but could also either cancel the registration or take no action. EPA could also initiate a reclassification for a restricted use or call in for additional field or laboratory data. After receiving public comments, the EPA must initiate a notice of intent to cancel or reclassify the pesticide for restricted use if it wants to cancel the pesticide or change the label of the pesticide. This notice is then reviewed by the USDA and a Scientific Advisory Panel before action is taken.

The EPA Administrator can suspend a pesticide registration at any point during the SRP. This action will be taken if the Administrator determines that suspension is necessary to prevent the use of a substance that has a high likelihood of being hazardous. Beside cancellations and suspensions, registrants may voluntarily choose to "lose" a registration, change a label, or withdraw a product rather than meet certain requirements. Registrants often "voluntarily lose" a registration after EPA presents its risk assessment and discusses label changes, including cancellation. Additionally, EPA must approve all voluntary label changes.

10.3.7   Applicator and Worker Protection Standard

EPA requires adherence to the applicator standard because most pesticides are highly toxic shortly after application. This standard applies to any pesticide registered for use on agricultural plants on farms or in forests, nurseries, or greenhouses. Its provisions are designed to (i) eliminate or reduce exposure to pesticides for up to 72 h in all areas in which pesticide products are applied; (ii) mitigate exposures that do occur; and, (iii) provide information to applicators and workers about proper pesticide handling and the hazards of pesticides.

10.3.8 Pesticide Applicator Certification

The 1972 FIFRA legislation mandated that EPA must classify all pesticides for either general or restricted use and that only certified applicators or laborers under the direct supervision of a certified applicator could administer restricted use pesticides. Certified applicators, in turn, were classified as either private or commercial applicators. Private applicators could administer restricted use pesticides on either their own land or that of a second party if no compensation was received. Commercial applicators, on the other hand, were compensated for being an applicator or being an immediate supervisor of an applicator of restricted use pesticides. Since states are permitted to establish standards that exceed EPA requirements, state pesticide applicator standards vary considerably and are not summarized.

10.3.9 Regulation of Biological Pesticides

The biotechnology portion of EPA's pesticide regulatory program controls registration of natural biochemical materials, such as pheromones, insect hormones, products of fermentation, and genetically altered microbial materials, such as bacteria, fungi, and viruses.

Regulatory costs are highest for viruses because pathogenicity tests require large quantities of test substances. Additionally, the presence and viability of a virus in animal tissue cannot be measured using chemical or microbiological techniques, making *in vivo* bioassays necessary. Plant and fermentation extracts also can be costly to register because they are treated much like chemicals, yet testing requirements for chemicals do not necessarily apply to them. Pheromones, on the other hand, require only toxicity and few other tests, and self-contained pheromone traps need no registration at all.

EPA regards each strain of a microorganism as a separate active ingredient. It uses a tier system for nontoxic and nonpathogenic microorganisms in which pesticides that pass a short-term test of toxicity to humans and fish and wildlife, pathogenicity, persistence, and replication in mammals gain waivers against additional human health tests. The EPA also gives regulatory priority to biologicals over chemicals and waives some of the 19 tests required of chemicals.

10.3.10 Regulation of Transgenic Crops

Both the USDA and the EPA regulate field research in plant biotechnology, and the FDA becomes involved if there is a well-defined difference in food safety or health risks between the original product and the one produced by biotechnology. The Animal Plant Health Inspection Service (APHIS) of USDA regulates plant biotechnology field-testing and transport. It requires that organizations receive an EUP with a duration of one year before conducting field-testing with regulated articles.* APHIS makes two site visits to ensure compliance.

EPA has regulatory authority over all genetically modified plants with pesticidal properties, such as insect resistance, but not herbicide tolerance, which is not considered a pesticidal property. EPA thoroughly scrutinizes insect resistant plants, but gives less scrutiny of virus-resistant plants because EPA believes that it is highly unlikely that there are adverse environmental or human health effects (conversation with Bob Torla of EPA, March 20, 1998).

As of the end of 1997, all insect-resistant plants approved by the EPA carried the Bacillus Thuringiensis (Bt) toxin, which has been used as an insecticide for several years and has well-known properties. In the registration process, EPA determined plant's digestibility and toxic residue absorption by the body; examined the potential for cross-breeding with wild plants and effects on nontarget insects and either endangered species or the habitats of endangered species; and, considered the planting restrictions needed to prevent insects from developing immunity. For other less well-known insect resistance traits, the regulatory approval process is likely to be more complicated and costly.

Some states also regulate field-testing, and there has been public debate over which government level should have jurisdiction. For example, North Carolina, which has state regulations, bans local ordinances prohibiting outdoor releases.

10.4 THE ECONOMIC AND ENVIRONMENTAL EFFECTS OF PESTICIDE REGULATION

Pesticide regulation has direct and indirect economic effects. The direct effects include the impact of pesticide regulation on pesticide cancellations and suspensions; new pesticide registrations (pesticide innovations); and the environmental and human health effects of new pesticides. Pesticide cancellations can be examined in the context of the coincidence of legislative timing and pesticide cancellations. Prior to FQPA, regulators were required to consider pesticide benefits during cancellation procedures. In their analyses, regulators measured cost in terms of risks to human health and the environment and benefits in terms of productivity gains realized by continued pesticide use.

The effect of regulatory stringency on pesticide innovations and environmental and human health effects can be statistically analyzed by using human health and environmental testing (regulatory) costs as a proxy for regulatory stringency. Regulation should negatively affect pesticide innovations because pesticides that were marginally profitable under a former less costly regulatory regime may be too costly to develop under a more stringent regulation.

---

* APHIS has added a notification process to reduce the time required for the regulatory review period (7 CFR 340). Several crops may be imported, moved interstate, or field-tested without first applying for an APHIS permit as long as certain eligibility criteria and performance standards are met. These crops were chosen after significant test experience.

Two hypotheses exist for the impact of regulation on pesticide toxicity (both environmental and human health effects). Greene, Hartley, and West (1977) and Lichtenberg, Spear, and Zilberman (1993) suggest that greater testing costs cause firms to develop pesticides that will be used on at least one major crop and many minor crops. They also assert that these wide-spectrum pesticides may be highly toxic to humans and fish and wildlife because they must be effective against numerous pests. Alternatively, Ollinger and Fernandez-Cornejo (1998) propose that an increase in the number of tests required to gain regulatory approval reduces the number of pesticide candidates that can pass regulatory scrutiny. As a result, pesticides that formerly complied with regulatory standards may no longer meet new guidelines, and the ones that do meet regulatory standards are, on average, less toxic than their predecessors.

The indirect effects of pesticide regulation include the impact of regulation on minor crop registrations, industry composition, and the development of biological pesticides and genetically modified plants. To gain a crop use registration, a firm must conduct toxicological and environmental tests and supply the data to the EPA. Potential revenues for each use, however, vary from less than a million dollars for some minor crops, such as many fruits and vegetables, to over a billion dollars for corn. Since many minor crops have low or no potential profits, a rise in regulatory related testing costs should cause a drop in the number of new minor use registrations.

A rise in testing costs makes it necessary for firms to generate greater revenues to remain profitable. Firms that realize substantial sales from each new pesticide or have more diverse sources of income are favored over marginally profitable firms with low sales. As a result, higher testing costs favor large, multinational pesticide firms over small, domestic ones.

Some chemical pesticide alternatives, such as biological pesticides and recently approved genetically modified plants, have substantially lower testing requirements than chemical pesticides. A rise in chemical pesticide regulatory costs, therefore, should encourage the development of these chemical pesticide alternatives by lowering relative development costs.

All economic research on the effects of pesticide regulation was based on pre-1992 legislation. Since FQPA is important, we discuss this Act, pesticide availability, pesticide regulation and Integrated Pest Management (IPM), and organic food regulation.

10.4.1   The Direct Effects of Pesticide Regulation

10.4.1.1   Regulation and Existing Pesticides: Restrictions, Cancellations, and Suspensions

The 1972 FIFRA amendments had a profound impact on pesticide development and use. By changing pesticide policy toward the protection of human health and the environment, this legislation put substantial limits on the uses of pesticides thought to cause cancers. With outright bans on most uses for DDT and other organochlorines, regulation contributed to the decline in organochlorine use from 70% of total insecticides applied in 1966 to about 5% in 1982. Although replacements, such as organophosphate and carbamate insecticides, did emerge, overall insecticide use declined from a peak during the mid 1960s (Osteen and Szmedra, 1989). Besides outright pesticide bans, EPA has also either severely restricted or canceled the registrations of 22 pesticides. Additionally, registrants have voluntarily restricted pesticide uses by changing pesticide labels or withdrawing pesticides from the market.

As indicated earlier, before a pesticide can be restricted, it must pass through a SRP in which EPA considers both the health dangers and the pesticide benefits (Cropper, Evans, Beradi, Dulca-Soares, and Portney, 1992). Benefits are estimated as the efficiency loss of switching to the next best alternative. If alternatives are less effective or more expensive or both, the cost per unit of output will increase. Benefits are then defined as the change in the cost per unit times the number of units. Estimates of pesticide benefits can readily be made for the major field crops, but not for many fruits and vegetables. Generally, experts must estimate yield differences and pesticide use, alternative controls, and the extent of alternative use.

The strategy of companies with pesticides undergoing an SRP is to avoid the loss of the most profitable crop use registrations. This use registration is typically the crop with the highest pesticide sales volume. For example, if an insecticide has 1% of the corn market, 50% of the kiwi fruit market, and a choice of losing one or the other market because of possible harm to human health or the environment, the company would choose to lose a registration on the kiwi fruit because the size of the market (and potential profits) is much higher for corn.

The requirement of the 1988 FIFRA amendment that all pesticides approved prior to 1984 be brought up to current standards necessitated pesticide companies to conduct extensive field tests for some pesticides approved during periods of less stringent regulation. Additionally, the amendment imposed yearly maintenance fees and a pesticide registration fee. Combined, the additional testing costs and the new maintenance and registration fees dramatically increased the costs of maintaining a crop use registration. Revenues, on the other hand, were either very low or did not exist for most pesticide uses (those for many minor crops). As a result, reregistration costs exceeded potential revenues, causing firms to voluntarily cancel about 25,000 of 45,000 pesticide uses (Gianessi and Puffer, 1992). Although only 5000 of the 25,000 canceled pesticides were still in use and the acreage on which they were used was small, there were substantial effects on pesticide availability for some crops in some states. For example, Gianessi and Puffer (1992) identify six herbicides that lost their registrations and adversely affected the New York onion, Ohio strawberry, Michigan pumpkin, Maryland spinach, Pennsylvania green pea, and the Georgia pepper crops. Additionally, Ferguson (1989) reports that registration expenses forced firms to cancel pendamethalin for garlic, glyphosate for broccoli, and prometryn for parsley. Korson (1993) indicates that the reregistration requirements caused a 30 to 80% drop in the availability of post-harvest apple pesticides and a 22 to 77% drop in pesticides for cherries. Other studies (Council for Agricultural Science and Technology, 1992 — and Gianessi and Puffer, 1992) indicate that firms may not renew registrations on 600 to 1000 existing pesticide crop uses because of high regulatory and liability costs.

The distributional effects of pesticide cancellations (voluntary and forced) on farmers and consumers depend on price changes in the commodity. As indicated by

Osteen and Szmedra (1989), if crop price does not change, then the farmer bears all of the costs associated with an increase in pesticide costs or crop yield losses. The distributional effects are far more complicated if prices rise. A rise in price induces a drop in quantity demanded of the affected crop and, thus, a rise in demand for substitute crops. The net result is still an increase in prices, causing consumers, particularly the poor, to bear at least some of the cost of pesticide cancellations (Zilberman et al., 1991).

There are also distributional impacts among farmer groups (Osteen and Szmedra, 1989). Some farmers in some regions may have no substitute pesticide and, thus, may suffer a much greater yield loss than a farmer in a different region that has substitute pesticides available. For example, Lichtenberg et al. (1988) found that users of ethyl-parathion on almonds, plums, and prunes would lose about $2 million in producers' surplus while nonusers would gain about $0.5 million if the pesticide became no longer available. Additionally, farmers in areas of heavy pest infestation will place a higher value on chemical use, and thus may suffer a greater loss if a pesticide is canceled, than farmers in areas free of the pest (Fernandez-Cornejo and Jans, 1998). If a chemical becomes unavailable, production may shift geographically as growers in pest-free areas or those with lesser pest pressure may expand production while growers in more heavily infested regions may exit the sector or shift to other crops. Finally, a price increase would likely increase revenues of farmers of substitute crops because demand for substitute crops would rise.

10.4.1.2 Regulation and the Number of Innovations

Beginning during the pesticide discovery stage and persisting throughout the development process, the effects of regulation on pesticide innovation are more subtle than the effects of pesticides on existing pesticides. Scientists discover pesticides by examining various compounds for their effectiveness against pests. Since compounds related to chemicals previously banned by the EPA would not likely gain EPA registration, regulation limits the number of viable chemical compounds and, thus, increases the number of chemical searches needed to make a pesticide discovery. NACA surveys indicate that the number of chemicals screened per new registered pesticide rose from 8500 over 1970–73 to about 21,600 for 1986–87.

After compounds with pesticidal properties are discovered, technicians synthesize them in larger quantities, and then examine their effectiveness in laboratory and field trials; determine whether the chemical is toxic to humans and fish and wildlife; and estimate production costs. Pesticide candidates that pose a risk to either the environment or human health are eliminated from consideration. Chemicals that may be able to pass regulatory scrutiny and also generate sufficient revenues to compensate for future development costs then pass through a series of more demanding field tests. First, technical personnel use small-scale field testing to determine the effectiveness of the chemical compound relative to existing pesticides. They also evaluate the impact of soil, sunlight, microbes, and climate on its effectiveness. During either step, further consideration of a pesticide may be terminated because of possible environmental or health risks. However, if the pesticide candidate fares well against existing pesticides, the firm obtains an experimental use permit (EUP).

An EUP allows the company to conduct larger field tests and are granted only if the EPA believes the evidence reveals no adverse environmental effects will occur. If the EPA does not grant a permit, the company must either specify a new field test that meets EPA specifications or abandon pesticide development.

In larger field tests, technicians conduct metabolic, environmental, residue, and toxicology studies in order to determine the impact of the compound on humans, mammals, fish, and wildlife. Simultaneously, chemical and production personnel develop formulation techniques and production methods. If metabolic, environmental, residue, and toxicology studies indicate that the pesticide may pass EPA scrutiny and other studies suggest the pesticide will be effective, then a firm applies for pesticide registration.

Pesticides must generate substantial revenue in order to generate sufficient revenue required to pay for the fixed costs associated with the more than 70 different tests of pesticide environmental and human toxicity characteristics and the associated administrative expenses. This high revenue hurdle also adds to development costs because it limits the number of viable pesticide candidates, thereby increasing the number of chemical compounds examined as possible pesticides. Since increases in either research or regulatory costs cause the gap between potential revenues and costs to narrow, the net effect of regulation is to make some marginally profitable pesticides under an old regulatory regime unprofitable under a new regime and, thus, reduce the number of pesticide innovations.

Research expenditures and the new pesticide development cycle rose substantially over the 1972–91 period. Deflated average industry research costs in 1982 dollars rose from $91.3 million per year over the 1972–76 period to $340.8 million over the 1987–91 period. Our own estimates indicate that new pesticide development costs, which includes the costs of pesticides that do not reach the market due to either regulatory or technical failures, rose from $16 million per new pesticide in 1972 to $42.5 million in 1987 in 1982 dollars (Table 10.1). Meanwhile, the pesticide development cycle (the time required to bring a pesticide from initial screening to market) rose from 5 years in 1967 to 10 years in 1987.

Both the increase in pesticide development costs and the new pesticide development cycle can deter firms from developing certain types of chemical pesticides. The increase in pesticide development time is costly because companies gain patent protection during the development process. Thus, a longer development time gives a pesticide company less time to sell a pesticide as a proprietary product. Higher development and environmental testing costs discourage innovation because a product must then realize greater revenue to be profitable.

Higher environmental and human toxicity testing costs were a major contributor to the increase in development costs and may have increased new pesticide development time. New pesticide testing costs, as a percentage of research expenditures, rose from 9.7 to 25%. Test costs of new and existing pesticides rose from 18% of total research expenditures in 1972 to 47% in 1989 (Table 10.1). New and existing pesticide testing costs rose more than new testing costs because existing pesticide regulatory costs include testing for reregistration requirements and the registration of pesticides for additional crop uses, and does not include novel research.

Table 10.1 New Chemical Pesticides, Industry Research, and Regulatory Effort for 1972–91

| Year | New chemical pesticide registrations | | Industry Research Million 1982 dollars | New Pest. Reg. Cost[b] Percent | All Pest. Reg. Cost[c] | Prod. Devel.[d] Years |
	Major Firms Number	All Firms[a]				
1972	12	16	98.5	9.7	18.0	7
1973	4	13	87.9	11.4	19.0	7
1974	11	22	65.7	17.2	18.0	8
1975	12	25	85.7	17.7	20.0	8
1976	7	8	107.7	21.2	33.0	8
1977	1	2	114.2	20.0	31.0	8
1978	0	5	127.3	18.1	29.0	8
1979	9	17	127.7	18.5	35.0	8
1980	9	11	129.2	18.7	29.0	8
1981	5	14	174.3	17.3	27.0	8
1982	7	18	204.4	18.9	30.0	9
1983	8	15	216.4	19.8	31.0	9
1984	7	11	245.5	18.7	28.0	9
1985	4	10	370.4	20.8	34.0	10
1986	8	9	290.3	25.9	39.0	10
1987	4	6	284.9	24.2	40.0	10
1988	4	9	305.9	24.5	41.0	10
1989	10	15	330.4	25.1	47.0	10
1990	3	6	392.8	25.0	55.0	11
1991	3	7	390.1	25.0	60.0	11
1972–76	46	84	91.3	13.0	22.0	7.6
1977–81	24	49	134.5	12.6	30.0	8
1982–86	36	63	265.4	14.2	32.0	9.6
1987–91	24	43	340.8	24.8	49.0	10.4

[a] Major firms are ranked among the top 20 companies at least once over the 1972–91 period.
[b] New pesticide health and environmental testing costs as percent of research for new pesticides.
[c] All pesticide health and environmental testing costs as percent of all industry research.
[d] Years required from discovery of the chemical compound to EPA registration of the pesticide.
Source: Ollinger, M. and J. Fernandez-Cornejo, *Regulation, Innovation, and Market Structure in the U.S. Pesticide Industry*, U.S. Dept. Agr. Econ. Res. Serv., Washington, D.C., AER-719, June 1995. No permission necessary.

Firms would undertake some human toxicity and environmental testing in the absence of regulation because farmers want pesticides that are safe to use and do not adversely affect the environment (Beach and Carlson, 1993). Nonetheless, EPA cost analyses for the 1978, 1982, and 1994 indicate that required testing costs doubled between 1972 and 1993 and strongly correlate with industry-reported toxicological and environmental testing costs (regulatory costs).

Now consider human toxicity and environmental testing costs and pesticide innovations. As the regulatory costs rose over the 1972–89 period, the number of chemical pesticides registered for use on its first crop (pesticide registrations) by major pesticide companies dropped from 46 to 24, and pesticide innovations by all companies dropped from 84 to 43 between the 1972–76 and 1987–91 periods (Table 10.1). Although suggestive, these trends alone do not reveal the precise contribution of regulation to the decline in new pesticide registrations.

One empirical study (CAST, 1981) found that EPA regulation correlated with a rise in research expenditures; a decline in the number of new pesticides registered per year; a delay in the time required to register and reregister pesticides; and a shift in the allocation of research expenditure from synthesis, screening, and field testing to administration, environmental testing, and residue analysis. Additionally, using annual data and the time required to develop a pesticide as a measure of regulatory stringency, Hatch (1982) found that increased regulatory costs led to a 7–9% decline in pesticide innovations. In a more rigorous examination using firm-level data and several alternative regulatory variables, Ollinger and Fernandez-Cornejo (1998) found that a 10% increase in regulatory costs led to a 15–20% decline in new pesticide registrations.

10.4.1.3 Regulation and the Environment

The 1972 FIFRA Amendment greatly stiffened regulatory control of pesticides with adverse environmental effects. Surprisingly, there has been very few studies of regulation and environmental effects. Ollinger and Fernandez-Cornejo (1998) considered the effect of regulation on fish and wildlife. Table 10.2 shows that the number of registered pesticides harmful to fish and wildlife dropped by about 33% between the 1972–76 and 1986–89 periods. Since many of the pesticides introduced during the 1970s were not later reexamined under far more rigorous current testing methods, some pesticides thought to have no effects during the 1970s may have had minor impacts. Hence, trends in changes in pesticide fish and wildlife effects are biased toward a less dramatic change.

10.4.1.4 Regulation and Human Health

Many pesticides are highly toxic to humans prior to application, but degrade rapidly after exposure to the environment, permitting no possibility of inducing cancer. Other pesticides do have the potential to cause cancer, however. It was these pesticides that 1972 FIFRA affected.

There have been very few studies of the effect of pesticide regulation on human toxicity. Ollinger and Fernandez-Cornejo (1998) used human toxicity information stated on pesticide labels to examine changes in pesticide toxicity over the 1972–89 period. These data include an acute toxicity rating (I, II, III, or IV), chronic human effects, and harm from inhalation, skin absorption, or other side-effects. An acute toxicity rating of I is the most toxic and is based on LD 50 values, which are the dose of a toxicant necessary to kill 50% of the test animals studied within the first 30 days following exposure.

Table 10.3 shows that the number of newly reregistered pesticides with a Class I acute toxicity rating changed very little over the 1972–89 period. The number of chemical pesticides with chronic health effects, however, dropped by 83% between 1972–76 and 1985–89. As with testing for fish and wildlife effects, testing technology for detecting chronic or other long-term human health effects has improved

Table 10.2 Number of New Pesticides Harmful
to Fish and Wildlife, 1972–89

Year	Number of Harmful New Pesticides
1972	3
1973	5
1974	4
1975	3
1976	3
1977	1
1978	0
1979	5
1980	2
1981	2
1982	3
1983	4
1984	3
1985	3
1986	2
1987	3
1988	2
1989	2
1972–76	18
1977–81	10
1980–84	13
1985–89	12

Source: Ollinger, M. and J. Fernandez-Cornejo,
*Regulation, Innovation, and Market Structure in the
U.S. Pesticide Industry,* U.S. Dept. Agr. Econ. Res.
Serv., Washington, D.C., AER-719, June 1995. No
permission necessary.

markedly, suggesting that some pesticides introduced during the 1970s that were reported to have no long-term human health effects may have had then-undetectable levels of long-term impacts on human health. The difference between the changes in chronic health effects and acute toxicity over the 1972–89 period relates to the nature of the 1972 FIFRA amendments. Congress designed the FIFRA amendments to restrict the sale of pesticides with chronic or other long-term human toxicity or fish and wildlife effects. Congress did not, however, substantially alter the way EPA regulated the sale of pesticides with high acute toxicity.

Many chemical pesticides introduced from 1972–89 were not highly toxic to humans and fish and wildlife. The difference in the human toxicity and fish and wildlife effects of pesticides allows pesticides to be classified as "more" or "less" toxic. "More toxic" pesticides are defined as those that have a Class I acute toxicity rating or have chronic or other human health effects or adversely affect fish and wildlife. "Less toxic" pesticides are all others.

Table 10.3 shows that between the 1975–78 and 1986–89 periods, sales of new pesticides with either chronic or fish and wildlife effects as a share of total pesticide

Table 10.3   New Pesticide Toxicity to Humans, 1972–89

Year	Toxic Pesticides Class 1/ acute	Chronic	Other	New Sales as Share of Total Chronic/ Fish/Wildlife	Total More Toxic	More Toxic New Sales as a Share of all New Sales
	Number			Percent		
1972	3	1	4	n.a.	n.a.	n.a.
1973	1	2	2	n.a.	n.a.	n.a.
1974	2	2	2	n.a.	n.a.	n.a.
1975	1	1	1	0.96	1.06	41.0
1976	0	1	1	4.21	4.38	64.0
1977	1	1	0	7.03	7.28	68.0
1978	0	0	0	5.21	5.39	92.0
1979	2	1	2	3.52	3.52	94.0
1980	2	2	1	2.31	2.31	85.0
1981	1	0	1	0.44	0.44	67.0
1982	1	1	3	0.31	0.31	35.0
1983	1	2	0	0.61	0.61	27.0
1984	0	2	0	2.18	2.19	54.0
1985	1	1	2	0.52	0.52	42.0
1986	1	0	0	0.78	0.83	41.0
1987	3	0	3	0.53	0.62	11.0
1988	2	0	2	0.65	0.75	10.0
1989	1	0	1	0.86	1.17	20.0
1972–76[a]	7	6	8	4.35	4.53	66.0
1977–81[b]	6	4	4	1.65	1.65	70.0
1980–84[c]	5	7	5	0.91	1.04	39.0
1985–89[d]	7	1	8	0.71	0.84	21.0

n.a. = not available.
[a] Covers the 1975–78 period for the percent sales categories.
[b] Covers the 1979–82 period for the percent sales categories.
[c] Covers the 1982–85 period for the percent sales categories.
[d] Covers the 1986–89 period for the percent sales categories.
Source: Ollinger, M. and J. Fernandez-Cornejo, *Regulation, Innovation, and Market Structure in the U.S. Pesticide Industry*, U.S. Dept. Agr. Econ. Res. Serv., Washington, D.C., AER-719, June 1995. No permission necessary.

sales fell from 4.35 to 0.71% of sales, a drop of 84%. More generally, "more toxic" new pesticide sales as a share of total pesticide sales declined from 4.53 to 0.84% of sales, a drop of 82% and "more toxic" new pesticide sales as a share of total new pesticide sales dropped from 66 to 21% of new pesticide sales. Hence, there has been a significant shift to "less toxic" pesticides, both in terms of number of new registrations and sales of new chemical pesticides.

Econometric analyses (Ollinger and Fernandez-Cornejo, 1998) confirm that regulation encourages the development of "less toxic" chemicals. They show that a 10% increase in regulatory costs (about $1.5 million per pesticide) would result in a 5% increase in the number of pesticides classified as "less" rather than "more" toxic. They also show that results do not change if high acute toxicity is included in the definition of "more toxic" pesticides.

10.4.2   Indirect Effects of Regulation

10.4.2.1   *Regulation and Pesticide Registrations for Minor Crops*

Chemical pesticides must be registered with the EPA for each crop use. The number of uses for which a pesticide is registered may give an idea of how much revenue the pesticide can generate. For example, a pesticide that is registered for 50 crops can realize greater potential revenue than a pesticide registered for 20 crops if all crop markets are of equal size and pesticides are of equal effectiveness. Crop market sizes differ, however. In 1994, corn and soybean pesticide sales exceeded $3 billion, while in 1991 pesticide sales for over 200 minor crop uses reached only $620 million (1994 prices). Hence, if pesticide development costs are constant for all types of pesticides, then differences in potential revenues suggest that firms have a much stronger incentive to develop pesticides strictly for major uses rather than pesticides only for minor uses.

Firms have historically developed pesticides for either only major or only minor crops as long as development costs were less than revenues. Firms also have registered major use pesticides for minor uses in order to increase revenues and profits for that pesticide. An increase in regulatory costs, however, narrows the gap between pesticide revenues and pesticide development and regulatory costs, thereby reducing the rate of return for each pesticide use. Since the decline in the rate of return is much greater for smaller pesticide markets than for larger markets, regulation should have a greater negative impact on pesticides strictly for minor uses than on pesticides only for major uses, and should adversely affect the number of minor use registrations for major use pesticides. Hence, a rise in regulatory testing costs should cause the number of new pesticide registrations to decline and the ratio of major use pesticides to all pesticides to rise. In addition, if firms focus their development effort on pesticides for major crops, then average pesticide sales per new pesticide should rise.

As regulatory costs rose over the 1972–89 period (Table 10.1), pesticides for major field crops and nursery and other crops dropped by about 50% between the 1972–76 and 1985–89 periods (Table 10.4). More dramatic was the 72% decline in crop registrations for minor field crops, 92% drop in registrations for vegetables, and 75% reduction in crop registrations for fruits and nuts.

Since major crops generate most of the pesticide sales in the U.S. market, if development costs are identical, a firm may be more likely to cover research and testing costs by developing a successful pesticide for major crops than for any other crop market. As an example, Table 10.5 shows that over the 1972–89 period, there was almost no decline in the number of herbicide uses for major field crops, and for pesticide use nursery crops and other crops with low regulatory costs. By contrast, herbicide uses dropped by about 59% for minor field crops; 90% for vegetables; and 50% for fruits and nuts. Even more dramatic was the contrast between herbicide uses for major field crops and insecticides for all markets, which dropped by 74%.

Econometric analyses (Ollinger, Aspelin, and Shields, 1998) indicate that a 10% increase in the EPA-estimated cost of regulation encourages a reduction of 10 to

Table 10.4   Number of New Pesticide Registrations by Crop Type, 1972–89

Year[a]	Major Field[b]	Minor Field[c]	Veg[d]	Fruit/Nut[e]	Nurse/Other[f]
1972	8	19	20	69	14
1973	5	9	23	42	4
1974	12	25	48	74	9
1975	12	15	4	16	10
1976	8	9	5	22	5
1977	0	0	0	0	5
1978	0	0	0	0	0
1979	13	20	99	101	22
1980	4	6	1	0	9
1981	2	4	3	13	4
1982	6	2	1	2	8
1983	5	11	24	73	12
1984	3	0	2	19	9
1985	0	0	2	0	4
1986	8	8	2	0	7
1987	4	4	0	0	7
1988	4	3	2	0	1
1989	6	7	2	50	4
1972–76	45	77	100	203	42
1977–81	19	30	103	114	40
1980–84	20	23	31	107	42
1985–89	22	22	8	50	23

[a] One pesticide can have multiple uses.
[b] Major field: corn, cotton, sorghum, soybean, and wheat.
[c] Minor field: 18 field crops, such as barley, clover, peanuts, potatoes, rice, rye, sugarcane.
[d] Veg.: 11 major vegetables with more than 100,000 acres planted and 35 with less than 100,000.
[e] 12 fruits and nuts with over 100,000 acres planted and 51, with less than 100,000 acres planted.
[f] Nurse/other: forage and pasture, forestry, eight nursery uses, and six other non-crop uses.
Source: Ollinger, M. and J. Fernandez-Cornejo, *Regulation, Innovation, and Market Structure in the U.S. Pesticide Industry,* U.S. Dept. Agr. Econ. Res. Serv., Washington, D.C., AER-719, June 1995. No permission necessary.

15% in registrations per pesticide. In this analysis, one registration was defined as one of eight major crop categories. Additionally, Ollinger and Fernandez-Cornejo (1995), hypothesizing that higher testing costs may cause companies to target major crop markets and reduce development of minor U.S. pesticides, found that a 10% increase in testing costs caused a 2% increase in the proportion of pesticides for major crop markets to pesticides for all crop markets. Other results (Ollinger, Aspelin, and Shields, 1998) show that regulation encourages increased sales per new pesticide. One criticism of empirical research by Ollinger and Fernandez-Cornejo (1995) concerns crop use data. The database was constructed in 1992; thus, the most recently registered pesticides had a much shorter time period to obtain additional crop use registrations than did pesticides registered earlier.

Table 10.5   Number of New Pesticide Registrations by Crop and Pesticide Type, 1972–89[a]

Year	Major field h,i,o[b]	Minor field h,i,o	Veg h,i,o	Fruit/nut h,i,o	Nurse/other h,i,o
1972	2,3,0	4,3,1	3,5,0	4,4,2	3,2,1
1973	1,1,1	2,1,2	4,0,2	3,1,2	0,2,0
1974	3,2,1	4,2,1	3,4,2	3,2,2	2,2,2
1975	4,1,0	5,1,1	2,1,0	4,3,0	7,1,1
1976	2,0,0	2,0,0	3,0,0	3,0,0	2,0,2
1977	0,0,0	0,0,0	0,0,0	0,0,0	2,0,1
1978	0,0,0	0,0,0	0,0,0	0,0,0	0,0,0
1979	1,4,2	1,3,3	2,6,6	2,5,6	3,5,2
1980	3,1,0	2,0,0	0,1,0	0,0,0	1,3,0
1981	0,0,2	0,0,1	0,0,2	0,0,4	0,0,3
1982	2,2,1	1,0,0	0,1,0	0,1,0	1,0,2
1983	3,0,0	4,0,0	5,0,2	4,0,2	3,2,1
1984	0,2,0	0,0,0	0,2,0	0,3,2	2,2,3
1985	0,0,0	0,0,0	0,2,0	0,0,0	2,2,0
1986	6,0,0	3,0,0	1,0,0	0,0,0	3,1,0
1987	3,0,0	2,0,0	0,0,0	0,0,0	2,1,0
1988	3,1,0	2,1,0	0,2,0	0,0,0	1,0,0
1989	2,1,1	2,0,2	0,0,2	2,2,4	2,0,1
1972–76	12,7,2	17,7,5	15,10,4	17,10,6	14,7,6
1977–81	4,5,4	3,3,4	6,7,8	2,5,10	6,8,6
1980–84	8,5,3	7,0,1	5,4,4	4,4,8	8,7,9
1985–89	14,2,1	9,1,2	1,4,2	2,2,4	10,4,0

[a] one pesticide can have multiple uses. See Table 10.2 notes for definition of crop categories.
[b] h is herbicides, i is insecticides, and o is other pesticides, such as fungicides, miticides, and others.
Source: Ollinger, M. and J. Fernandez-Cornejo, *Regulation, Innovation, and Market Structure in the U.S. Pesticide Industry*, U.S. Dept. Agr. Econ. Res. Serv., Washington, D.C., AER-719, June 1995. No permission necessary.

10.4.2.2   *Industry Composition*

As discussed above, higher testing costs favor large, multinational pesticide firms over small, domestic ones. Table 10.6 illustrates the impact. As human health and environmental testing costs rose from 17.5 to 47% of total research costs over the 1972–89 period, the number of pesticide firms undertaking research and development dropped from 33 to 19; the number of small firms undertaking research and development declined much more (from 16 to 6) than did large firms (17 to 13); the U.S. market share held by foreign-based companies rose from 18 to 43%; and the percent of U.S. firm sales from foreign markets rose from 23 to 60% over 1974–89.

Structural change during the 1970s and 1980s resulted in little change in industry concentration because most sales were of large domestic firms with large pesticide operations to pesticide firms with an even larger international but a smaller U.S. presence. Among the most newsworthy of these sales were those of the pesticide operations of Shell, Stauffer, and Union Carbide to DuPont, ICI, and Rhone Poulenc, respectively.

Table 10.6   Market Structure Trends in the Pesticide Industry, 1972–89

Year	All Firms	Small Firms	Large Firms	4-Firm Concen. Ratio	Foreign Firm U.S. Market Share[2]	Foreign Firm Entrants	U.S. Firm Sales Abroad as Share of Total Firm Sales
	Number			Percent		Number	Percent
1972	33	16	17	0.496	18	0	n.a.
1973	34	17	17	0.501	16	1	n.a.
1974	34	17	17	0.484	20	1	23
1975	36	18	18	0.487	20	2	18
1976	36	18	18	0.478	21	2	25
1977	36	18	18	0.441	20	2	25
1978	36	18	18	0.421	22	2	17
1979	36	18	18	0.407	21	3	20
1980	34	16	18	0.394	21	3	25
1981	34	16	18	0.378	21	3	24
1982	33	15	18	0.372	21	3	29
1983	32	14	18	0.392	21	3	33
1984	29	10	19	0.402	23	3	25
1985	28	9	19	0.385	28	4	31
1986	26	8	18	0.380	29	4	32
1987	23	8	15	0.454	36	4	33
1988	23	8	15	0.466	38	4	30
1989	19	6	13	0.483	43	6	n.a.

n.a = not available

[a] Firms with one or more new products or among top 20 firms at least once over 1972-89. Entry date is first year in top 20 or 4 years before first new product in Aspelin and Bishop (1991).
[b] Share of production includes the production of foreign-based plants in the U.S. and imports into the U.S. market by foreign-based companies.
Source: Ollinger, M. and J. Fernandez-Cornejo, *Regulation, Innovation, and Market Structure in the U.S. Pesticide Industry*, U.S. Dept. Agr. Econ. Res. Serv., Washington, D.C., AER-719, June 1995. No permission necessary.

Econometric results (Ollinger and Fernandez-Cornejo, 1998) suggest that regulation negatively affects the number of pesticide firms, with a greater negative impact on smaller firms than larger ones; and encourages the U.S. growth of large foreign-based firms. Their empirical results suggest that a 6% rise in pesticide product regulation costs causes two small firms and one large one to exit the pesticide industry, and expands the share of the U.S. market held by foreign-based firms by 4%.

10.4.2.3 Regulation and Biological Pesticides

Regulatory control over biological pesticides varies considerably in its intensity. Properties such as nontoxicity and low application rates generally encourage rapid EPA approval of pheromones (*Pesticide Regulation Handbook*, 1991). Similarly, the EPA favors bacteria spores and crystals because they are immobile, occur naturally in the environment, require little protective applicator equipment, and generally do not harm beneficial organisms. Other types of bacteria, such as those able to reproduce and move between plants, however, have never even been tested because of regulatory concerns that the bacteria would colonize nontarget plants.

Recognizing that many biological pesticides posed fewer regulatory-defined risks to either the environment or human health, the EPA established a tier approval system in which biological pesticides that pass a short-term test of toxicity to humans and fish and wildlife, pathogenicity, persistence, and replication in mammals can gain waivers against additional tests. The EPA also gives regulatory approval priority to biologicals over chemicals and waives some of the 19 tests required of chemical pesticides that do not apply to biologicals. As a result, biological pesticides have far lower development and regulatory costs than chemical pesticides. For example, Tantillo (1989) reports that one biological pesticide cost only $5 million to develop, had $500,000 in regulatory costs, and was registered in 6 months. By comparison, chemical pesticide development costs were about $42.5 million with 25% for regulatory costs in 1987.

The pace of biological pesticide development illustrates their lower regulatory costs when compared with chemical pesticides. Over the 1980–85 period, there were four biological pesticides registered by the EPA. From 1987–92, the EPA registered 13 new biological pesticides, about 26% of all EPA registered pesticides. Additionally, pesticide programs in which biological pesticides are the only pest control agents have been established for citrus, nut, and apple crops, and have been used as components of IPM in cotton, citrus, rice, nuts, soybeans, vegetables, and deciduous fruit crops. Despite these recent gains, biological pesticides have been applied against relatively few economically significant pests and have less than 2% of the pesticide market.

Perhaps the biggest reason for the low market share of biological pesticides is their narrow host range. Plapp (1993) indicates that biological herbicides have not been as successful because target weeds are often replaced by other weeds not affected by the biological herbicide. As a result, of the 13 biological pesticides registered from 1986–92, five were insecticides, six were fungicides, two were other types, and none were herbicides.

10.4.2.4 Regulation and Genetically Modified Plants

Plant scientists have long recognized that some plants have natural resistances to some herbicides and produce Bt and other naturally occurring pest control agents. With the advent of plant biotechnology techniques, it became possible to incorporate herbicide-tolerance (ability to tolerate certain types of herbicides), insect-resistance (ability to withstand attacks by certain insects), and virus- and other pest-resistant characteristics (immunity to some types of plant viruses and other plant pests) directly into plants. As of September 1994, several genetically modified plants had been commercialized and had elicited optimism that genetically modified plants would become an important new approach to controlling pests.

The cost of genetically modifying plants relative to the cost of developing new chemical pesticides is a prime factor motivating plant biotechnology research. Recent estimates suggest that genetically modified plants cost about $10 million and require about 6 years to develop, while chemical pesticides cost $42.5 million and take 11 years to develop. In addition, Krimskey and Wrubel (1993) report that it is becoming more costly for firms to develop chemical pesticides that are both harmless to the crops and sufficiently toxic to kill all or most pests.

Growth in the number of environmental release permits indicates the increased pace of technology adoption. From November 1987 to November 1988, USDA's Animal Plant Health and Inspection Service (APHIS) issued 20 environmental release permits for testing genetically modified plants. The number of permits increased to 30 for the 12 months preceding November 1989, 50 in 1990, 86 in 1991, and 149 in 1992. By April of 1998, over 2400 field test permits and notifications have been approved at over 9600 sites in the United States, and 23 plant lines have been proven safe and no longer need to be regulated by APHIS.

Many applications of plant biotechnology have been for pesticide-related characteristics. Of the 3200 environmental release permits issued from 1987 to April 30, 1997, 902 (28.2%) were for herbicide tolerance, 784 (24.5%) were for insect resistance, 334 (10.4%) were for virus resistance, 130 (4.1%) were for fungal resistance, 827 (25.8%) were for changes in product enhancement qualities, such as delayed product spoilage and research needs, and 223 (7.0%) were for other purposes (APHIS, 1997). Herbicide-tolerant plants can withstand applications of a particular type of herbicide and, thus, are designed to encourage farmers to switch pesticide brands (Krimskey and Wrubel, 1993). Pest and insect resistances are designed to replace chemical pesticides if they are targeted at crop markets in which chemical pesticides are used for an identical purpose.

10.4.2.5 Regulation and Pesticide Availability

Chemical pesticides have been approved for use on many crops, but many registrations have also been withdrawn or canceled. Moreover, the tendency of pests to develop resistance to pesticides further reduces the number. Eichers (1980), for example, indicates that insect resistance had caused a drop in sales of DDT, chlordane, and heptachlor before the EPA banned these organochlorine insecticides; and that weed resistance to the herbicide 2,4-D led to the decline in its market share from 32 to 4% over 1966–76. Similarly, Plapp (1993) reports that 447 species of insects and mites, 100 species of plant pathogens, and 55 species of weeds are known to be resistant in one location to one or more pesticides used for their control. The decline in the number of new chemical pesticide registrations for minor crops, coupled with a rise in the number of pests resistant to chemical pesticides, suggests a decline in the availability of effective chemical pesticides.

Biological pesticides and genetically modified plants may compensate for some of the drop in pesticides. However, biological pesticides are effective only for single-pest infestations and in conjunction with chemical pesticides for multi-pest infestations. Genetically modified plants are currently being developed only with insect and virus resistance, and with no characteristics that can protect against harmful weeds. Moreover, regulatory approval for the currently used insect resistant characteristic (Bt) was swift because Bt has been used as biological pesticides for many years and is well known. If pests develop resistances to Bt, as suggested by Krimskey and Wrubel (1993), then there will be a need for other varieties of insect-resistant plants. Yet insect-resistant properties other than Bt would likely encounter much more regulatory scrutiny (conversation with Bob Torla of EPA, March 20, 1998).

Fears of a decline in the number of pest control alternatives are not new. Similar fears have encouraged Congress to reduce regulatory requirements for some minor use pesticides and, during the early 1960s, led the USDA to establish the Interregional Research Project No. 4 (IR-4) to help register minor uses of pesticides. Currently, IR-4 funds some field tests and works with the EPA to reduce minor use pesticide field test requirements. Demand for its services has grown because the 1988 FIFRA amendment required pesticide manufacturers to reregister all pesticides approved prior to 1984 by 1997. Rather than comply, many manufacturers canceled pesticides that were currently in use. As a result, the number of requests to gather data to support the pesticides about to be canceled was 5082 minor uses in August 31, 1991. IR-4 identified 1324 of these requests as pesticides for which the company was willing to register the pesticide use if funding were available. However, budget limits have prevented IR-4 from meeting all requests (Gianessi and Puffer, 1992).

Some private mechanisms have addressed the needs of farmers of some minor crops. For example, a state-based hops growers' association met to ensure registration of seven pesticides needed for hop production. After gaining the approval of the pesticide company to seek registration, it used funds from members and breweries to register the pesticides. Private sector and IR-4 responses have not met all the needs of farmers of minor crops, however. Gianessi and Puffer (1992), for example, cite six cases in which farmers had no pest control products after the commonly used chemical pesticide was dropped.

10.4.3 The Effects of the FQPA

Although just recently enacted, the FQPA has caused substantial controversy over its possible effects. Previous economic research on the effects of pesticide regulation provides implications for the effect of the FQPA on pesticide development and usage. The immediate impact of the 1972 FIFRA Amendments was a ban on most uses of organochlorine insecticides. The longer term effect was to negatively affect the number of pesticide innovations, particularly pesticides for minor crops, and discourage the development of pesticides with chronic or environmental effects. Additionally, it encouraged industry consolidation and the development and use of biological pesticides and genetically modified plants. The 1988 FIFRA legislation led to the voluntary cancellations of over 25,000 pesticide use registrations (about 55% of all use registrations). Most cancellations were for inactive pesticide uses, but of the actively used pesticides, most were for minor crops and some these had no substitutes.

Mintzer and Osteen (1997) provide an excellent overview of the FQPA, discussing the regulatory issues, old provisions, changes made by legislation, and economic implications. Provisions of the FQPA deal with a common tolerance standard for processed and unprocessed foods; aggregate pesticide exposure; consideration of the diet of infants and children; use of benefits for establishing pesticide tolerances; state right to set tolerances stricter than that of the EPA; harmony with international standards; pesticide suspensions; minor-use pesticide regulatory costs; petitions for tolerances; estrogenic screening; pesticide reregistrations; how to handle existing stocks of suspended or canceled pesticides; and a right-to-know provision.

The provisions relating to common tolerance standard for processed and unprocessed foods, aggregate pesticide exposure, and consideration of the diet of infants and children have been briefly described. In terms of economic impacts, establishing a uniform health-based standard that applies to both processed and unprocessed foods and the treatment of cancer and noncancer risks eliminates the dual treatment of carcinogenic residues. For farmers, this means there is no immediate need to seek substitutes for pesticides threatened by the Delaney clause.

The use of an aggregate pesticide exposure standard and the consideration of the diet of infants and children will likely increase the cost of developing new pesticides because the failure rate for potential pesticide candidates should rise. Previous research (Ollinger and Fernandez-Cornejo, 1998) shows that an increase in regulatory restrictiveness reduces innovation. An aggregate pesticide exposure standard will also likely make it more difficult to register new pesticide uses if the risk standard is met or exceeded and provides an incentive to develop "new chemistry" to avoid risks from previously registered uses and common mechanisms of toxicity.

There could be a much more dramatic impact on existing pesticides. Although EPA has not yet specified the measures it will use to enforce FQPA, the Environmental Working Group (1998) recently published a report indicating that, under one scenario, organophosphate pesticides have an aggregate exposure rate for children that greatly exceeds that permitted by the EPA. If this proves to be true, there could be a substantial impact on agriculture. Gianessi (1997) reports that organophosphates currently account for about 65% of the non-oil poundage of insecticides and have over 50% of the U.S. pesticide market (in acreage treated) for corn, alfalfa, canola, mint, oats, rye, safflower, seedcrops, sod, sorghum, sugar cane, sunflowers, wheat, wild rice, and 17 fruits and vegetables. Historically, firms have canceled minor crop use registrations first. With the health of infants and children provision, the first crop uses dropped would likely be for minor crops consumed heavily by infants and children. In cases where no alternatives exist, loss of pesticide registrations could be very costly to farmers and increase consumer prices.

Gianessi (1997) adds that the effects could be even more serious, suggesting that EPA may be compelled to restrict or ban all organophosphate and carbamate insecticides. These insecticide types include 30 of the 59 existing insecticide classes and account for 84% of the non-oil insecticide poundage used in the U.S. In terms of affected crops, organophosphates and carbamates combined were used on over 50% of all treated acres for 11 fruits, 22 vegetables, 12 small field crops, 5 major field crops, and 6 other crops.

Another provision of FQPA lowers the cost of registering minor use pesticides, particularly important ones, by granting additional time for submission of data or in some cases waiving data requirements altogether; extending the period of exclusive use; requiring EPA to expedite review; and directing EPA to establish a matching fund to develop data supporting pesticides. A minor use pesticide was defined as a pesticide used on less than 300,000 acres or a case in which there is insufficient incentive to register a pesticide that is important to crop production.

Petitions for tolerances, estrogenic screening, and pesticide reregistrations all could raise production costs and food prices. Petitions for tolerances would permit anyone, rather than only the pesticide registrant, to petition to establish, modify, or revoke a tolerance. Registrants must provide a summary of the data and authorize publication of it and health information. This provision would make it much easier for public interest, grower, and environmental groups to initiate proceedings to revoke a tolerance, thereby increasing the likelihood of pesticide cancellations.

The estrogenic screening provision would increase testing costs, and thus, may negatively affect innovation. More significant is the reregistration provision. The 1988 Amendment required reregistration of all pesticides registered before 1984 and resulted in cancellation of 25,000 pesticide uses. If FQPA has a similar impact, then many minor use pesticides would likely be canceled, leaving some crops with no effective way to control some pests, and result in higher production costs and food prices, especially for fruits and vegetables.

The use of benefits for establishing pesticide tolerances, state rights to set tolerances stricter than those of the EPA, harmony with international standards, suspensions, how to handle existing stocks of suspended or canceled pesticides, and a right-to-know provision have modest economic implications. Since EPA has not previously used benefits for setting tolerances, revocation of the right to use benefits to set tolerances codifies an existing practice and will have little effect. Similarly, most states do not have stricter tolerances than those of the EPA; thus, this provision also would have a modest impact. Since some FQPA standards are stricter than the CODEX international system, there may be some trade effects. Under the FQPA, the EPA could order a suspension prior to issuing a notice of intent to cancel for pesticides suspected of having adverse health effects. This provision accelerates the rate at which a pesticide could be withdrawn from the market and would address a potential hazard, but would have almost no economic effect. The provision for existing stocks of suspended or canceled pesticides grants EPA authority to continue an existing practice, while the right-to-know provision includes directives to improve consumer access to dietary information.

10.4.4   FQPA and Integrated Pest Management

Integrated Pest Management (IPM) had received scant attention under FIFRA until the enactment of FQPA. Even under FQPA, however, it is mentioned only in one section. That section requires the Secretary of Agriculture to implement research, demonstration, and education programs to support the adoption of IPM. Under this mandate, IPM was defined to be a sustainable approach to managing pests by combining biological, cultural, physical, and chemical tools in a way that minimizes economic, health, and environmental risks.

The use of nonchemical pest-control alternatives in an IPM program could become particularly important if FQPA results in the widespread loss of pesticide registrations, as suggested by Gianessi (1997) and other observers of the pesticide industry. The chemical pesticide alternatives include biological controls (beneficials); cultivation, mulching, crop rotation and other cultural controls; planting location and other strategic controls; and host plant resistance, such as insect- and disease-resistant plant varieties and root stock.

10.4.5 Organic Farming and Pesticide Use

Federal pesticide legislation makes no reference to organic farming; yet, it could have a major impact on pesticide use. Growing at an average of 20% annually over the past six years, sales of organically grown foods have reached $2.8 billion. Of the total organic food market, fruits and vegetables account for 10 to 15% of sales.

As a way to standardize the definition of "organic" from among 11 state organic certifying boards and 33 private certifying groups, USDA proposed a set of standards that would govern the definition of organically produced foods. Although the proposed standards do not permit the direct use of synthetic chemical pesticides, there are several controversial sections, particularly those allowing the use of irradiation, seeds from genetically modified plants, sludge from city waste treatment plants, and intensive confinement of livestock. Additionally, the standards do not permit the use labels that differentiate production methods. Having received almost 100,000 comments, most of which came from organic food producers and consumers who think the standards are too permissive, the USDA definition of an organically grown food will not likely include the use of genetically modified plants, irradiation, and sludge from city waste treatment plants and may be even more restrictive. With only a small fraction of the total fresh produce and grain markets, the likely impact of either the current or future version of a standard definition of organic on pesticide use is likely to be small compared to the provisions of FQPA.

REFERENCES

Aspelin, A. and F. Bishop, *Chemicals Registered for the First Time as Pesticidal Active Ingredients Under FIFRA*. U.S. Environmental Protection Agency, Office of Pesticide Programs, Washington, D.C., May 1991.

Beach, E. D. and G. A. Carlson, Hedonic Analysis of Herbicides. *American Journal of Agricultural Economics*. 75, pp. 612-23, 1993.

Carson, R., *Silent Spring*. Fawcett Crest, New York, 1962.

CAST (Council for Agricultural Science and Technology), *Impact of Government Regulation on the Development of Chemical Pesticides for Agriculture and Forestry*. Council for Agricultural Science and Technology, Ames, IA, Number 87, 1981.

CAST (Council for Agricultural Science and Technology), *Pesticides: Minor Uses/Major Issues*. Council for Agricultural Science and Technology, Ames, IA, June 1992.

Cropper, M., William N. Evans, S. J. Berardi, M. M. Ducla-Soares, and P. R. Portney, The Determinants of Pesticide Regulation: A Statistical Analysis of EPA Decision-Making. *Journal of Political Economy*. 100, pp. 175-97, 1992.

Eichers, T.R., *The Farm Pesticide Industry*. U.S. Dept. Agr., Econ. Res. Serv., Washington, D.C., AER-461, Sept., 1980.

Farm Chemicals Handbook '95, Meister Publishing Company, Willoughby, OH, 1995.

Ferguson, M., "The Team Approach to Successful Minor Crop Registration," presented at the 41st California Weed Conference, 1989.

Fernandez-Cornejo, J. and S. Jans, "Issues in the Economics of Pesticide Use in Agriculture: A Review of the Empirical Evidence" Forthcoming in the *Review of Agricultural Economics*.

Gianessi, L. P. and C. A. Puffer, Reregistration of Minor Use Pesticides: Some Observations and Implications, in *Agricultural Resources Inputs Situation and Outlook Report*, Ar-25-4, U.S. Dept. Agr., Econ. Res. Serv., Washington, D.C., Feb. pp. 52-60, 1992.

Gianessi, L. P., *The Uses and Benefits of Organophosphate and Carbamate Insecticides in U.S. Crop Production*, National Center for Food and Agricultural Policy, Washington, D.C., 1997.

Greene, M. B., G. S. Hartley, and T. F. West, *Chemicals for Crop Protection and Pest Control*. Pergamon Press, Oxford, England, 1977.

Harper, C. R., and D. Zilberman, Pest Externalities from Agricultural Inputs. *American Journal of Agricultural Economics*. 71, pp. 692-702, 1989.

Hatch, U., *The Impact of Regulatory Delay on R&D Productivity and Costs in the Pesticide Industry*. Ph.D. dissertation, University of Minnesota, St. Paul, MN., 1982.

Korson, P., The Minor Use Pesticide Situation, in annual report of the Horticultural Society of Michigan, East Lansing, MI, p. 54, 1993.

Krimskey, S. and R. Wrubel, *Agricultural Biotechnology: An Environmental Outlook*. Department of Urban and Environmental Policy, Tufts University, Medford, MA., 1993.

Lichtenberg, E., R. C. Spear, and D. Zilberman, The Economics of Reentry Regulation of Pesticides. *American Journal of Agricultural Economics*. 75, pp. 946-58, 1993.

Lichtenberg, E., D.D. Parker, and D. Zilberman, Marginal Analysis of Welfare Costs of Environmental Policies: The Case of Pesticide Regulation. *Amer. J. Agr. Econ.* 70, pp. 867-74, 1988.

Mintzer, E.S. and C. Osteen, New Uniform Standards for Pesticide Residues in Food. *Food Review*, USDA, Economic Research Service, 20, pp. 18-26. January-April, 1997.

NACA (National Agricultural Chemicals Association), *Pesticide Industry Profile Study*. National Agricultural Chemicals Association, Washington, D.C., various issues, 1971-89.

NAS (National Academy of Sciences), *Regulating Pesticides in Food: The Delaney Paradox*. National Academy Press, Washington, D.C. 1987.

NAS (National Academy of Sciences), *Pesticides in the Diets of Infants and Children*. National Academy Press, Washington, D.C., 1993.

Ollinger, M., A. Aspelin, and M. Shields, Regulation and New Pesticide Registrations and Sales in the U.S. *Agribusiness: An International Journal*. Forthcoming, 1998.

Ollinger, M. and J. Fernandez-Cornejo, *Regulation, Innovation, and Market Structure in the U.S. Pesticide Industry*. U.S. Dept. Agr. Econ. Res. Serv., Washington, D.C., AER-719, June 1995.

Ollinger, M. and J. Fernandez-Cornejo, Innovation and Regulation in the U.S. Pesticide Industry. *Agricultural and Resource Economics Review*. pp. 1-13, 1998.

Ollinger, M. and J. Fernandez-Cornejo, "Sunk costs and regulation in the U.S. Pesticide Industry." *International Journal of Industrial Organization*. 16, pp. 139-68, 1998.

_____. and L. Pope, *Plant Biotechnology: Out of the Laboratory and Into the Field*. U.S. Dept. Agr., Econ. Res. Serv., Washington, D.C., AER-697, Apr. 1995.

Osteen, C. and P. Szmedra, *Agricultural Pesticide Use Trends and Policy Issues*. U.S. Dept. Agr., Econ. Res. Serv., Washington, D.C., AER. 622, Sept. 1989.

Pesticide Regulation Handbook. McKenna and Cueno and Technology Sciences Group, Inc., 1991.

Plapp, F., The Nature, Modes of Action, and Toxicity of Insecticides, in *CRC Handbook of Pest Management in Agriculture*, Pimentel, D., Ed., New York State College of Agriculture and Life Sciences, Ithaca, NY, 1993.

Ribaudo, M. and D. Gadsby, (1997) State Water Laws That Affect Agriculture: A Summary, unpublished manuscript, U.S. Dept. Agr., Econ. Res. Serv, Washington, D.C.

Tantillo, L., The Growing Field of Biopesticides. *Chemical Week*. pp. 46-47, June 28, 1989.

Torla, Bob, personal conversation, March 20, 1998.

USDA (U.S. Dept. of Agriculture), *Agricultural Outlook*. U.S. Dept. Agr. Econ. Res. Serv., Washington, D.C., June, 1997.

USEPA (U.S. Environmental Protection Agency), *Registration of Pesticides in the United States: Proposed Guidelines; Economic Impact Analysis*. Environmental Protection Agency, Washington D.C., Sept. 6, 1978 Part II.

USEPA (U.S. Environmental Protection Agency), *Regulatory Impact Analysis Data Requirements for Registering Pesticides Under the Federal Insecticide, Fungicide, and Rodenticide Act*. Environmental Protection Agency, Washington D.C., August 1982.

USEPA (U.S. Environmental Protection Agency), *Changes to Part 158*. Environmental Protection Agency, Washington D.C., mimeo, June 22, 1994.

Wiles, R., K. Davies, and C. Campbell, *Overexposed: Organophosphate Insecticides in Children's Food*. Environmental Working Group, Washington D.C.: 1998.

Zilberman, D., A. Schmitz, G. Casterline, and J.B. Siebert, The Economics of Pesticide Use and Regulation. *Science*. 253, pp. 518-22, 1991.

Index

A

Abamectin, 133

Acanthoscelides obtectus, 71

Acaricides selective toxicity, 134

Acarina, *see* Mites

Acceptable Daily Intake (ADI), 346

Accessory proteins in Bti, 253–254

Achroia grisella (lesser wax moth), 71

Acorus calamus, 108

Acremonium, 138

Action threshold (AT) for pest assessment, 119

Aculus schlechtendali (apple rust mite), 134

Acyrthosiphon pisum (aphid), 153

Adalia bipunctata (twospotted lady beetle), 32

Adelophocoris lineolatus (alfalfa plant bug), 65

ADI (Acceptable Daily Intake), 346

Adoxophyes orana (summerfruit tortrix), 139

Aedes aegypti (mosquitos)

    Bacillus used against, 247

    ITU established using, 274

Aerial invertebrates and insecticide use, 110–113

Agamermis unka in natural farming environments, 160

Agmenellum quadruplicatum (cyanobacterium), 266

Agreement on Sanitary and Phytosanitary, *see* Sanitary and Phytosanitary Measures (SPS)

Agricultural imports, regulations concerning, 308

Agroecology definition, 108

Ahasverus advena (foreign grain beetle), 73

AIM (alfalfa integrated management), 191–192

Aldrin/dieldrin phase-out, 342

Alfalfa crops

    integrated management, 191–192

    pest control

        looper, 139

        plant bug, 65

        seed chalcid, 17, 47

        weevil, 61, 65

    refuge creation, 153

Algae, aquatic, and insecticide use, 110

Almond moth (*Cadra cautella*), 39, 40

ALP (appropriate level of protection) in SPS, 331

Amaranthus (pigweed), 18

Amblyseius spp. (mite), 155, 158

Ampelomyces quisqualis (fungus), 139

Amphasia sericaea (beetle), 161

Amphibians and insecticide use, 114–115

Amyelois transitella (navel orangeworm), 52

Anabaena (cyanobacterium), 266

Anagrus spp.

    abundance management, 151

    multicropping techniques using, 16–17

Anastrepha ludens (Mexican fruit fly)

    Bti used against, 248

    color traps used to control, 31

    pulsed electric fields used to control, 62

Anastrepha suspensa (Caribbean fruit fly)

    heat and steaming used to control, 64

    spinosad toxicity to, 134

Androctonus australis (scorpion), 139

Angoumois grain moth (*Sitotroga cerealella*), 61

Animal and Plant Health and Inspection Service (APHIS)

    analysis process overview

        initiating points, 318–319

        pathway analysis, 319–322

        stages of risk analysis, 317–318

    environmental release permits issued, 364

    export regulations, 311

    legal basis, 308

    risk analysis principles, 316–317

    risk assessment, 317

    risk assessment and SPS, 315

    transgenic crop regulations, 349–350

Annelid susceptibility to insecticides, 113

Anopheles (mosquitos)

    sinensis and irrigation, 50

    stephensi and *Bacillus* use, 247

Anthonomus grandis grandis, *see* Boll weevil

Antibiosis, 121

Ant (*Pheidole megacephala*), 38

I